CITRUS
Biotechnology and Development

CITRUS
Biotechnology and Development

Nalini Sharma

2025
Agri Horti Press
New Delhi-110 084

ISBN: 978-93-92969-12-6 (Hardbound)

Published By

Agri Horti Press

New Delhi

E-mail: agrihortipress@gmail.com

Digitally Printed at : Replika Press Pvt. Ltd.

Preface

Welcome to the captivating world of **"Citrus: Biotechnology and Development."** This book serves as a comprehensive exploration of the multifaceted realm of citrus cultivation, where the convergence of biotechnology and traditional agricultural practices holds immense promise for the future of citrus production and global food security.

Citrus fruits, cherished for their vibrant flavors, nutritional value, and cultural significance, occupy a central place in agriculture and culinary traditions worldwide. However, the citrus industry faces myriad challenges, including emerging pests and diseases, climate variability, and shifting consumer demands. In response to these challenges, the application of biotechnological innovations has emerged as a potent tool for enhancing citrus resilience, productivity, and sustainability.

The primary objective of this book is to provide readers—ranging from researchers and practitioners to students and policymakers—with a comprehensive overview of the latest advancements in citrus biotechnology and their potential applications in agricultural development. Through a synthesis of cutting-edge research, practical insights, and case studies from around the globe, we aim to illuminate the diverse facets of citrus biotechnology and its transformative impact on citrus cultivation and value chain.

Within these pages, esteemed experts and scholars share their expertise on a wide range of topics, including citrus genomics, genetic engineering, tissue culture, disease management, post-harvest technologies, and market dynamics. By delving into these diverse areas of inquiry, we hope to inspire collaboration, innovation, and sustainable development within the citrus industry.

As we embark on this intellectual journey together, let us remain cognizant of the profound implications of citrus biotechnology for agricultural development, food security, and environmental sustainability. By harnessing the power of biotechnology and fostering interdisciplinary collaboration, we can unlock new pathways for citrus cultivation that are resilient, productive, and equitable.

We extend our sincere gratitude to the contributors, reviewers, and supporters who have contributed to the realization of this book. It is our fervent hope that "Citrus: Biotechnology And Development" serves as a beacon of knowledge and inspiration for all those engaged in the quest to cultivate a brighter future for citrus agriculture and beyond.

Nalini Sharma

Contents

SECTION 1
CITRUS GENEALOGY, CLASSIFICATION, AND BIOTECHNOLOGY

Chapter-1

Introductory Chapter: Citrus for a Healthy Life

1. An overview

Citrus is an extensively produced fruit crop and is cultivated predominantly in tropical and subtropical regions of the world. The genus Citrus and related genera (Fortunella, Poncirus, Eremocitrus, and Microcitrus) belong to the family Rutaceae. Of these genera, Citrus is the widely grown genus and is well known for fruits like oranges, mandarins, lemons, limes, and grapefruits [1]. The classification of citrus is complexed however, the genus Citrus consists of more than 100 species. The number of species is variable and this species variation in a single genus is due to the admixture of wide morphological diversity, intra- and interspecific sexual compatibility, apomixis, and spontaneous mutations. However, intergenic hybrids such as citranges, citrumelos, and citrandarins, citremons, citradias, and citrumquats are also reported and are getting increasing importance. Several indigenous varieties are developed and consumed locally in specific regions. The citrus fruits are tangy with pleasant flavor and taste, a combination of sweet and sour flavors. Oranges and mandarins are predominant species of genus Citrus, marketed as fresh or processed juice [2].

The citrus plants in the orchards are confronted worldwide with increasing biotic and abiotic factors due to the changing climate. Amongst abiotic factors, fluctuating temperature and unexpected frosts are the main limiting factors whereas, bacteria, viruses, viroids, nematodes, fungi, and phytoplasmas are major biotic factors. Some factors result in a massive reduction in production and quality while others may destroy altogether the citrus industry. Citrus improvement through conventional approaches is discouraged due to the genetic and reproductive characteristics of the plant. The omics and biotechnology-based interdisciplinary interventions may allow combating such external factors and improving the health, nutritional quality of the fruit. The book, Citrus: Research and Development describe the citrus plant, the biotic as well as abiotic challenges, nutrients, and nutritional value, and nutraceutical applications to improve human health. Citrus production, management, detection, and documentation of citrus pathogens and their management, fruit nutritional quality, and potential use as nutraceutical is an interdisciplinary endeavor; therefore, it is difficult to cover all aspects of this subject in a single book. The editor of the book is conscious of the fact that there is considerable scope for improving citrus production and controlling the diseases and benefitting from the availability of chemicals of nutritional and nutraceutical importance, novel approaches for detection of such chemicals, enriching the genetic information through next-generation sequencing and improving the genome by incorporating new genes through genetic engineering and knocking out genes using CRISPR/Cas technology, and hence the information relevant to the topics is covered in the book.

2. Citrus genealogy

Citrus domesticated in Southeast Asia started several thousand years ago and was distributed to different regions of the world through ancient land and sea routes. The genealogy of the modern cultivated citrus is controversial because these are either selections from or hybrids of wild progenitors (Figure 1). The biological features and cultivation of Citrus have further complicated the lineage of modern species. Citrus is being cultivated either through clonal grafting or by asexual means of propagation to maintain the identified superior traits however, diversity within such populations is due to somatic mutations. Further, spontaneous mutants have been reported and are occasionally selected from limb sports or nucellar seedlings. The genetic diversity in citrus is the major impediment in the classification of the ever-increasing number of varieties. Several methods have been used to classify the citrus varieties however the methods proposed by Swingle [3] and Tanaka [4] are commonly adopted. Swingle's classification is based on native varieties rather than cultivated, and he placed two subgenera Papeda with six species and Citrus with ten species in the genus Citrus [3], and the rest as natural hybrids of native species. Whereas Tanaka [4] classified both indigenous and cultivated varieties as species and placed two subgenera namely; Archicitrus with 111 species and Metacitrus with 48 species, the list of species was swelled to 159 [5]. The intervention of biotechnology approaches like RAPD, RFLP, AFLP, SSR, SRAP, and more recently the next-generation sequencing and pan-genomics has made it easier to determine the genealogy of citrus, extensively reviewed elsewhere [6]. The genomes of clementines, mandarins, pummelos, sweet oranges, and sour oranges have been sequenced using Sanger whole-genome sequencing method [7]. The sequence data together with earlier similar work was a step forward in elucidating the phylogenetic history of citrus domestication and highlights the genetic basis of the diversity in the colors, flavors, sizes, and aromas of citrus fruits, which could be introduced in novel varieties [8]. Another exciting and challenging work has been published, again by Wu et al. [9], where they worked out that distant genera i.e., Fortunella, Eremocitrus, and Microcitrus are a part of the citrus monophyletic group, whereas a related genus Poncirus, originally believed to be a part of Citrus group, is declared as a distinct clade based on whole-genome phylogeny. These findings have challenged the earlier taxonomic and phylogenetic developments and have warranted reformulation of the genus Citrus.

3. Citrus production and management

Worldwide, citrus is cultivated for consumption as fresh or processed fruit. Major citrus-producing countries are; China, Brazil, USA, Mexico, India, Spain, Iran, Italy, Nigeria, and Turkey. Production and consumption trends are diverse in different regions and countries however, variably 147 million tones citrus is produced, annually [10]. As citrus is grown from subtropical to tropical and the Mediterranean regions of the world, hence, its production is dependent on soil and climate conditions. The global orange production is projected to grow 3.6 million metric tons from the previous year due to favorable weather in Brazil and Mexico. Only a slight increase in production, as well as consumption, is expected for mandarins/tangerines. Unexpectedly, global grapefruit consumption and exports will rise to their highest levels in three years due to favorable weather conditions and expanded areas in China and Mexico. More than 80% of the fruit is being processed for juice production in developed countries, and the juice demand is increasing day by day. The market and consumption trends of major types of citrus fruit as well as of juice are not affected even by the COVID-19 pandemic, this is perhaps due to the perception of the consumers that citrus fruits are immunity boosters being rich in vitamin C.

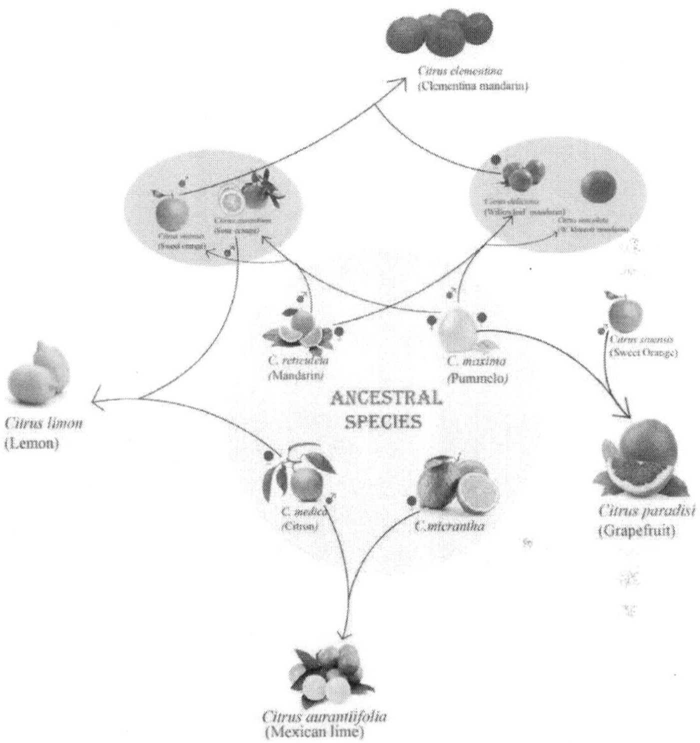

Figure 1. The genealogy and evolution of citrus fruits. Four ancestral species namely; Citrus reticulata, Citrus maxima, Citrus medica, and Citrus micrantha, depicted in the central circle, have contributed to the evolution of cultivated species. The figure is derived from Velasco and Licciardello, 2014 [8] and other published articles.

In addition to climate, plant–soil interaction affects citrus production by affecting the availability of nutrients to the plants [11]. For efficient nutrient availability, the practices of controlled release of fertilizers are preferred [12]. The published data confirm that the nitrogen uptake is greater by using controlled means, compared to conventional approaches, resultantly, the plant growth is improved [13]. Hence, technology-based precision management of orchards including the application of balanced fertilizers, herbicides, and pesticides is required for sustainable citrus production. However, there are several biotic and abiotic factors including diseases that affect the citrus industry.

Amongst diseases, bacterial, fungal, and viral are constant threats and cause substantial economic impact in all growing areas around the world. Citrus greening is an extremely dangerous disease, caused by different species of the bacterium Candidatus Liberibacter including Candidatus Liberibacter asiaticus, Candidatus Liberibacter americanus, and Candidatus Liberibacter africanus. Of these three species, the Candidatus Liberibacter asiaticus (CLas) has become a serious threat to the citrus industry. The disease was reported from Brazil in 2004, from Florida, the USA in 2005, in 2007 from Cuba, and molecularl detection of the disease has been reported in 2007 from Pakistan, in 2008 from the Dominican Republic; in 2010 from Mexico [14, 15, 16, 17, 18]. However, the disease has been reported and described in China since 2019. The disease is transmitted by the psyllid Diaphorina citri Kuwayama (Hemiptera: Psyllidae), commonly called Asian citrus psyllid. As CLas is difficult to culture on artificial media hence its detection is possible through polymerase chain reaction (PCR) and particularly by quantitative real-time polymerase chain reaction (qPCR) targeting the 16S rDNA gene [14]. Recently, chloroplasts provide opportunities for pathogens to directly or indirectly target 'chloroplast immunity' as these organelles are the main sites for the synthesis of precursors of phytohormones, hence are coordinating plant defense responses. As hormonal crosstalk between host and pathogens is well established. The identification of chloroplast genes targeted by the CLas effectors will open the window to control this disease. The pathogen genome has been sequenced from Diaphorina citri from America [19] and the genome is composed of 1231639 bp with GC 36.5% contents. The pathogen genome has been sequenced from two strains from Pakistan. The CLas genomes of two strains (PA19 and PA20) were sequenced that is comprised of 1224156 bp and 1226225 bp, respectively with an average GC content of 36.4% [20]. The genome sequence of CLas from Thailand has also been reported with total GC contents of 36.4% and 1230623 bp genome. Several genes from the Candidatus spp. have been identified that interact with the genes of the plant defense system and scientists are working to identify the plant genes regulated by these pathogen effectors. The chloroplast being the main site of phytohormone precursor synthesis provides opportunities for pathogens to target, directly or indirectly, the 'chloroplast immunity'. As hormonal crosstalk between host and pathogens is now well established hence identification and editing of chloroplast genes using CRISPR/Cas technologies hold promise to control the disease. Similarly, the canker susceptibility gene, CsLOB1, of Duncan grapefruit has been knockout. The infection by Xanthomonas citri was significantly reduced with no disease development on plants [21].

4. Citrus nutritional and nutraceutical importance

Citrus fruits, diverse in color and size, are highly nutritious. They are beneficial for health, due to the presence of bioactive compounds such as carotenoids, flavonoids, and ascorbic acid [22]. These compounds have antioxidant, antitumor, anti-inflammatory and blood clotting inhibiting characteristics [23, 24]. These fruits are also a rich source of vitamins and minerals like vitamin C, A, and B-complex [25, 26, 27], vitamin A benefits skin and vision whereas vitamin B-complex like thiamin, folates, and pyridoxine are required as external sources to replenish. Of minerals, potassium, magnesium, calcium, and sodium are present in citrus from very high to low levels in citrus fruits [28]. However, zinc, iron, and manganese are present in trace amounts (Figure 2).

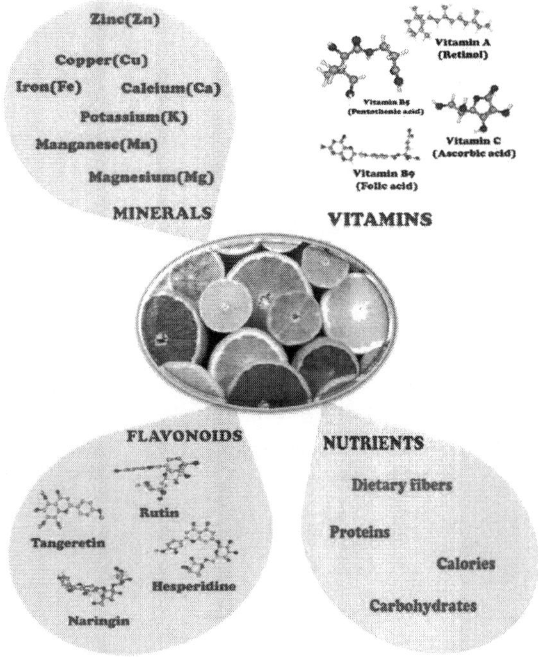

Figure 2. The citrus, a source of nutrients and nutraceuticals. Citrus species are a rich source of carotenoids, flavonoids, minerals, and vitamins that function as antioxidants, antitumor, and anti-inflammatory compounds.

Nutraceutical is a combination of two words; 'nutrition' and 'pharmaceutical', hence the word infers that the nutraceuticals could be regulated as dietary supplements, medicine, and food ingredients. The word, 'nutraceutical' was first coined by Stephen L. DeFelice in 1989 who explained nutraceuticals as; "Food, or parts of food, that provide medical or health benefits, including the prevention and treatment of disease" [29, 30]. However, the government of Japan started approving foods with proven benefits for the general public in the 1980s, reviewed elsewhere [31]. Nutraceuticals protect against diseases developed due to nutrient deficiencies and also have physiological benefits. These are used as dietary supplements and food ingredients. Being dietary supplements, these may contain vitamins, minerals, botanical extracts, essential amino acids, Poly Unsaturated Fatty Acids (PUFA), and enzymes that are probably deficient in most of our diets. Another category of nutraceuticals is nutrient-fortified food. For example, the iron-fortified wheat flour protects the wheat-dependent population from diseases that develop due to iron deficiency. A salient example is iron deficiency anemias. Other examples include purple cauliflower and purple potatoes, having additional anthocyanin content. Golden rice and Golden potatoes as well as Provitamin-A-fortified maize are the rich sources of Provitamin-A and carotenoids. These crops could reduce the disease development in humans caused by Vitamin A deficiency diseases e.g. night blindness, xerophthalmia, prevalent in Africa and Asia. Similarly, Quinoa is of great nutritional value. Laden with fiber, vitamins, and minerals, this plant also is rich in lysine, hence its proteins are nutritionally more complete than many vegetables [32]. Thus, quinoa holds great potential as a Nutraceutical, to curb the malnutrition rampant in many third-world countries, the likes of Pakistan, India, Nepal, Bangladesh, and several African nations.

Several plants, including food crops, have been reported that contain compounds to prevent diseases. Even in this era of rapid medicine evolution, cancer remains a major threat to the population and a leading cause of mortality in developed nations. Introduction of plants into lifestyle at an early age could reduce cancer risk up to 33%. For instance, blue maize is useful in preventing different types of cancers, such as colon cancer [33]. Several chemotherapeutic agents such as Taxol, Vincristine, and Vinblastine are derived from plants such as *Taxus brevifolia* and alkaloids of Vinca species. Nutraceuticals have also been shown to reduce the toxic effects of chemotherapeutic agents and radiation therapies [34].

Bacterial infections and the growing resistance to synthetic antibiotics is a serious concern. It has been proven experimentally that medicinal plants are effective against bacterial infections. In this era of technology development, the introduction of medicinal and nutritional traits transgenically into food crops is on the top priority of biotechnologists, hence engineering the citrus genome will be a better choice.

5. Citrus improvement through biotech approaches

Citrus fruits are highly nutritious and are beneficial for health, due to the presence of bioactive compounds such as carotenoids, flavonoids, and ascorbic acid. However, citrus production and quality are challenged with several biotic and abiotic problems. Biotech interdisciplinary interventions including transgenesis, genome editing, and OMICS could offer solutions to the issues of this fruit crop. Genetic transformation has been established in many citrus species thereby transgenic plants have been developed against bacterial, viral, and fungal pathogens. Equally, OMICS approaches; genomics, transcriptomics, proteomics, metabolomics, interactomics, and phenomics are exploited to improve the citrus fruits. Since, first attempt to manipulate the citrus genome remained unsuccessful hence, the protocols for efficient regeneration from explants like seeds, embryogenic cells, epicotyls, callus, nodal stem segments, and protoplasts, followed by transformation and selection have been optimized for different citrus species. Maximum regeneration potential has been observed in explant 'epicotyl' hence, the epicotyl has been used as an explant for the genetic transformation of citrus plants. Transgenically stable plants were recovered from Agrobacterium treated Duncan grapefruit epicotyls. The recovered plants were confirmed for transgene presence using PCR and Southern blotting techniques. Similarly, transgenic plants using epicotyl tissues as explants were developed from sweet orange, and citrange. The transformation efficiency remained as high as 93%. Hence, transgenic technology is proven as one of the most reliable interventions to genetically improve tolerance/resistance to abiotic/biotic factors in citrus [35, 36].

Using the technology, the nutrition and medicine-related traits have been successfully tailored in citrus fruits. For example, the expression of genes that encode enzymes like phytoene synthase, lycopene-β-cyclase, and phytoene desaturase of the carotenoid biosynthesis pathway have been modulated to supplement human nutrition with vitamin A and antioxidants. The Valencia orange is majorly grown for its juice but the quality of the juice is deteriorated due to the degradation of an enzyme, named thermostable pectin methylesterase (TSPME). Hence, the gene (CsPME4) that encodes TSPME was downregulated to improve the juice quality [37]. Further, an environmentally friendly technology named, 'chloroplast transformation' is available to develop transgenic plants [38, 39, 40, 41, 42, 43, 44, 45, 46]. This technology offers several superior advantages like overexpression of transgenes up to 70% due to polyploidy at organelle and genome (plastome) levels, accumulation of functional proteins, and natural containment of transgenes since plastids are transmitted to the next generation through ovary, rather than pollens that cause horizontal gene transfer, in most of the cultivated plant species. The chloroplast genome of several citrus species has been sequenced [47, 48, 49, 50, 51], thus necessary genetic information of the subcellular organelle is available to develop chloroplast transformation vectors and achieve chloroplast transformation, successfully. Therefore, the development of transgenic plants through chloroplast genome engineering is a promising way forward for cost-effective production of nutraceuticals.

6. Conclusions

Conventional research has played a pivotal role in the improvement of citrus. Enhanced heterozygosity has helped in the development of genetically diverse germplasm in most of the citrus species and numerous varieties have been released for commercial cultivation. However, with the advent of modern biotechnological tools, the period involved in crop improvement through indirect mutagenesis and polyploidization could be further reduced and enhance cost-effectiveness. Transgenic technology and OMICS have great potential to improve this fruit crop.

References

1. Scora RW. On the history and origin of Citrus. Bull. Torrey Bot. Club. 1975; 102(6): 369-375

2. Raveh E, Goldenberg L, Porat R, Carmi N, Gentile A, La Malfa S. Conventional breeding of cultivated Citrus varieties. In: Gentile A., La Malfa S., Deng Z. (eds.) The Citrus Genome. Compendium of Plant Genomes. Springer, Cham. 2020; p 33-48 DOI:10.1007/978-3-030-15308-3_4

3. Swingle WT. The botany of Citrus and its wild relatives. In: Reuther W., Webber H. J., Batchelor L. D. (eds.). The Citrus Industry. California: University of California; 1967; p. 190-430

4. Tanaka T. Species problem in citrus: a critical study of wild and cultivated units of citrus, based upon field studies in their native homes (Revisio Aurantiacearum IX). Tokyo: Japanese Society for the Promotion of Science. 1954; p 152

5. Tanaka T. Misunderstanding with regards citrus classification and nomenclature. Bull Univ Osaka Prefect Ser. B, Agriculture and biology. 1969; 21:139-145

6. Shimizu T, Kitajima A, Nonaka K, Yoshioka T, Ohta S, Goto S, et al. Hybrid Origins of Citrus Varieties Inferred from DNA Marker Analysis of Nuclear and Organelle Genomes. PLoS ONE. 20 16;11: e0166969. DOI:10.1371/journal.pone.0166969

7. Wu GA, Prochnik S, Jenkins J, Salse J, Hellsten U, Murat F, et al. Sequencing of diverse mandarin, pummelo and orange genomes reveals complex history of admixture during citrus domestication. Nat Biotechnol. 2014;32: 656-662

8. Velasco R, Licciardello C. A genealogy of the citrus family. Nat Biotechnol. 2014;32: 640-642

9. Wu G, Terol J, Ibanez V. et al. Genomics of the origin and evolution of Citrus. Nature. 2018;554: 311-316

10. FAOSTAT. Food and Agriculture Data. Available online: http://www.fao.org/faostat/en/#data/QC

11. Mattos-Jr D, Kadyampakeni DM, da Silva JR, Vashisth T, Boaretto M. Reciprocal effects of huanglongbing infection and nutritional status of citrus trees: A review. Trop. plant Patho. 2020;45: 586-596

12. Vashisth T, Kadyampakeni D. Diagnosis and Management of Nutrient Constraints. 2020; p 723-737

13. Dou H, Alva A. Nitrogen uptake and growth of two citrus rootstock seedlings in a sandy soil receiving different controlled-release fertilizer sources. Biol Fertil Soils. 1998;26: 169-172

14. Teixeira DC, Dane JL, Eveillard S, Martins EC, Jesus Junior WC, Yamamoto PT, et al. Citrus huanglongbing in São Paulo state, Brazil: PCR detection of the "Candidatus" Liberibacter species associated with the disease. Mol. Cell. Probes. 2005;19:173-179

15. Halbert S. The discovery of huanglongbing in Florida. In: Gottwald, T.R., Graham, J., Dixon, W.N., LaVigne, A. and Berger P. (eds.) Proceedings of the 2nd International Citrus Canker and Huanglongbing Research Workshop. Orlando, FL. 2005; p 50

16. Gottwald TR, Parnell S, Taylor E, Poole K, Hodge J, Ford A, Therrien L, Mayo S, Irey M. Within-tree distribution of Candidatus Liberibacter asiaticus. In: Gottwald TR, Graham JH (eds.) Proceedings of the 1st International Research Conference on Huanglongbing. Florida Citrus Mutual, Orlando, FL, USA. 2008; p 310-313

17. Matos L, Hilf ME, Camejo J. First report of 'Candidatus Liberibacter asiaticus' associated with citrus huanglongbing in the Dominican Republic. Plant Dis. 2009;93:668 DOI:10.1094/PDIS-93-6-0668

18. Yaqub MS, Khan IA, Usman M, Rana IA. Molecular detection of Candidatus liberibacter asiaticus, the causal organism of huanglongbing (Citrus greening) in Faisalabad, Pakistan for Huanglongbing management. Pak. J. Agri. Sci. 2017;54:21-26

19. Wang Y, Kondo T, He T, Zhou Z, Lu J. Genome sequence resource of Candidatus Liberibacter asiaticus' from Diaphorina citri Kuwayama (Hemiptera: Liviidae) in Colombia. Plant Dis. 2021;105: 193-195

20. Liu Z, Ge XX, Wu XM. et al. Genome-wide analysis of the citrus B3 superfamily and their association with somatic embryogenesis. BMC Genomics. 2020;21:305 DOI:10.1186/s12864-020-6715-9

21. Jia H, Zhang Y, Orbović V, Xu J, White FF, Jones JB, Wang N. (2017). Genome editing of the disease susceptibility gene CsLOB1 in citrus confers resistance to citrus canker. Plant Biotechnol J. 15:817-823. DOI:10.1111/pbi.12677

22. Ma G, Zhang L, Sugiura M, Kato M. Citrus and health. In: Talon M. et al. (eds.) The Genus Citrus. 2020; p 495-511 DOI:10.1016/B978-0-12-812163-4.00024-3

23. Cha JY, Cho YS, Kim I, Anno T, Rahman SM, Yanagita T. Effect of hesperetin, a citrus flavonoid, on the liver triacylglycerol content a phosphatidate phosphohydrolase activity in orotic acid-fed rats. Plant Food Hum. Nutr. 2001;56(4):349-358

24. Rehman SU, Abbasi KS, Haq NU, Ahmad Q, Taran NU, et al. Comparative nutraceutical properties of seeds of eight citrus varieties grown in Rawalpindi region. Pure and Applied Biology. 2020;9:1-9 DOI:10.19045/bspab.2020.90001

25. Etebu E, Nwauzoma AB. A review on sweet orange (Citrus sinensis L Osbeck): Health, diseases and management. American Journal of Research Communication. 2014; 2: 33-70

26. Tsuda H, Ohshima Y, Nomoto H, Fujita KI, Matsuda E, Iigo M, Moore MA. Cancer prevention by natural compounds. Drug Metab Pharmacokinet. 2004;19: 245-263

27. Lee SE, Hwang HJ, Ha JS, Jeong HS, Kim J. Screening of medicinal plant extracts for antioxidant activity. Life sci. 2003;73:167-179

28. De Moraes Barros HR, de Castro Ferreira TAP, Genovese MI. Antioxidant capacity and mineral content of pulp and peel from commercial cultivars of citrus from Brazil. Food Chem. 2012;134:1892-1898

29. Kalra EK. Nutraceutical-definition and introduction. AAPS PharmSci. 2003;5:2-3

30. De Felice L Stephen. The nutraceutical revolution, its impact on food industry. Trends in Food Sci. and Tech. 1995;6:59-61

31. Pandey M, Verma RK, Saraf A. Nutraceuticals: New era of medicine and health. Asian Journal of Pharmaceutical and Clinical Research. 2010;3:11-15

32. Maradini-Filho AM. Quinoa: Nutritional aspects J Nutraceuticals Food Sci. 2015;2:1

33. Guzmán-Gerónimo RI. Blue maize as a potential Nutraceutical in the prevention of Colon Cancer and metabolic syndrome. J Nutraceuticals Food Sci. 2017;2:2

34. Martinotti S, Ranzato E. 3 – Natural compounds and anticancer effects: The whole is greater than the sum of its parts. In: Kabir Y. (ed.) Functional Foods in Cancer Prevention and Therapy. 2020;p 47-58

35. Cervera M, Ortega C, Navarro A, Navarro L, Pena L. Generation of transgenic citrus plants with the tolerance-to-salinity gene HAL2 from yeast. J. Hort. Sci. and Biotechnol. 2000;75:26-30

36. Caserta R, Teixeira-Silva NS, Granato LM, Dorta SO, Rodrigues CM, Mitre LK, Yochikawa JT, Fischer ER, Nascimento CA, Souza-Neto RR, Takita MA. Citrus biotechnology: What has been done to improve disease resistance in such an important crop?. Biotechnol. Res. Innov. 2020;3:95-109

37. Guo WW, Duan YX, Olivares-Fuster O, Wu ZC, Arias CR, Burns JK and Grosser JW. Protoplast transformation and regeneration of transgenic Valencia sweet orange plants containing a juice quality related pectin methylesterase gene. Plant Cell Rep. 2005;24:482-486

38. Svab Z, Maliga P. High-frequency plastid transformation in tobacco by selection for a chimeric aadA gene. Proc Natl Acad Sci USA. 1993;90:913-917

39. Khan MS, Maliga P. Fluorescent antibiotic resistance marker for tracking plastid transformation in higher plants. Nat Biotechnol. 1999;17:910-915

40. Daniell H, Khan MS, Allison L. Milestones in chloroplast genetic engineering: An environmentally friendly era in biotechnology. Trends Plant Sci. 2002;7:84-91

41. Bock R, Khan MS. Taming plastids for a green future. Trends Biotechnol. 2004;22:311-318

42. Khan MS, Joyia FA, Mustafa G. Seeds as economical production platform for recombinant proteins. Protein and Peptide Letters. 2019;27:1-16

43. Khan MS. Utilizing heterologous promoters to express green fluorescent protein from jellyfish in tobacco chloroplasts. Pak J Bot. 2001;33:43-52

44. Khan MS. Engineering photorespiration in chloroplasts: A novel strategy for increasing biomass production. Trends in Biotechnology. 2007;25:437-440

45. Khan MS. Engineered male sterility. Nature. 2005; 436:783-785

46. Mustafa G, Khan MS. Transmission of engineered plastids in sugarcane, a C4 monocotyledonous plant, Reveals that Sorting of Preprogrammed Progenitor Cells Produce Heteroplasmy. Plants. 2021;10: 26 DOI:10.3390/plants10010026

47. Kiwoung Yang, Jonghoon Lee, Sang-Choon Lee, Tae-Jin Yang, Hoyteak Kim, Mi-Young Chung, Ill-Sup Nou. The complete chloroplast genome of Korean popular Citrus hybrid Hallabong mandarin [(Citrus unshiu×C. sinensis)×C. reticulate] (Rutaceae), Mitochondrial DNA Part B. 2016;1(1):29-30 DOI:10.1080/23802359.2015.1137805

48. Su H-J, Hogenhout SA, Al-Sadi AM, Kuo C-H. Complete chloroplast genome sequence of Omani lime (Citrus aurantiifolia) and comparative analysis within the Rosids. PLoS ONE. 2014;9(11):e113049 DOI:10.1371/journal.pone.0113049

49. Shi-Rong Xu, Chang-Yue Huang, Ya-ting Deng, Lin-Li Zhou, dong-Ming Pan, He-Li Pan. The complete chloroplast genome sequence of Citrusmaxima (Burm.) Merr. 'Guanximiyou', Mitochondrial DNA Part B 2020;5(1):482-483 DOI:10.1080/23802359.2019.1704661

50. Bausher MG, Singh ND, Lee SB, Jansen RK, Daniell H. The complete chloroplast genome sequence of Citrus sinensis (L.) Osbeck var 'Ridge Pineapple': Organization and phylogenetic relationships to other angiosperms. BMC Plant Biol. 2006;30(6):21 DOI:10.1186/1471-2229-6-21

51. Lee M, Park J, Lee H. et al. Complete chloroplast genomic sequence of Citrus platymamma determined by combined analysis of sanger and NGS data. Hortic. Environ. Biotechnol. 2015;56:704-711 DOI:10.1007/s13580-015-0061-x

Chapter-2

Horticultural Classification of Citrus Cultivars

1. Introduction

Generally, there is a strong demand for citrus varieties of superior eating and processing quality. A shortage of supply of consumer-preferred varieties and high prices are the dominant market forces responsible for the revitalisation of the fresh citrus sector. The general demand is for sweet, low acid fruit, with an aromatic flavor. The shortage of supply has meant the acceptance of a range of varieties, some with marginal quality, which it is expected will have a limited commercial potential. Growing citrus in your own backyard and field can be both enjoyable and rewarding! Beautiful green foliage, fragrant blossoms, and delicious, healthful fresh fruit readily available at your doorstep make citrus excellent garden trees. You can choose a citrus variety according to the climate in your area. While this chapter provides ripening season information for worldwide, farmers/gardeners have had success with citrus in many different regions of world where tropical/subtropical climatic conditions occur. In general appearance and other respects, the citrus fruits of principal commercial importance fall into four, reasonably-well-defined horticultural groups: the Sweet oranges, the mandarins, the grapefruits and the pummelos and the common acid members. The true *or 'biological'* citrus fruit trees, including species of *Citrus (C. reticulata, C. maxima and C. medica)*, share certain characteristics, however, these are clearly differentiated according to the morpho- taxonomic key of Swingle [1, 2]. Wu et al. [2] found that several named genera (*Fortunella, Eremocitrus* and *Microcitrus*) are in fact nested within the citrus clade. These and other distinct clades that they have identified are therefore more appropriately considered species within the genus *Citrus*. The pulp vesicles contain droplets of oil, which are more abundant in Poncirus, Microcitrus, and the papedas. Fruits of the true citrus species are segmented and fruits of the genera other than *Citrus* are smaller than those of Citrus itself.

Fortunella and Eremocitrus have ovaries with three to five locules, each of which has only two ovules, whereas Citrus, Microcitrus, and Poncirus have ovaries with six to eight locules, each of which contains many ovules. Members of the true citrus fruit trees are generally cross and graft compatible with other members of the group [3, 4, 5]. Fortunella (Kumquat) trees, leaves, flowers, and fruits are generally smaller than those of Citrus. Kumquats are adapted to climates that are marginally cool for most of the other members of the subfamily Aurantioideae, they require less heat to achieve fruit maturity and have a certain level of winter dormancy [1]. Eremocitrus and Microcitrus are both endemic to the Oceania region. Both differ from Citrus in having dimorphic foliage and free stamens; however, Microcitrus has an ovary with four to eight locules, whereas Eremocitrus has an ovary with three to five locules. The cold hardiness of Eremocitrus stated in Swingle [1] and Swingle and Reece [6] is in error; Eremocitrus can probably tolerate temperatures as low as −5.5 °C, consistent with the original description of the genus in 1914 [3, 7, 8]. Microcitrus, on the other hand, is considered semixerophytic and able to withstand prolonged periods of drought [1, 6]. Trifoliate orange was considered as a mono-typic genus for many years, represented by *Poncirus trifoliata* [1], with distinctive trifoliate leaves (unique among the true citrus fruit trees) and deciduous growth habit. This gives to trifoliate oranges the highest degree of cold hardiness among the true citrus fruit trees, surpassing that of kumquats. The adaptation of Poncirus to cold conditions led Swingle [1] to speculate that the remote ancestor of the true citrus fruit trees originated in a tropical or semitropical climate. While the other genera of the true citrus fruit trees remained in these climates, *Poncirus* (or its ancestors) "migrated" to the temperate climate of Northeastern Asia, during which time it developed the adaptations to colder winters mentioned previously. In addition to cold tolerance, Poncirus exhibits many other characteristics that have been and continues to be used in citrus rootstock breeding, notably disease tolerance (including citrus tristeza virus immunity) and dwarfing. For a more complete utilization of Poncirus, the reader is referred to Krueger and Navarro [3, 9, 10]. Relatively recently, a new species, *Poncirus polyandra*, was published [11, 12], which differs from *P. trifoliata* by its larger leaves, some floral differences, and most notably, being evergreen. Perhaps, this latter characteristic is related to its habitat in Yunnan, the southernmost province of China. Clymenia is a very distinctive member of the other true citrus fruit trees. Clymenia was separated from Citrus by Swingle [1] based upon the structure of the pulp vesicles, which are short, plump, blunt, ovoid or sub-globose, sessile or very short stalked, and attached to the side walls of the 14–16 locules. Table 1 summarizes the correspondence between the proposed classification and the former most important ones of Tanaka [13], Swingle and Reece [6], and Mabberley [22] revised by Zhang and Mabberley [14]. Commercially grown citrus trees such as the varieties discussed in this publication are not grown from seed but are grafted or budded onto a seedling of a rootstock variety. Varieties that are used as rootstocks provide a number of important qualities to the entire tree such as disease tolerance, cold hardiness, soil adaptation, and, to a certain degree, tree size. In the world, most citrus nurseries do not label or identify the rootstock of a tree, but they do select rootstocks that protect trees from important diseases of commercial citrus and are adapted to a range of regions and soil conditions.

Table 1. Correspondences between the new phylogenomic classification and the former classifications of Tanaka, Swingle and Reece, and Mabberley revised by Zhang and Mabberley.Reference: [20, 21].

Phylogenomic classification	Tanaka (1961) [13]	Swingle and Reece (1967) [6]	Zhang and Mabberley (2008) [14]	Common names (examples)	Phylo-genomic references
C.cavaleriei H. Lev. ex Cavalerie	*C. ichangensis* Swingle	*C. ichangensis*	*C. cavaleriei*	Adsae	[2]
C. maxima (Burm.) Merr.	*C. maxima*	*C. maxima*	*C. maxima*	Pummelos (Pink, Deep Red, Timor, …)	[2, 15]
C. medica L.	*C. limonimedica* Lush.	*C. medica*	*C. medica*	Etrog citron	[2, 15, 16]
	C. medica			Citrons (Corsican, Diamante, Buddha's hand, Humpang)	[2, 15, 16]
C. micrantha Wester	*C. micrantha*	*C. micrantha*	*C. hystrix* DC.	Small-flowered papeda, small-fruited papeda	[2, 15, 16]
C. reticulata var. austeraSwingle		*C. reticulata* var. austere	*C. reticulata* Blanco	Sun-Chu-Sha-Kat mandarin	[2]
C. reticulata var. tachibanained.	*C. tachibana* (Makino) Tanaka	*C. tachibana*	*C. reticulata*	Tachibana mandarin	[2]
C. × *amblycarpa*	*C. amblycarpa*	*C. reticulata* hybrid		Nasnaran mandarin	[15]
C. × *aurantiifolia* var. aurantiifolia	*C. aurantiifolia*	*C. aurantiifolia*	*C.* × *aurantiifolia*	Mexican, Key, West Indies limes…	[2, 16, 17]
C. × *aurantiifolia* var. macrophyllained.	*C. macrophylla* Wester	C . *aurantiifolia* (Christm.) Swingle		Alemow	[16]
C. × *aurantiifolia* var. auratained.	*C. aurata*Risso	*C. limon* (L.) Burm. f.	*C.* × *aurantium* L.	Adam's apple	[16]
	C. excelsa Wester	*C. aurantiifolia*		Excelsa and Nestour lime	[16]
C. × *aurantium* var. L. var. aurantium	*C. aurantium*	*C. aurantium*	*C.* × *aurantium*	Sour orange, Bouquetier	[2, 15, 16, 18, 19]
	*C. myrtifolia*Raf.			Myrtle-leaf orange, Chinoto	[15]

Phylogenomic classification	Tanaka (1961) [13]	Swingle and Reece (1967) [6]	Zhang and Mabberley (2008) [14]	Common names (examples)	Phylo-genomic references
C. × *aurantium* var. clementinained.	*C. clementina* hort. ex Tanaka	*C. reticulata*		Clementine	[2, 15, 18, 19]
C. × *aurantium* var. deliciosained.	*C. deliciosa* Ten.	*C. reticulata*	*C. reticulata*	Willowleaf, Chios mandarins	[2, 15, 18, 19]
C. × *aurantium* var. erythrosained.	*C. erythrosa* hort. ex Tanaka	*C. tachibana*	*C. reticulata*	Fuzhu and San huhongchu mandarins	[19]
C. × *aurantium* var. kinokuniined.	*C. kinokuni* hort. ex Tanaka	*C. tachibana*		Kinokuni, Kishu, Huanglingmiao mandarins	[2, 19]
C. × *aurantium* var. nobilisined.	*C. nobilis* Lour.	*C. reticulata* hybrid	*C.* × *aurantium*	King mandarin	[2, 15, 19]
C. × *aurantium* var. paradisiined.	*C. paradisi*Macfad.	*C. paradisi*	*C.* × *aurantium*	Star Ruby, Marsh, Duncan, etc.	[2, 15, 17, 19]
C. × *aurantium* var. paratangerinained.	*C. paratangerina* hort. ex Tanaka	*C. reticulata*		Ladu Mandarin	[19]
C. × *aurantium* var. sinensis L.	*C. sinensis* (L.) Osbeck	*C. sinensis*	*C.* × *aurantium*	Sweet oranges (Valencia, Washington Navel, Tarroco, etc.)	[2, 15, 18, 19]
C. × *aurantium* var. suhuiensisined.	*C. suhuiensis* hort. ex Tanaka	*C. reticulata*	*C. reticulata*	Szibat and Se Hui Gan mandarins	[19]
C. × *aurantium* var. tangerinained.	*C. tangerina* hort. ex Tanaka	*C. reticulata*	*C. reticulata*	Dancy, Beauty mandarins	[2, 15, 19]
C. × *aurantium* var. temple ined.	*C. temple* hort. ex Yu. Tanaka	*C. sinensis*		Temple tangor	[19]
C. × *aurantium* var. unshiuined.	*C. unshiu* Marcow.	*C. reticulata clone*	*C. reticulata*	Satsuma mandarins	[2, 15, 17, 19]
C. × *latifolia* var. nov. 1				India lime	[16]
C. × *latifolia* var. nov. 2				Kirk lime	[16]

Phylogenomic classification	Tanaka (1961) [13]	Swingle and Reece (1967) [6]	Zhang and Mabberley (2008) [14]	Common names (examples)	Phylo-genomic references
C. × *latifolia* var. latifolia	*C. latifolia*	*C. aurantiifolia*	*C.* × *latifolia*	Bears, Tahiti, Persian limes	[16]
C. × limon var. bergamiained.	*C. bergamia*Risso and Poit.	*C. aurantiifolia*	*C.* × *limon*	Fantastico, Femminello, Castagnaro bergamots	[16, 17]
C. × *limon* var. meyeriiined.	*C. meyerii* Yu. Tanaka	*C. limon*	*C.* × *limon*	Meyer lemon	[16]
C. × *limon* var. limettioidesined.	*C. limettioides* Tanaka	*C. aurantiifolia*		Palestinian and Brazil sweet limes and Butnal sweet lemon	[16, 17]
C. × *limon* var. limettained.	*C. limetta* Risso	*C. limon*		Marrakech limonette	[16]
C. × *limon* var. limon (L.) Burm. f.	*C. limon* (L.) Burm. f.	*C. limon*	*C.* × *limon*	Lemons (Lisbon, Eureka, Verna, Luminciana, Interdonato, etc.)	[2, 16]
C. × *limonia* var. nov. 1				India sweet lime, Indian lemon	[16]
C. × *limonia* var. jambhiriined.	*C. jambhiri* Lush.	*C. limon*	*C.* × *taitensis*Risso	Rough lemon	[2, 16, 17]
C. × *limonia*Osbeck var. limonia	*C. limonia*	*C. limon*		Rangpur lime	[2, 16]
	C. karna Raf.			KhattaKharna lime	[16]
C. × *limonia* var. nov. 2				Voangiala	[16]
C. *x* *limonia* var. volkameriana Pasquale	*C. limonia* Osbeck	*C. limon*		Volkamer lemon	[16]
C. × *lumia* var. nov. 1				Bitrouni lime	[16]
C. × *lumia* var. nov. 2				Fourny hybrid	[16]
C. × *lumia* var. lumia	*C. lumia* Risso and Poit.	*C. limon*		Jaffa lemon	[16]
C. × *lumia* var. pyriformis ined.	*C. pyriformis* Hassk.	*C. limon*	*C. maxima*	Ponderosa lemon	[16]
C. × *microcarpa*	*C. madurensis* Lour.	*C. reticulata* hybrid	*C.* × *microcarpa*	Calamondin, Calamansi	[16]
C. × *pseudolumia*ined.				Borneo, Barum, Baboon lemons	[16]

2. Varietal groups

Four main varietal groups are distinguished in the international market:

2.1 *Citrus sinensis* (L.) Osbeck

Sweet orange is the main group which is used both for fresh fruit and processing. It probably originated in China but its major center of diversification is the Mediterranean Basin (Figure 1). Major cultivars in this group are classified as navel oranges (Washington Navel, Navelina, Navelate, Powell, Rhode Navel, Cara Cara), blonde oranges (Shamouti, Valencia Late, Hamlin, Pineapple, Trovita, Salustiana, Delta Valencia, Pera), and blood oranges (Tarocco, Moro, Sanguinelli, Maltese).

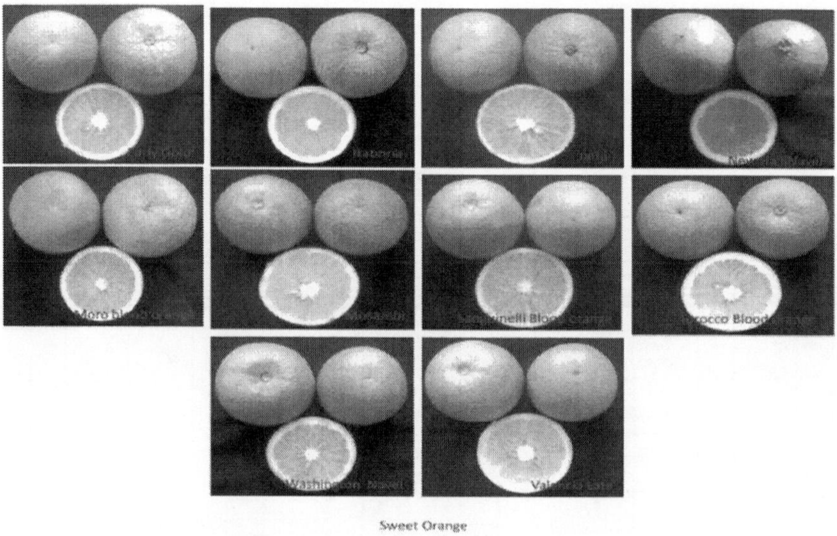

Figure 1. Sweet orange varieties.

2.1.1 Navel oranges

- **Washington (Riverside, Bahia, Baia or Baiana):** Is considered to be a limb sport of a variety 'Selecta' in Bahia, Brazil. Tree medium in size and vigor, crown round topped, anthers are without pollen. Rind medium thick tender flesh deep orange, firm less juicy, rich in flavor and taste. Processing quality poor, ships and stores well. Seedless.
- **Navelencia:** In growth characters is less vigorous than Washington. Flesh light in color (as rind) firm more juicy, flavor good more than Thompson. Fruit matures earlier than Washington and after Thompson, and hangs well on tree, almost seedless.
- **Thompson:** Limb sport of Washington. Tree compact semi dwarf type, less in vigor than Washington. Rind and flesh less colored, rind smooth glossy pitted finely, flesh firm more juicy, taste and flavor good. Ripens early, seedless.

2.1.2 Blonde oranges

- **Shamouti (Palestine Jaffa, Jaffa, Chamouti):** Tree upright moderate in growth and vigor, branches thick and thornless, petioles with narrow wing. Rind thick leathery smooth pitted, oil glands faint, flesh light orange, firm juicy fragrant and sweet in taste, peels easily. Shipping and storage very good. Mid-season cultivar. Seedless or nearly.

- **Valencia (Valencia Late, Hart's Tardiff, Hart Late):** Has wide adaptation with alternate bearing, tendency to be heavy cropper. Tree large upright vigorous. Matures late. Storage and transportation qualities are very good and hangs well on trees. Seeds few to none.
- **Hamlin (Norris):** Originated as chance seedling and named after the owner of the orchard AG Hamlin. Tolerate cold better than most oranges, productive. Tree medium large with moderate vigor. Matures very early in the season, seeds few to none.
- **Pineapple:** Chance seedling very sensitive to frost, very productive and excellent for processing. Tree medium large with moderate vigor. A mid-season ripening cultivar. Has two limb sport which are seedless. Seedless Pineapple and Blaquemines.

2.1.3 Pigmented or blood oranges

This group of oranges differ from the common sweet orange in that the fruit generally has pink or red coloration on the rind, in the flesh and juice and also has distinct flavor.

- **Ruby (Ruby Blood):** This cultivar was introduced from Mediterranean region. Plants are compact, not very large, moderate in vigor and production. Ripens in mid-season. Rich in flavor and few seeds.
- **Spanish Sanguinelli (Syn. Sanguinelli, SanguinellaNegra):** Originated as a limb sport of Doblefina. Tree foliage light green, spineless, productive small medium in size. Fruit large very attractive with persistent style, late to mid-season in maturity, shipping and storage qualities very good.
- **Torocco (Tarocco dal Muso, Taracco di Francofonte):** An Italian cultivar. Tree medium in size, irregular in bearing but moderate production. Fruits are quite large, variable in shape. Shipping and storage quality good and has mid-season maturity, does not retain quality if left for long on trees. Seeds few to none.
- **Doblefina (Oval Sangre, Blood Oval):** Tree small poor in growth, branches spreading, crown open, foliage light green. Precocious and heavy bearer. Shipping, storage quality good but does not hang well on tree, almost seedless.
- **Blood Red (Blood Red Malta):** Origin unknown probably came from Mediterranean basin. Grown widely and commercially in India and Pakistan. Good coloration is generally attained in the submountain region.
- **Maltaise Sanguine (Portugaise):** Origin of this variety is uncertain but the Egyptian variety Baladi Blood introduced from Malta is said to be Maltaise Sanguine. Likewise Maltase Blood and Bloodred Malta seems to be a clone of this cultivar.Trees are very productive, medium large moderate in vigor. Rind soft easy to peel, taste and flavor excellent, seeds few, storage and shipping quality poor.

2.1.4 Sugar or Acidless oranges

- **Maltaise Meski:** Originated in Tunisia, a non-acid orange cultivar. For all plant, foliage, flower and fruit characters it is like the parent cultivar Maltaise Blonde.
- **Shamouti Maski or Shamouti Moghrabi:** It is a Labenese cultivar which is indistinguishable from Shamouti (Palestine Jaffa) baring that it is acid less (insipid in taste) and more seedy in nature.

2.1.5 Other sweet oranges

- **Jaffa (Florida Jaffa):** It is a clone of Palestine beledi seedling group. Flesh pale orange, tender juicy with pleasant flavor, seeds few. Has good shipping quality but does not hang well on tree.
- **Joppa:** Resembles Jaffa for number of characters and is different in that it starts bearing early and prolifically, branches are stiff and thornless, branchlets are stout. Flesh light orange, soft juicy with rich aroma. Mid-season cultivar, seeds few.
- **Foster:** Tree medium in growth semi spreading foliage dense. Seeds large oval, maturity mid to late season.
- **Marrs (Marrs Early):** Precocious and heavy bearer matures very early. Flesh orange well colored juicy, sweet in taste, acid very low. Fruits hang well on tree and maintain the quality. For good quality and high juice picking should be delayed. Seeds moderate in number.
- **Parson (Parson Brown):** Originated as a chance seedling. It is an early maturing and relatively more seedy. Trees are large vigorous and productive. Rind pitted and pebbled moderately, flesh orange, firm juicy highly flavored.

2.1.6 Indian continents

In India, Mosambi and Sathgudi are invariably placed under sweet oranges as the acid content is very low.

- **Mosambi (Mosambique):** A very popular variety grown commercially in India, early in maturity. Rind quite thick, stripes faint with longitudinal ridges and grooves. Flesh light yellow juicy, acid low, tastes insipid, seeds few.
- **Sathgudi:** Origin is unknown and a popular cultivar grown extensively in India. Tree vigorous produces moderately. Flesh orange, juicy flavor fair sweet in taste, acid very low, mid-season cultivar, moderately seedy.

2.1.7 Sequenced oranges

- **Chinese box orange** (*Severinia buxifolia* (Poir.) Tenore) is native to China and grows as a compact tree or a small shrub. Among the trees related to citrus is the hardiest one. It produces small fruits that have no commercial value and it is used as an ornamental species (IVIA147-).
- **Amber sweet orange**, [(An unnamed hybrid of Clementine mandarin x Orlando tangelo) x unnamed midseason sweet orange seedling], is a variety released by the USDA because of its resemblance to sweet orange, early maturity and deep flesh color. It is the only such hybrid ever legally designated as a "sweet orange", so that its

juice could be used to blend with true sweet orange juice, according to juice industry regulations in Florida. All other known sweet oranges are derived only by somatic mutation, not by sexual hybridization, so Amber sweet is not a true sweet orange [23, 24]. There is one true sweet orange (*C. sinensis*) from which many somatic mutants are derived, including Washington navel and blood orange. The Amber sweet orange is a mandarin x sweet orange hybrid, and not a true sweet orange, as noted above.

2.1.7.1. Sour oranges

Wu et al. [2] reserved the name "sour orange" (*C. aurantium*) to refer to the genome from which cultivar Seville and other somatic mutants are derived. It is the maternal parent of lemon (*C. limon*). The two sour oranges from South China [24] represent two different genomes both unrelated to sour orange (*C. aurantium*).

2.2 Lemon and lime

Lemon and limes are included in the second group. Two main types of limes are distinguished: the small diploid and seedy lime (Mexican) and the big seedless triploid lime (Tahiti, Bears). Several lemon cultivars having major contribution in the world production include Lisbon, Verna, Eureka, Feminello, Fino and Primofiori.

2.2.1 Lemon (C. limon)

There are some distinct fruits in which lemon characteristics are evident, however the differences are to such a magnitude which warrants their separate characterization and classification. In this group, the most important are the karna, the galgal or hill lemon and jambhiri or rough lemon and all of them are widely grown in India. Meyor lemon and the limettas are also lemon like fruits, described below.

- **Hill lemon (Galgal) C. pseudolimon**: An ancient Indian citrus whose origin is unknown and is commonly called as hill or kumaon lemon, grown extensively in the submountain areas along the foothills of Himalayas and in Punjab as a substitute for lime or lemon. It is indigenous to North India grown in sub-Himalayan region. Plant is tall, vigorous, upright and spreading with an irregular and loose crown, fruit ovate oblong, yellow, apex nipped, base rounded or nippled, rind medium, axis hollow, segments 7–11, seed 25–50. Ripening during October–December.
- **Karna** (*C. karna* Raf): Karna Khatta, Karna Nimbu, Khatta Nimbu. A very old Indian citrus fruit of unknown origin, moderately polyembryonic. Considered to be a natural hybrid between rough lemon and sour orange, as the characters exhibited resemble the two species. Widely employed as a rootstock in northern India, second only to rough lemon. Flowers and fruits only once a year. Rind quite thick adhere tightly, golden yellow to deep orange, smooth or ribbed. Flesh orange, texture coarse, semidry, acidic in taste, flavor sour orange like. Seedy, cotyledons white.
- **Rough Lemon** (*C. jambhiri* Lush.), Jatti Khatti, Lemon, Citronelle [Red rough lemon (*C x jambhiri* (Lush)) Wu et al. [2]. The species is regarded to be native of Himalayan foothills in India, where even today it grows wild. It was thought to be a natural hybrid between citron and lemon. However, Wu et al. [2] was reported that it originated from an F_1 cross *C. reticulata* x *C. medica* by whole genome sequence

comparison and is not a true lemon. It is presumed that Portuguese while returning home introduced it in the southeast Africa. Later towards the end of the fifteenth or early sixteenth century it was brought to Europe from where it reached new world. Fruits are acidic, medium sized, shape variable, usually oblate to elliptic oblong. Rind lemon yellow to brownish orange in color, medium thick, surface typically deeply pitted, bumpy (sunken oil glands) deeply pitted or ribbed, separates readily. Flesh pale yellow to pale orange, acidic in taste, juice moderate, segments 10, hollow and large. Seeds many, small highly polyembryonic, cotyledons light green in color.

- **Eureka** (*C. x limon* L. (Burm. f.)): Carvalho et al. [25] were sequenced cultivar Eureka and related somatic mutants. Its seed parent is sour orange and pollen parent is an unknown citron. It is one of the most important commercial varieties around the world [2]. It is seedling selection of Sicilian lemons. Tree is medium, spreading and having few thorns. Its fruit color is lemon yellow, surface rugose, pitted, shape obovate, size medium, apex round, rind medium thin axis small, solid, segments 8–10, juice acidic with excellent flavor and quality. Eureka is heavy yielder and begins bearing at early age. It has tendency of tip bearing. Rind semi thick pitted, oil glands sunken. Fruiting more in winter, spring and early summer. Seeds few to none.
- **Lisbon:** Originated in Portugal. Its appearance and yield is superior to Eureka. It is resistant to frost, heat and high wind velocity. Tree is large and vigorous with spreading shoots. It has upright thorny growth, lemon yellow fruit color, smooth surface, medium size, pitted rind, small axis, solid, 10–6 segments with 8–0 seeds. Rind quite thick adherence tight, pitted surface smooth and less ribbed than Eureka.
- **Lucknow seedless:** It is hardy, medium, vigorous, spreading drooping, dense foliage, thorny, fruit color yellow, smooth, nippled apex, base round, thin rind, hollow axis, segments 12–10 maturity during November–February.
- **Plant lemon:** Fruit size medium, juicy, heavy fruiting, tolerant to pests and diseases.
- **Villafranca lemon (CRC 280)**: It belongs to Eureka group and was introduced into Florida from Europe in about 1875. Originated in Sicily. Commercially not very important compared to Eureka and Lisbon. Tree characters resembles to that of Lisbon, but the plants are more erect or upright in growth, foliage less and with few thorns. In fruit characters, it is similar to Eureka but fruits more during winter like Lisbon.
- **Nepali oblong (Assam, Nimber or Pat Nebu)**: An ever bearing lemon cultivar. Plant medium sized, hardy, spreading drooping with irregular crown, fruit shape oblong, color lemon yellow, segments 10–12, the fruit ripen during December–January. Rind relatively thick, greenish yellow, glossy and smooth. Flesh fine grained, greenish yellow, juicy, not too acidic, seeds none to few.
- **Meyer lemon**: Flowers throughout the year, but more so in spring. Tree semi dwarf, thornless, spreading, cold resistant, fruit color light orange, smooth surface, finely pitted, shape obolate or oblong base rounded, rind thin, axis small, 8–10 segments and 8–12 seeds. Rind thin, adhere tightly, surface smooth yellowish orange to orange, flesh light orange yellow, soft, very juicy and typical lemon flavor.

2.2.2 Limes

Like the citron and lemon, the limes likely have originated in north eastern India. Limes are generally of two forms- Small fruited acid limes (*C. aurantifolia*), West Indian lime and the large fruited acid limes (*C. lantifolia*). In its natural habitat, several forms are recognized which differs markedly in size, form, shape, spine and seediness character etc. The West Indian or Mexican lime is the Kagzinimbu and has number of vernacular names. Australian desert lime (*Eremocitrus glauca* (Lindl.) Swingle, *C. glauca* (Lindl.) Burkill) is native to Australia and produces fruits of sour taste that can be used as condiment. It is drought tolerant and has very few soil requirements (UCR12-B01-38-). Eremorange, Australian desert lime hybrid (*Eremocitrus glauca* x *C. sinensis*) (SRA-871). Australian finger lime (*Microcitru saustralasica* (F. Muell.) Swingle, *C. australasica* F. Muell), native to Australia, develops elongated finger-shaped fruits of different colors. Juice vesicles that can be broken down and separate very easily are of sharp acid flavor. It is used as a food seasoning (UCR18-B04-16-). Australian finger lime is an accession that we find has Australian round lime admixture. BC2 backcross. (SRA1002-). Australian round lime (*Microcitrus australis*, (Swingle), *C. australis* (A. Cunn. ex Mudie)) native to Australia produces rounded green fruit although at full maturity they become yellow. The pulp has low cohesive juice vesicles as the Australian finger lime. It is used as a food seasoning (UCR18-A01-32-).

Acid lime (C. aurantifolia (Christm.) Swingle.):

West Indian (Mexican sour lime): It is native of India and widely cultivated in the tropics. Rouiss et al. [26] reported that it is a natural hybrid between micrantha and citron. It is tenderest species among all the citrus species. Tree medium sized, hardy, semi vigorous, upright growth, thorny, fruit round to oblong, yellow apex rounded and slightly nippled, base round, rind thin, papery, 8–10 segments and seeds both. Rind thin, surface fine, smooth, adhere tightly, greenish yellow. Polyembryonic with few seeds, aroma characteristic, flesh greenish yellow, fine, soft juicy, very acidic.

Tahiti lime (Persian lime) C. lantifolia.

Cold resistant, same as that of lemons. It is large fruited acid lime. Flowers throughout the year more in spring. Purple pigmentation present on shoot and flowers. The plants are large, spreading, cold resistant, thornless, fruit large, seedless, triploid and produce non-viable pollen. It is considered as hybrid between lime and lemon. Fruit color orange yellow, smooth surface and 8–10 segments. It is late variety. Rind thin lemon yellow, adherence tight, flesh greenish yellow, tender juicy acidic. Usually 10 segments.

Rangpur lime (C. limonia Osbeck*):*

Rangpur lime (*C. x limonia* (Osbeck)) produces non-commercial small and very acidic fruits of orange color. It is indigenous to India and is commonly used as rootstock. Rangpur lime is mainly grown for home consumption and ornamental purpose [2]. Fruits are used for making limeade. It is also known as Marmalade orange. It has loose rind, easily separable segments and pulp is light orange yellow. It is very cold hardy.

Sweet lime (C. limettioides):

In north-eastern India to which it is native it is said that sohsynteng of Assam is the acid form of this fruit. Like acid limes, the Indian sweet lime is the mithanimbu and number of its forms differing in fruitfulness, fruit shape, size and with or without nipple etc. are easily recognized. Generally, sweet lime is grown as a rootstock and for its non-acidic fruits.

- **Vikram:** It was developed at MAU, Parbhani, fruit medium size, heavy fruiting, fruit color golden.
- **Pramilini:** It was also developed at MAU, Parbhani, high yielder, golden fruit color, tolerant to canker.
- **Sai Sarbati:** It was developed at MPKV, Rahuri, high yielder, suitable for summer cropping, tolerant to canker and tristeza.
- **Jai devi:** It was developed at FRS, Periakulum, high yield, juicy, thin peel and pleasant aroma.

2.3 Mandarins

The easy peeling mandarins are becoming more important in the fresh fruit market. Principal in importance in the Orient are the Mandarins, a large, distinctive, and highly varied group that includes some of the finest and most highly reputed citrus fruits. These fruits are commonly referred to as loose-skin oranges. Clementines are the most important mandarins in the Mediterranean Basin, while Satsumas predominate in Japan. Other commercial mandarins include intraspecific or interspecific hybrids such as Fortune, Kinnow, Minneola and several chance seedlings such as Ponkan, Ellendale, Ortanique, Murcott, and Nadorcott (Figure 2). In the United States, where the name tangerine first came into common usage, mandarin and tangerine are used interchangeably to designate the whole group. Since mandarin is the older and much more widely employed name, its use is clearly preferable. Presumably because of the orange-red color of the Dancy variety, which originated in Florida and was introduced in the markets as the Dancy tangerine, horticulturists have tended to restrict the use of term tangerine to the mandarins of similar deep color. Tangerine is applied more strictly to those varieties which produce deep orange or scarlet fruits. Mandarin is known as the mikan of Japan, the suntara or sangtra (numerous modifications) of India, mandarino of Italy and Spain and the mandarine of French-speaking countries.

Figure 2. Mandarins varieties.

Due to remarkable diversity of mandarins and the writer's lack of firsthand knowledge of many of the Oriential members, considerable difficulty was experienced in developing a satisfactory horticultural classification for this group. Webber (1948) has separated the mandarin oranges into (a) King group (b) Satsuma group (c) Mandarin group (d) Tangerine group (e) Mandarin-Lime group (f) Mitis group. In this treatment, therefore, the mandarins are presented as the following classes.

2.3.1 The Satsuma mandarins

The Satsuma mandarins (unshiu) mandarin, cv. Owari (UNS) (*C. unshiu* [(Mak.) Marc]; *C. reticulata* (Swingle)) is a commercial midseason, sterile and parthenocarpic, easy peeling mandarin. Satsumas are a group of commercial varieties with relatively high tolerance to low winter temperatures. It is a Japanese variety that was introduced in Florida in 1876. It is also resistant to canker, gummosis, scaly bark and melanose. Plant is thornless having spreading growth habit, orange fruit color, rough surface, oblate to spherical shape, medium size, thin and easily separable rind, flavor rich and seedless. Ripens early than any other oranges as its heat requirement is very low. Number of segments is 12–10, axis hollow and capillary membranes are loose. Fruits should be picked quickly when mature otherwise quality deteriorates and it stores well.

2.3.2 The King mandarins

The King mandarins (*C. nobilis* (Lour.), *C. reticulata* (Swingle)) is thought to be a natural tangor, i.e., a hybrid between mandarin and orange, that originated in Vietnam. However, this conventional wisdom is evidently wrong, as Wu et al. [2] was reported that whole genome sequence analysis shows that sweet orange is not a direct parent of King mandarin. King mandarin was first introduced from Cochin China into California in 1882. The king is a prolific bearer, frost resistant and produces high quality fruit. Fruits have had much commercial interest since they are large, develop good flavor when ripe and are of late harvest.

2.3.3 Tachibana mandarin

Tachibana mandarin (*C. tachibana* (Mak.) Tanaka, *C. reticulata* (Blanco)) is thought to be native to Japan and surrounding islands. It develops easy peeling, small fruits of pale-yellow-orange color and acid flavor. Although taste is not completely unpleasant the fruit is not palatable.

2.3.4 Sunki mandarin

Sunki mandarin (sour mandarin, suanju) (*C. sunki* (Hayata, Hort. ex Tanaka, *C. reticulata* (Blanco)) produces easy peeling, very acidic small fruits, of an attractive orange color. Its fruits are not palatable and the plants are used as rootstocks.

2.3.5 Cleopatra mandarin

Cleopatra mandarin (*C. reshni* (Hort. ex Tanaka), *C. reticulata* (Blanco)) is native to India. It produces unpalatable, small and very acidic fruits. It is widely used as a salt tolerant rootstock and also as an attractive ornamental because of the deep red color of the peel. Plant is thornless with dense top. Fruits are produced singly or in bunch, fruit color is dark orange red, shape oblate, flattened at both ends, size is small with15–12 segments.

2.3.6 The Mediterranean mandarins

The Mediterranean mandarins (*C. deliciosa* Tenore), which is of principal importance in the Mediterranean Basin. It differs from other mandarins as seeds are plumb and spherical, leaves are small narrow lanceolate and aroma of oil glands and juice is very aromatically flavored.

Willow leaf mandarin (C. deliciosa): The tree is willowy in growth, almost thornless, fruits usually born singly at the tip of slender branches. Fruit color orange, surface smooth, glossy, frequently slightly lobed, necked base, apex depressed, wrinkled, rind thin with 12–10 segments. It is an early variety. Trees are cold hardy but at the ripening time rind separates rapidly and lacks storage ability.

2.3.7 The common mandarins

The common mandarins (*C. reticulata* Blanco), which have worldwide importance and are represented by numerous varieties.

2.3.8 The small-fruited mandarins

The small-fruited mandarins, which are of considerable importance in the Orient and consist of many varieties.

2.3.9 *Clementine (Algerian Tangerine)*

Clementine (Algerian Tangerine) C. clementina Hort.ex Tanaka: It is a tangerine and is probably an accidental hybrid of the mandarin and sour orange which originated in Algeria. Fruit color is deep orange, shape globose to elliptical, size medium with depressed apex, rind thick and segments are 12–8, adhered slightly. It is an early variety. Cotyledons are green in color and seeds are monoembryonic.

2.3.10 Kishu mandarin

Kishu mandarin (Kinokuni mandarin) (*C. kinokuni* Hort. Ex Tanaka). The seeded form of this small tangerine grows in southern China and also in Japan, where it was introduced. We sequenced the seedless mutant known in Japan as Mukakukishu; sweet, juicy, and easy to peel, it is appreciated because of its pleasant taste and wonderful aroma. Whole genome sequence comparison shows that it has the same base genotype (i.e., is a somatic mutant of) Huanglingmiao1 mandarin.

2.3.11 Dancy mandarin

Dancy mandarin, Dancy tangerine (*C. tangerine* (Tanaka), *C. reticulata* (Swingle)) is an easy peeling commercial late harvesting variety of excellent color and good size and perdurability on the tree. Originated in 1867 from a chance seedling. In USA, dancy is the best known and highly prized of all the mandarin oranges. Tree is large, nearly thornless and has upright growth. It has tendency towards alternate bearing and seeds are polyembryonic.

Fallglo [Hybrid of Bower mandarin (Clementine mandarin x Orlando tangelo) x Temple tangor, a presumed mandarin-sweet orange hybrid of unknown parentage], a seeded, early maturing and large fruited mandarin hybrid, developed by the USDA and produced primarily in Florida, USA [24].

2.3.12 *Calamondin (C. madurensis* (Lour.))

Calamondin (C. madurensis (Lour.)): Tanaka has recognized it as a loose skinned orange group. It has less value as a fruit but hangs well on tree and is widely used as an ornamental fruit. It is very cold resistant true citrus fruit and as hardy as Satsuma. Fruit color is orange to deep orange, smooth and glossy surface, pitted shape oblate, size small with flattened base having 7–10 segments.

- **Sun Chu Sha Kat mandarin** (*C. reticulata* (Blanco), *C. reticulata* var. austere (Swingle), *C. erythrosa* (Tanaka)) is characterized by small flowers, small but narrow leaves and small fruits. These are broader than long, peel color may change from yellow to deep red and taste is acidic or acidic-sweet. It is used as rootstock.
- **Changsha mandarin** (*C. reticulata* (Blanco)) produces small, juicy, puffy, brilliant orange-red and seedy fruit. The taste is sweet or acidic-sweet. The tree is rather tolerant to frost and yields heavy crops. It is also grown as an ornamental (UCR-12B07-23-).
- **Wilking** [It is a sister hybrid of Kinnow mandarin both having King mandarin x Willowleaf mandarin parentage], developed by the University of California, Riverside in 1915. Fruit are small, quite fragrant and richly aromatic. Because it produces monoembryonic (zygotic) seeds, it has been used in breeding programs, but not grown commercially to any great extent [24].
- **Feutrell's early:** It is an old variety of New South Wales. Its parents are unknown. The fruit characteristics indicate that it may be a natural tangore and those of the three suggest the possibility that medaterian or Willow leaf might have been the mandarin parent.
- **Coorg orange:** It is an important variety of South India particularly in Coorg and Wynad tracts. Fruits are medium to large, bright orange color, oblate to globose in shape, finally papillate and winkled, glossy with 11–9 segments.
- **Deshi mandarin (Pathankot):** This variety is mainly grown in Punjab hills. The tree is large with semi-upright growth habit with compact foliage and spineless. Fruit is ovoid to sub globose, color uniformly cadmium, surface pitted, semi glossy and finally wrinkled, rind medium, adherence slight with 7–10 segments.
- **Khasi mandarin:** It is commercial variety of Assam. Fruit is depressed, globose to oblate, orange yellow to bright orange, surface smooth, glossy, base even or obtuse, rind thin, soft and 10–8 segments.
- **Nagpur santra (Ponkan, Warnucro):** This variety occupies premier position in Indian market and is one of the finest mandarins grown in the world. It is also known as Ponkan. Tree is large, vigorous, spineless with compact foliage. Fruit size is medium, cadmium color, smooth surface, glossy, rind thin, soft, and slightly adhered with 10–12 segments.
- **Kinnow mandarin:** It is first generation hybrid between the King and Willow leaf mandarin that was developed by H.B. Frost at the California Citrus Experiment Station in 1915. It was introduced into Indian Punjab from USA. Tree is vigorous, large, top erect, dense symmetrical with few scattered thorns. Fruit color resembles

of King, deep yellowish orange, surface smooth, glossy, very shallow pitted, shape slightly oblate, size medium with flattened base, rind thin, peel tough and leathery, 10–9 segment easily separable and 24–12 seeds. It is a late variety, cold resistant, has strong alternate bearing tendency and seeds are polyembryonic.

- **PAU Kinnow-1:** It is bud mutant of Kinnow mandarin and is low seeded compared to Kinnow.
- **Tengors:** The mandarin-like fruits include the synthetic tangors; the so-called natural tangor, Temple. Kiyomi [Hybrid of Miyagawa-wase satsuma mandarin x Trovita sweet orange], developed by the Okitsu Branch Fruit Research Station, now known as the Okitsu Citrus Research Station, National Institute of Fruit Tree Science. This is a large fruited juicy tangor, with aroma closely resembling sweet orange, and is seedless in the absence of cross pollination. It produces abundant monoembryonic (zygotic) seeds when cross pollinated and has been used as a scion breeding parent in Japan and elsewhere [24].
- **Temple mandarin:** It is a hybrid between Tangerine and sweet orange. Temple mandarin is most beautiful and highly flavored fruit of the citrus group. Tree is medium, thorny, spreading with deep orange to reddish fruit color, rugose glossy surface, medium to large, depressed or nearly flat apex, loose rind, solid axis with 12–10 segments and orange pulp. It is late in maturity.
- **Dweet:** Evolved from a cross between Mediterranean Sweet Orange and Dancy Tangerine. Late in maturity. Tree character intermediate of the two parents. Fruits do not hang well on tree. Fruit reddish-orange in color, globose-oblate, neck distinct, medium to large, surface pebbled. Rind adherence tight and peels very poorly, also puffs, flesh orange, firm very juicy, flavor rich, seeds numerous.
- **Mency:** Originated from a reciprocal cross of Dweet. Fruits ripen early, do not hang well on tree, susceptible to sunburn, small, reddish orange, somewhat oblate and necked, surface pebbled due to oil glands. Rind adherence is not tight and peels easily. Flesh is orange colored and juicy with acidic flavor. Seeds are many.
- **Tangelos:** The mandarin-like fruits include the synthetic tangelos; the so-called natural tangelo, Ugli. Hybrids of mandarin, grapefruit and pummelo are designated as tangelos. They exhibit characters of both the parents.
- **Clement:** It is a hybrid between Duncan grapefruit and Clementine mandarin. Tree and foliage characters are intermediate and productive. Fruit is subglobose to oblong, light orange yellow in color and medium large. Rind is quite thick, pebbled and peels easily. Flesh is dull yellow soft and moderately sweet. Fruit is early to medium in ripening.
- **Minneola:** It is hybrid of Duncan grapefruit and Dancy tangerine. Less cold hardy than Orlando, requires cross pollination for regular and higher yields. Maturity midseason. Rind deep reddish orange, smooth, medium thick, adherence moderate, surface pitted.
- **Orlando:** A hybrid of Duncan grapefruit x Dancy tangerine. Early in maturity. Fruit broad oblate to subglobose without neck, medium large. Rind, thin orange in color,

adherence quite tight, does not peel easily, pebbled. Flesh orange, very juicy tender, somewhat sweet seedy.

- ❖ **Seminole:** Parentage is same as that of Orlando and Minneola. More fruitful. Late in maturity, trees are productive, medium large, leaves small to medium, rounded and cupped. Fruit broad oblate deep reddish orange in color. Rind adhere moderately, thin, pebbled, peelable, axis usually hollow. Flesh dark orange, juicy soft, somewhat acidic in taste, seedy.
- ❖ **Page:** An early ripening cultivar obtained from a cross between Minneola tangelo x Clementine mandarin.

2.4 Grapefruit and Pummelo

The last group is grapefruit which is divided into the yellow flesh cultivars (Marsh, Duncan) and the red flesh cultivars ('Hudson', 'Star Ruby', 'Ray Ruby' 'Rio Red'). In the Southeast Asia and the Pacific, pummelo (*C. maxima*) and many traditional local mandarin cultivars are still important in the domestic market.

2.4.1 *Citrus paradisi* Macf. Grapefruit

It is closely related to pummelo, originated in Barbados (West Indies), as old records refer to 'forbidden fruit' (Figure 3). It has become popular as a breakfast fruit because of its typical flavor and mild bitterness of the juice. Grapefruit is regarded to be either an interspecific hybrid of pummelo and sweet orange or a hybrid or a mutant of pummel Wu et al. [2]. Seeds are polyembryonic. The fruit is also called as small shaddock, obtained from pummelo, and the name has been derived from the fact that it bears in clusters, and flower resemble that of grape. Like pummelo, grapefruit is also of two types- the common and pigmented grapefruits. The varieties obviously differ in a number of characters i.e., maturity time, seedless or seedy and flavor etc. Important cultivars grown worldwide are described briefly. It is a hybrid between a pummelo and sweet orange. The Cocktail grapefruit is not a true grapefruit.

Figure 3. Grapefruit varieties.

Common Grapefruit Varieties:

- **Marsh:** (Syn. White Marsh, Marsh seedless, (*C. x paradisi* (Macfadyen)). It is most extended varieties of grapefruit, originated as a chance seedling around 1860 in Lakeland, Florida. It is a late-ripening, self-incompatible variety that shows long tree storage capability and very good behavior during postharvest. It comes under pallid pulp group of grapefruit. It is bud sport from cultivar Marsh. Fruit color light yellow, surface smooth, shape oblate to globose, size medium to large, basal area small, rind thin, segments 12–14 and 2–5 seeds. It is a late cultivar. Cultivars more significant value is its seedlessness character and its late ripening. Fruit medium sized, spherical or oblate, light yellow, areole ring almost absent. Rind surface even, smooth semi thick, flesh light yellow, soft very juicy, with pleasant flavor, but less than that in seedy cultivars. Has very good storage quality and fruit hangs for long on the tree. Seeds few to none.
- **Duncan:** It was developed as chance seedling in Florida. It is the hardiest variety for cold, fruit color yellow, surface smooth, shape oblate to globose, size large, basal area depressed, apex round, rind medium thick, firm, axis medium in size segments –12 14 and 50–25 seeds. Plants are large, very vigorous and productive early to medium in ripening and also among the few which are cold hardy. Rind surface even/ smooth, glossy, medium in thickness. Flesh light yellow/ buff, soft very juicy, flavor very characteristics strong and pleasant.

Pigmented Grapefruit Cultivars

- **Foster (Foster Pink):** It belongs to pink or red pulp group and originated as bud sport of Walters grapefruit by R.B. Foster in 1907–1906. Fruit color is light yellow, surface smooth, oblate or globose shape, size medium large, base rounded, apex round, rind medium thick with 15–12 segments and 50 or more seeds. It matures during November–December. Cultivar is said to be the first pigmented grapefruit variety that ripens in mid-season, originated as a limb sport of cv. Walters. Fruit medium large in size, round flat to spherical, light yellow overlaid by pink blush, which is also seen in the albedo (mesocarp), areole inconspicuous, furrows at base are small and diverging. Rind smooth, medium thick, flesh light yellow, pink under favorable conditions, soft, flavor good.
- **Red blush (Syn. Ruby, Red Marsh, Red Seedless):** It belongs to pink or red pulp group. It originated as bud sport from Thompson. Deep red color which uniformly distributed throughout pulp. Fruit resembles Thomson for most of the characters except intense pigmentation of the flesh, albedo/ mesocarp also pink with bright red blush on the fruit surface. Hangs well on the tree as the parent cultivar.
- **Thompson (Pink Marsh):** It originated from bud sport to marsh. Fruit color light yellow, surface smooth, 12–10 segments with 5–2 seeds. Plants are very productive large with vigorous growth. Fruits medium sized, spherical-oblate, areole indistinct light yellow, rind relatively thick, smooth and tough. Has excellent storage and shipping quality, fruits hang well for long period of time. Seeds few to none.
- **Saharanpur Special:** Fruit round to oblate, empire yellow, surface small, segments 15–10 and 100–50 seeds. Fruits ripen in November to February.

2.4.2 *Citrus grandis* (L.) Osbeck. Pummelo, shaddock

Seeds are large yellowish and ridged, monoembryonic. It is native of Polinasia and Malaysia and commonly grown in South China. Fruit is pyriform, largest fruit size among citrus fruits, rind thick, juice is acid bitter, juice sacs easily separable. Seeds are monoembryonic. Fruits are of two types (a) elongated pear shaped with neck (b) Oblate or globose, flattened and neckless. In India there is no improved cultivar except Nagpur chakotra.

In all likelihood is indigenous to the Malayan and East Indian archipelago. In respect of fruit characters, the pummelos fall into two major groups the pigmented and the non-pigmented types. The pigmentation in the former types is caused by carotenoid lycopene and varies from pink to deep red some of which are very attractive with very high in flavor. The non-pigmented (common pummelo) are very variable in number of characters, have moderate to high acid contents and mostly seedy. Description of some known pummelo cultivars is described briefly.

Non-Pigmented Varieties

- **Banpeiyu:** The variety is considered to have originated in Malayan region and is very popular in Japan and Taiwan. Rind thick tightly adhering to vesicles, surface smooth. Flesh pale yellow, soft and juicy, segments many (18–15) central axis solid and large, carpellary membranes thin and tough. Mid-season ripening cultivar, flavor excellent and pleasant taste of acid and sugar, storage quality very good, seedy.
- **Hirado (Hirado Buntan):** Originated as a chance seedling at Nagasaki Prefecture, Japan and rated very highly resistant to cold. Fruit large bright yellow, glossy, oblate both ends slightly depressed, surface smooth, rind medium thick, flesh light green, yellow soft semi dry, flavor pleasant with balanced blend of acid and sugar with some bitterness, segments many, carpellary membranes thin and tough. Seedy.
- **Kao Phuang:** One of the most famous cultivars grown in Thailand. Mid to late in ripening, fruits hang well on tree for long. Tree upright very vigorous. Rind semi thick adherence medium with the vesicles which are firm large and separates easily, juicy, flavor good, capillary membranes semi thick, axis small and compact. Seedy.

Pigmented Varieties

In its place of origin-Orient a number of pink and red fleshed types are present some of which are pomologically described below.

- **Chandler:** Is a hybrid cultivar obtained from a cross between Siamese Sweet x Siamese Pink. Flesh in pink and in other characters it is intermediate between the parents. Rind smooth medium thick, vesicles moderate in juice flesh firm later turns soft, flavor pleasant sub acidic in taste, seeds many.
- **Ogami (Syn. Egami):** This cultivar has performed very well in Florida and other parts of the world. Fruit quite large broadly round flat (oblate), rind smooth shiny and moderate in thickness, pink to deep pink in color, flesh deep pink, pigmentation extends far into the albedo (mesocarp), seedy.
- **Siamese Pink (Siam):** Considered to be one of the finest cultivars in respect of its strong flavor, matures late. Rind adheres tightly medium thick, smooth and glossy.

Segments many, carpellary membranes quite thick and tough but open at axis when mature. Almost seedless.

- **Siamese Sweet:** Commonly called as sweet pummelo practically non-acid. Plants are dwarf type, branches drooping, leaves distinctly round pointed. Vesicles separate easily, are large crisp, lack in juice, sweet in taste.

3. Some wild and semi wild species

As north eastern region of India has been recognized as one of the major centre of citrus origin, a number of species are said to be native to this place Table 1.

- **C. indica** (Indian wild orange): Specie is found growing in many parts of Assam, Nagaland, Meghalaya and other north eastern parts of India in wild form. Fruit small broad obovoid or subpyriform, appear singly on terminal twigs, about 2 cm in diameter, pedicel very small. Rind very thin, red in color, segments few, orange red in color, inedible, vesicles spindle shaped, very soft, pulp slimy, acidic in taste with unpleasant flavor. Seeds are very large, smooth, monoembryonic and occupy major portion of fruit.
- **C. assamensis** (Adajamir): A very distinct specie known for its peculiar aroma that resembles to ginger or eucalyptus smell that emits from the crushed leaf or from the fruit. A new specie which was indigenous to Assam region. Fruit medium small, spherical to round in shape, surface smooth, very acidic juice hence has very limited use as fresh fruit.
- The other species *C. latipes*, *C. ichangensis* and *C. macroptera* are usually placed under subgenus Papeda which has number of true wild species of citrus.
- **C. macroptera** (Malanesian Papeda): Almost all the species of Papeda group, this species is most promising as a rootstock. Widely distributed in Indo-China, Thailand, Philippines, New Guinea, New Caledonia and Polynesia. All plants parts are with amber colored glands, spine axillary straight. Flowers are like that of orange. Fruit globose, pale yellow, small (2.5 in dia), smooth, glands small, segments 12–10 with 2–1 seed/vesicle. Juice scanty, highly acidic.
- **C. latipes** (Khasi Papeda): Cold hardy. Native of northeastern Khasi hills (India), and Northern Burma. A small spiny shrub or a small tree (15–10 feet), twigs angular with sharp stout spines (2.5–1.2 cm), reduced or absent on flowering twigs. Fruit medium sized, globular, rind somewhat thick, leathery, segments 9, quite large pulp, vesicles few, spindle shaped, well developed. Seeds many 7–5 in each vesicle, round in shape.
- **C. ichangensis** (Ichang Papeda): Native of south western and west central China. This specie is the most cold hardy of all evergreen types. Differs from other Papeda in that it has large flowers, stamens are connate. Small shrub or a tree (12–15 feet), twigs angular, spines stout. Fruit small (3–4 cm dia) glossy, peel rough, bumpy, medium thick, segments 7–9, each with many seeds large and thick, blunt at both ends, monoembryonic.
- **Micrantha, Biasong** (*C. micrantha* (Wester)) it is thought to be native of the Southeast of the Philippines. It produces small, bitter and inedible fruit with a skin comparatively thick and broadly winged leaves (SRA1114-).

Citron (*C. medica* L).

Citron is considered as native of India having probable region of origin in South Western Asia. Citron is often ranked as the first specie to be cultivated in the western world and in China.

Major citron cultivars are divided into two groups:

- **Acid cultivars:** Diamante is commercially grown in Italy and Etrog is the main cultivar of Israel.
- **Nonacid cultivars**: Corsican-well established in California. Sarcodactylis is well known in Japan and China, often referred as fingered citron, because fruit split into number of segments (like finger) with very less pulp.

This citrus fruit is most likely to have originated in the north eastern India and the areas nearby. As the fruit is inedible and tree lacks ornamental value, one ponders why the Romans and Greeks liked it so much. This was the first citrus fruit which was known to the Romans, further the fruit fragrance is soft, penetrating and lasting. Features of some citron cultivars grown extensively are described.

- **Diamante (Cedro Liscio):** A very important and commercial cultivar of Italy, origin is unknown. New growth, buds and flowers are distinctly purple pigmented. Fruit is usually large, long oval to ellipsoid, basal cavity furrowed, apex pointed-nipple like and lemon yellow in color. Rind is very thick, fleshy surface lobed or ribbed and smooth. Flesh is crisp, dry, acidic in taste and seeds are many.
- **Etrog (Atrog, Ethrog):** New growth, flowers and flower buds all are purple pigmented. Trees are small, productive and moderate in growth. Fruit is small to medium, ellipsoid, neck distinct, nipple at apex sharp, style persistent and lemon yellow. Rind is fleshy, thick, surface bumpy/ribbed, flesh firm and crisp, acidic in taste and low in juice content. Citrus is a highly heterogeneous group and its various species hybridize freely with each other in nature.
- **Humpang citron** (*C. medica* L.) fruit is large, oblong or oval, of green color when growing but generally yellow when ripe. The surface usually is smooth, the rind and the albedo are very thick and the segments are filled with acidic pale greenish pulp-vesicles. Citrons were the first citrus fruit to reach the Mediterranean region and are cold sensitive, monoembryonic, unpalatable and very fragrant.
- **Mac Veu citron** (*C. medica* L., *C. lumia* Risso & Poit). Similar to Humpang citron.
- **Corsican citron** (*C. medica* L.) is an acidless citron of unknown origin.
- **Buddha's hand citron var. Sarcodactylus** (*C. medica* L. (Noot.) Swingle) produces a very characteristic fruit usually without pulp and split into a number of finger-like sections. This fingered citron is well-regarded because of its fragrance for perfuming rooms and clothing. It is also grown as a dwarf plant for ornamental purposes.

Fortunella margarita Fortunella/Kumquat.

Kumquats are part of genus *Fortunella*, and in Chinese language the word kumquat means 'Gold orange'. This distinct group of citrus has been named after Robert Fortune, the well-known English horticulture worker.

Other species of *Fortunella* are:

F. crassifolia Swingle (Meiwa kumquat)

F. margarita Swingle (Oval or Nagami kumquat)

F. japonica Swingel (Round or Marumi kumquat)

F. hindsii Swingle (Hongkong wild kumquat)

Pomological description of important Kumquats is briefly summarized below:

- **Meiwa- Large round Kumquat** (*F. crassifolia* Swing.): It is less cold hardy compared to Nagami but considered to be the best as fresh fruit. This specie is the Ninpo, Neiha or Meiwa Kinkan of Japan. A natural hybrid of oval and round Kumquats.
- **Nagami or Oval Kumquat** (*F. margarita* [Lour] Swing): Popularly called as Naga or Nagami Kinkan of Japan. The characteristic features are that fruits are oval, oblong or obovate in shape having fewer number of segments usually 5–4, rind deep orange, both flesh and rind are richly flavored. Tree and leaves are larger in size. Since trees are small and show slow, cold-tolerant growth it is also used as an ornamental. It produces fertile hybrids when crossed with species of the genus *Citrus*.
- **Marumi or Round Kumquat** (*F. japonica* [Thumb.] Swing): Maru or Marumi Kinkan of Japan. Tree is low in vigor with some thorns. Leaves are small and apex is not sharp pointed. Fruit is small round to slightly oblate-obovate. Rind is thin and sweet in taste, segments vary from 7–4.
- **Poncirus and hybrids.**
- Trifoliate orange, Poncirus Pomeroy (*Poncirus trifoliata* (L.) Raf.) shows trifoliate leaves and deciduous behavior, two dominant characters that are not present in citrus. The tree also has high resistance to cold. Its fruit has no commercial value and the plant is commonly used as rootstock like its hybrids, especially the citranges, Carrizo and Troyer.
- **Citranges:** It is hybrid between *Poncirus trifoliata* and *C. sinensis* and is hardy than sweet orange. It is nearly deciduous, fruit color varies from yellow to deep orange, surface rough, wrinkled, ribbed or smooth, rind thin, juicy pulp, highly acidic. It is used as dwarfing rootstock for grapefruit, Satsuma, sweet orange and lemon.
- **Varieties:** Coleman, Etonia, Mortan, Rusk, Cunnigham, Rustic, Sanford, Savage, Troyer.
- **Citrangequats:** It is trigeneric hybrid of Citrange (Sweet orange x Trifoliate orange) and Kumquat. Citranges are subjected to spring frost injury due to its tenderness while Kumquat is tolerant to the spring frost because of their winter dormancy.

4. Sequenced genome of *Citrus* spp. and their varieties

Xu et al. [27] reported Valencia sweet orange (*C. sinensis* cv. Valencia): A model of the sweet orange origin. With pummelo as the female parent crossed with mandarin, the interspecific hybrid was backcrossed with mandarin and produced the ancient sweet orange. Haploid Clementine, Clementine mandarin, Ponkan mandarin, Willowleaf mandarin, W. Murcott mandarin, Chandler pummel, Low-acid pummel, Sweet orange, Seville sour orange [18] Table 1.

5. Conclusion

Classification of the Citrus varieties has long been debated by taxonomists and botanists. The reticulate evolution combined with partial apomixis has led to very different classification systems. Earlier classification was very difficult due to lacking of genomic study but recently, phylogenomic data revealed the origins and admixtures of modern cultivars and wild types. Coupled with reproductive biology, phylogenomy supports the inclusion of all true citrus of the Swingle system plus Oxanthera in the genus Citrus. The variety rank is defined by the old independent reticulation events from which groups of cultivars were differentiated by asexual mechanisms. It provides an unambiguous conceptual framework for Citrus classification based on the phylogenomic and genetic data. However, today, the available genomic data remain available for further study and further WGS studies are needed to establish a definitive classification of the Citrus varieties. Commercially grown citrus varieties discussed in this text, are not grown from seed but are grafted and budded onto a seedling of a rootstock.

References

1. Swingle WT. The botany of citrus and its wild relatives of the orange family. In: Webber HJ, editors. The Citrus Industry: Berkeley and Los Angeles. University California Press. 1943; p. 129-474

2. Wu GA, Terol J, Ibanez V, López-García A, Pérez-Román E, Borredá C,Domingo C, Tadeo FR, Carbonell-caballero J, Alonso R, Curk F, Du D,Ollitrault P, Roose ML, Dopazo J, Gmitter Jr FG, Rokhsar DS, Talon M. Genomics of the origin and evolution of Citrus. Nature. 2018; 554: 311-316. DOI: 10.1038/nature25447

3. Krueger RR, Navarro L. Citrus germplasm resources. In: Citrus Genetics, Breeding and Biotechnology. CAB International, Wallingford, UK. 2007; p. 45-140

4. Siebert T, Kahn T, Krueger R. Observations of graft compatibility between Citrus spp and related Aurantioideae taxa. Acta Horticulturae. 2015; 1065:173-179

5. Siebert TJ. Cross-Compatibility, Graft-Compatibility, and Phylogenetic Relationships in the Aurantioideae: New Data From the Balsamocitrinae. University of California, Riverside, California. 2016

6. Swingle WT, Reece PC. The botany of Citrus and its wild relatives. In: Reuther W, Webber HJ, Batchelor DL. editors. The Citrus Industry. University of California Press, Berkeley, CA, USA, 1967.p. 190-430. ISBN/Record No. 19680306358

7. Hazarika TK. Citrus genetic diversity of north-east India, their distribution, ecogeography and ecobiology. Genetic Resources and Crop Evolution. 2012; 59 (6):1267-1280. DOI: 10.1007/s10722-012-9846-2

8. Swingle WT. Citrus and related genera. In: Bailey LH, editors. Standard Cyclopedia of Horticulture. The Macmillan CO., New York. 1914; p. 779-785

9. Matsukawa T, Nito N. Metabolic fingerprinting of Citrus cultivars and related genera using HPLC and multivariate analysis. Journal of Plant Studies. 2018; 7 (1):1-9. DOI: 10.5539/jps.v7n1p1

10. Nicolosi E, Deng ZN, Gentile A, La Malfa S, Continella G, Tribulato E. Citrus phylogeny and genetic origin of important species as investigated by molecular markers. Theoratical and Applied Genetics. 2000; 100:1155-1166. DOI: 10.1007/s001220051419

11. Ding SQ, Zhang XN, Bao ZR, Liang MQ. Poncirus polyandra. Acta Bot. Yunnanica. 1984; 6 (3): 292 p

12. Duan Z. Primary study of fuming evergreen trifoliate orange (P. trifoliata Raf). In: Proceedings of the International Citrus Symposium Guangzhou, China. 1990; p. 166-169

13. Tanaka T. Citologia: Semi-Centennial Commemoration Papers on Citrus Studies. Osaka, Japan, 1961. Citologia Supporting Foundation

14. Zhang Z, Mabberley DJ. Citrus Linnaeus. Sp. Pl.2: 782. 1753. Fl. China, 2008; 11:90-96

15. Curk F, Ancillo G, Ollitrault F, Perrier X, Jacquemoud-Collet JP, Garcia-Lor A, et al. Nuclear species-diagnostic SNP markers mined from 454 amplicon sequencing reveal admixture genomic structure of modern citrus varieties. PLoS One. 2015; 10 (5):e0125628

16. Oueslati A, Salhi-Hannachi A, Luro F, Vignes H, Mournet P, Ollitrault P. Genotyping by sequencing reveals the interspecific C. maxima/C. reticulata admixture along the genomes of modern citrus varieties of mandarins, tangors, tangelos, orangelos and grapefruits. PLoS One. 2017; 12 (10):e0185618

17. Penjor T, Mimura T, Kotoda N, Matsumoto R, Nagano AJ, Honjo MN, et al. RAD-Seq analysis of typical and minor Citrus accessions, including Bhutanese varieties. Breed. Sci. 2016; 66 (5):797-807

18. Wu GA, Prochnik S, Jenkins J, Salse J, Hellsten U, Murat F, et al. Sequencing of diverse mandarin, pummelo and orange genomes reveals complex history of admixture during citrus domestication. Nat. Biotechnol. 2014; 32 (7):656-662

19. Curk F, Ollitrault F, Garcia-Lor A, Luro F, Navarro L and Ollitrault P. Phylogenetic origin of limes and lemons revealed by cytoplasmic and nuclear markers. Annals in Botany. 2016; 117:565-583. DOI: 10.1093/aob/mcw005

20. Ollitrault P, Curk F, and Krueger R. Citrus taxonomy. In: Talon M, Caruso M, Gmitter Jr FG, editors. The Genus Citrus. 1st ed. Elsevier; 2020. p. 57-81 DOI: 10.1016/B978-0-12-812163-4.00004-8. Ch4

21. Shimizu T, Kitajima A, Nonaka K, Yoshioka T, Ohta S, Goto S, Toyoda A, Fujiyama A, Mochizuki T, Nagasaki H, Kaminuma E, Nakamura Y. Hybrid origins of citrus varieties inferred from DNA marker analysis of nuclear and organelle genomes. PLoS One. 2016; 11 (11): e0166969. DOI: 10.1371/journal.pone.0166969

22. Mabberley DJ. Citrus (Rutaceae): a review of recent advances in etymology, systematics and medical applications. Blumea. 2004; 49 (2/3): 481-498

23. Bayer RJ, Mabberley DJ, Morton C, Miller CH, Sharma IK, Pfeil BE, Rich S, Hitchcock R, Sykes S. A molecular phylogeny of the orange subfamily (Rutaceae: Aurantioideae) using nine cpDNA sequences. American Journal of Botany. 2009; 96:668-685. DOI: 10.3732/ajb.0800341

24. Wang X, Xu Y, Zhang S, Cao L, Huang Y, Cheng J, Wu G, Tian S, Chen C, Liu Y, Yu H, Yang X, Lan H, Wang N, Wang L, Xu J, Jiang X, Xie Z, Tan M, Larkin RM, Chen LL, Ma BG, Ruan Y, Deng X, Xu Q. Genomic analyses of primitive, wild and cultivated citrus provide insights into asexual reproduction. Nature Genetics. 2017; 49 (5):765-772. DOI: 10.1038/ng.3839

25. Carvalho R, Soares Filho WS, Brasileiro-Vidal AC, Guerra M. The relationships among lemons, limes and citron: a chromosomal comparison. Cytogenetic and Genome Research. 2005; 109 (1/3):276-282. DOI: 10.1159/000082410

26. Rouiss H, Bakry F, Froelicher Y, Navarro L, Aleza P, Ollitrault P. Origin of C. latifolia and C. aurantiifolia triploid limes: the preferential disomic inheritance of doubled-diploid 'Mexican' lime is consistent with an interploid hybridization hypothesis. Annals of Botany. 2018; 121 (3):571-585. DOI: 10.1093/aob/mcx179.molecular

27. Xu Q, Chen LL, Ruan X, Chen D, Zhu A, Chen C, et al. The draft genome of sweet orange (Citrus sinensis). Nat. Genet. 2013; 45 (1):59-66

Chapter-3

Citrus Biotechnology: Current Innovations and Future Prospects

1. Introduction

Citrus is one of the most diverse members of the family *Rutaceae* and is the leading tree fruit crop in the world. Citrus comprises different species of edible fruits like mandarins (*Citrus reticulata* Blanco), sweet oranges (*C. sinensis* Osbeck), grapefruit (*C. paradisi* Macf.), acid limes (*C. aurantifolia* Swingle), and sweet limes (*C. limettioides*), lemons (*C. limon* Burmf.) and their hybrids including tangerines, tangelos, tangors, etc. [1].

Citrus is being widely cultivated in the sub-tropical, tropical, and temperate regions of the world. Global citrus production is 157 million tons per annum from an area of 15 million hectares. About 50% of the area and production of citrus is being contributed by the northern hemisphere of the world. China (28%) and the Mediterranean regions (25%) are the major contributors to global citrus production followed by Brazil (13%). China is leading in grapefruit and mandarin production. Among Mediterranean countries, Spain is leading in global citrus production (6 million tons) including mandarins, oranges, limes, lemons and exports. Brazil is leading in global fresh sweet orange and its juice production. Mexico and India are major lime producers [2]. Pakistan's share in global citrus production is quite low (1.6%) which includes mandarin and sweet oranges as major species whereas limes, grapefruit, and lemons have less production and are dealt as minor species. The global citrus industry is facing many biotic (Citrus greening, Citrus tristeza virus, sudden death, citrus canker, and Phytophthora) and abiotic stress (salinity, drought, and temperature fluctuations) which have a direct impact on fruit crop production and yield [3].

2. Origin and diversification

Citrus and other genera including *Poncirus*, *Clymenia*, *Fortunella*, *Eremocitrus*, and *Microcitrus* belong to tribe Citreae and sub-tribe Citrineae and are considered as true citrus [4]. Classification in citrus has been controversial since ancient times due to vast morphological diversity, interspecific and intergeneric sexual compatibility. However, molecular biology tools have revealed four species including mandarins, citron (*C. medica*), pummelos (*C. grandis* Linn.), and wild cultivar of papeda (*C. micrantha* Wester) as the true parental species that have contributed to the development of other species during the process of evolution [5, 6]. Based on phylogenetic and genomic studies it is revealed that mandarin originated in China, Vietnam, and Japan whereas citron was originated in northeast India and China. Pummelo originated in Indonesia and Malay whereas *C. micrantha* was originated in the Philippines [6]. Other citrus species including sweet oranges, grapefruit, lime, lemon, sour oranges, and hybrids (tangelos and tangors) have developed from these ancestral species through random hybridization and natural mutation events [7].

Among citrus genetic resource centers, major collections are found in the USA, China, Spain, France, Japan, and Brazil where a large number of wild species, their relatives, old and new varieties, and breeding lines are conserved [8]. In Pakistan, citrus genetic resources are conserved mainly in the field as orchards or germplasm units in Sargodha, Faisalabad, and Sahiwal in different academic and research institutes.

3. Conventional approaches for crop improvement

Citrus breeders have been using different approaches for their improvement including conventional breeding, mutation breeding, polyploidization and *in vitro* culture tools particularly somatic hybridization which has played an essential role in developing new somatic hybrids. These techniques have contributed towards the selection and development of new potential cultivars and are still being used as important fundamental tools for the development of genetically diverse germplasm which could be further screened and characterized using modern breeding technologies.

3.1 Classical and mutation breeding

Though conventional breeding has limitations in citrus due to its complex reproductive behavior, nucellar embryony, long juvenility, sterility, sexual incompatibility, and endogametic depression [9, 10]. However, still, many hybrids have been developed by conventional breeding and recovered using *in vitro* tools.

Mutation breeding has played a pivotal role in fruit crop improvement including citrus and has developed several mutants with improved phenotypic and genotypic traits [11]. Spontaneous or induced mutants do not have intellectual property rights (IPR) related issues that have to be faced in the case of conventional breeding and transgenics [12]. Both spontaneous and induced mutations have enhanced genetic diversity in existing varieties and have provided the raw material for making selections for the novel horticultural traits [13]. About 3365 mutant varieties belonging to 170 plant species have been released including citrus and 20 other fruit species [14]. Among continents, Asia is leading with 2052 mutants released followed by Europe (960 mutants). Among countries, China (817), Japan (479), India (341) and the USA (139) are leading in mutant development whereas Pakistan has released 59 mutants in different crops [15]. In citrus, a total of 15 mutants have been released since 1970 including mandarins and clementine (6), sweet oranges (6), grapefruit (2), and lemon (1) [10]. Pakistan has registered a single mutant variety in citrus, a Kinnow mandarin induced mutant having less number of seeds and named it as "NIAB Kinnow" in 2017.

The rate of spontaneous mutations has been much higher in citrus compared with other fruit crops, however, due to random genetic alterations it has been difficult to identify and utilize such mutants [16, 17]. Induced mutations using different irradiation sources including gamma rays (physical mutagens) and various chemical compounds have enhanced the frequency of genetic variability. Physical mutagens or ionizing radiations have been more commonly used for inducing genetic diversity, chromosomal aberrations, and point mutations. About 70% of the mutant varieties have been developed using physical mutagens [18]. In fruit crops, physical mutagens have altered key horticultural traits like seedlessness, precocious bearing, and dwarfism [19, 20, 21]. Other traits include fruit ripening time, fruit skin and flesh color, fruit aroma, self-compatibility, pathogen resistance, and fertility restoration in sterile hybrids. Among physical mutagens, gamma rays have been most used for the development of mutants due to their shorter wavelength and greater penetration [22], however, the ion beam is getting more popular and is being widely used due to its greater efficiency and precision compared with gamma rays [23]. Among chemical mutagens, ethyl methanesulfonate (EMS), diethyl sulfate (DES), ethylenimine (EI), sodium azide (SA) has been most frequently used for reliable and gene-specific mutations. A comprehensive review of the role of mutation breeding in mandarins and lime crop improvement has been discussed [24, 25]. Irradiation and chemical mutagen treatment of seeds and budwood have been commonly used by breeders for inducing variation followed by selection and clonal propagation. Mutation breeding applications have been reported in different fruit crops including papaya, peach, pear, grapes, sweet and sour cherries, banana, plum, almond [26], apple [27], and rough lemon [28]. Natural bud mutants include Washington navel orange, most of the early grapefruit varieties including Marsh, Foster, Shamber, Salustiana sweet orange, and Shamouti orange have originated as bud sports. Now there are several commercial seedless varieties including Daisy SL, Kinnow SL, Fairchild SL, and Tango that have been developed from their seedy parents through mutation breeding and are being commercially cultivated [29]. Other commercial mutants in citrus include sweet orange varieties Jin Cheng [30], Kozan [21], and NIAB Kinnow mandarin [31]. In grapefruit, Rio Red and Star Ruby are two induced mutants that have obtained commercial significance due to their better fruit color and seedlessness, respectively [32]. These are leading grapefruit varieties in Texas, USA. Star Ruby is the leading variety in Turkey, South Africa, Australia, and Spain. Rio Red is the main cultivar in China, India, and Argentine [33]. In Pakistan, Shamber is the main grapefruit variety that needs to be replaced with other potential candidate varieties like Star Ruby, Rio Red, and Flame [10]. Conclusively mutation breeding has shown its enormous potential in citrus crop improvement particularly in economically important horticultural traits. However, it is a slow and long-term process and takes more time to detection of desirable phenotypic variability. Utilization of modern breeding tools including molecular markers, advanced methods for phenotypic screening like Targeting Induced Local Lesions IN Genomes (TILLING) [34], using targeted mutagenesis and genome editing technologies [35] like Clustered Regularly Interspaced Short Palindromic Repeats (CRISPR/Cas9) could enhance the efficiency and cost-effectiveness of variety development having novel traits in citrus.

3.2 Ploidy manipulations

Polyploid organisms have a greater number of chromosomes compared with their diploid progenitors. Breeders have utilized polyploidization for the investigation of inheritance patterns in genes of interest. Polyploids have shown tremendous success in nature due to higher heterozygosity, less inbreeding depression, and more tolerance to biotic and abiotic stress conditions compared with their diploid progenitors [36, 37, 38]. The duplicated genes may evolve new functionalization during evolution [39]. Polyploids have been reported in many fruit crops including grapes, apples, strawberry, and citrus [40, 41], however, the frequency of spontaneous polyploid events is quite low and breeders prefer to induce hyperploidy using different chemicals.

Among chemical mutagens, colchicine is mostly used for the induction of polyploids due to its more reliability, higher efficacy, and cost-effectiveness. Colchicine is an alkaloid derived from *Colchicum autumnale* (meadow saffron). It is used for inducing chromosome doubling or developing tetraploids by restricting the chromosomal segregation at metaphase in meiosis [42, 43]. Other methods of polyploid induction include interploid hybridization [44], unreduced gamete formation [45, 46], and endosperm culture [47, 48].

Members of the subfamily *Aurantioidae* including *Citrus*, *Fortunella*, and *Poncirus* are mainly diploid having chromosome number 2n = 18 [49]. The occurrence of spontaneous polyploids in citrus is known since the 1940s [50]. Important spontaneous polyploids include triploid Tahiti lime [51], Triphasia desert lime [52], *Clausena excavata* [53], tetraploid mandarins [54], sweet oranges [41, 42] and grapefruit [43]. In spontaneous polyploids, triploids and tetraploids are believed to be formed by doubling of chromosomes in nucellar cells and fertilization of the unreduced gametophytes [55, 56]. Polyploids have been induced using colchicine in several citrus species and tetraploids produced have been used for interploid crossing to develop triploid progenies that are usually seedless due to irregular distribution of the chromosomes during cell division particularly gamete formation and formation of unreduced gametes. In interploid crossing, the formation of tetraploids in addition to triploids indicates the predominant formation of the unreduced (2n) gametes which may be formed by the first division restitution (FDR) or second division restitution (SDR) during meiosis. Production of 2n gametes was predominantly via SDR in lemon [44, 45] and monoembryonic Orah mandarin [57]. Higher tetraploid: triploid ratio in the progeny of the interploid hybridization indicates greater production of the 2n megagemtophytes in that cultivar which is promising to produce a greater number of polyploids.

3.3 *In vitro* culture: somatic hybridization

Plant tissue culture tools offer advantages related to efficient regeneration, propagation, and crop improvement in citrus and other horticultural crops. Endosperm cultures have been used for the development of triploids in citrus [48]. In interploid and wide hybridizations, the progeny may be sterile or have underdeveloped or shriveled seeds with viable embryos. The embryo rescue technique has shortened the breeding cycle and many plants have been recovered from these embryos through *in vitro* culture in different citrus species [45]. Similarly, micrografting is another tool in which a miniature bud is grafted under aseptic conditions on *in vitro* raised rootstocks and micrografted plants have been reported in many citrus varieties [58]. Micrografting is also useful for the production of virus-free citrus plants.

Another highly promising and most widely used approach is somatic hybridization which is utilized to overcome sexual incompatibility and to enhance genetic variability by combining nuclear and organelle (chloroplast and mitochondria) genomes followed by their characterization for hybrid confirmation and variability assessment [59]. The organelle genomes are known to encode genes related to photosynthesis and male sterility and new hybrids could be developed having novel genetic recombinations. Somatic hybrids may be developed through electrofusion of plant embryogenic protoplasts predominantly with mesophyll protoplasts. The plant progeny having nuclear origin could be characterized and separated using flow cytometry and molecular markers [60].

Protoplast fusion of distantly related citrus species bypasses the biological barriers and develops allopolyploids that could not be obtained through classical breeding. Somatic hybridization is an important tool and has been widely used in citrus scion and rootstock breeding. The first intergeneric allotetraploid somatic hybrid of Trovita sweet orange and *Poncirus trifoliata* was reported by Ohgawara et al. [61] followed by several interspecific and intergeneric hybrids in citrus from the USA [62], Japan [63], and other citrus-producing countries. Triploids were also reported from interspecific and intergeneric somatic hybridization of Citrus species, kumquats (*Fortunella japonica*), and *Poncirus trifoliata* by protoplast fusion [64]. Fusion of protoplasts from the haploid lines and diploid cultivars may also yield triploids and hundreds of triploids and tetraploids were developed and planted for field evaluations [65]. Polyethylene glycol may also be used to induce regeneration in the fused protoplast cultures as reported in Willow leaf mandarin (embryogenic parent) and Duncan grapefruit and sweet orange (mesophyll parents). The regenerated plants were identified as alloplasmic cybrids [66].

Polyploids developed through somatic hybridization have also shown enhanced tolerance to abiotic and biotic stress conditions. Allotetraploids of cv. FlhorAG1 (FL-4x) developed by somatic hybridization of diploid *Poncirus* and *Citrus* showed greater tolerance to cold and higher light conditions compared with parents (diploid) and their tetraploids [36]. Kumquats (*Fortunella* species) chloroplast have demonstrated higher resistance to canker in diploid kumquats and their tetraploid somatic hybrids developed with other citrus species including grapefruit [37].

4. Innovative approaches/technologies

4.1 Transgenesis

Since the advent of recombinant DNA technology, transgenesis has proved its significance, and 190.4 million hectares of transgenic crops were grown in more than 29 countries in 2019. They have significantly contributed to food security, climate change mitigation, sustainability thus uplifted the lives of 17 million biotech farmers worldwide. The first transgenic plant was developed in the 1980s and was available as commercial food in the 1990s. More than 400 transformation events have been approved so far wherein 356 events have been approved for crop plants, 23 for ornamentals, 22 for fruit plants, and 2 for trees. Hence a wide range of plant species (maize, cotton, canola, papaya, rice, tomato, sweet pepper, squash, popular, petunia, sugarcane, alfalfa, and citrus) have been engineered for various valuable traits i.e. insect resistance, herbicide tolerance [67], abiotic stress tolerance, improved nutritional value, and disease resistance. In addition to nuclear transformation plastid genome has also been targeted and has proved to be of more value as multiple genes can be introduced at a specific target site, the transgene is contained owing to maternal inheritance, and hyperexpression of the transgene, etc. [68, 69].

Citrus is an economically important fruit crop worldwide. It not only provides essential minerals and vitamins but is also of great commercial importance. Conventional research has contributed a lot to the improvement of this fruit yet serious problems are evolving which are difficult to tackle with these conventional approaches [70]. Juvenility, sexual incompatibility, high heterozygosity, apomixes, large plant size, and nucellar polyembryony, and certain other biological limitations hinder the improvement of these plant species through conventional breeding. Genetic manipulation through advanced innovative techniques is a potential approach to improve crop plants as well as fruit species. Though citrus species are recalcitrant to transformation and subsequent rooting, yet consistent efforts by the researchers have resolved these bottlenecks and proficient protocols have been established. Likewise, various transformation methods i.e. Agrobacterium-mediated transformation [71], biolistic transformation [72], and chemically assisted uptake of recombinant DNA by protoplasts [73] have been attempted to introduce genes of agronomic value as well as to strengthen it against bacterial, viral, and fungal pathogens (Figure 1).

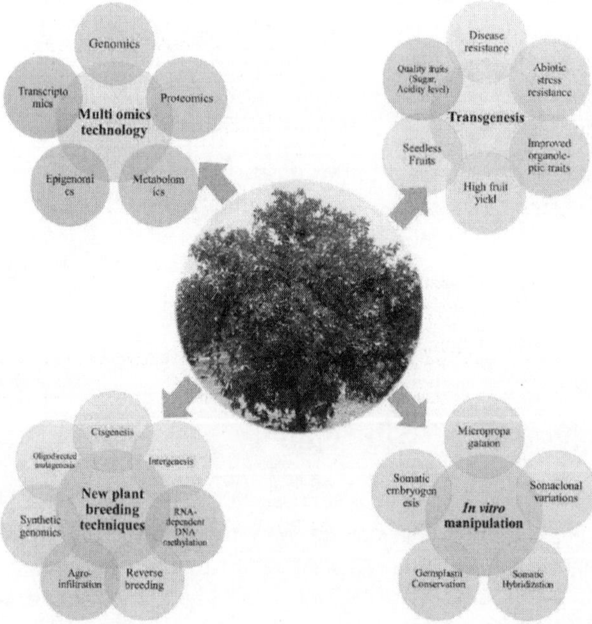

Figure 1. Schematic sketch showing the importance of conventional and advanced innovative approaches for the improvement of different citrus species.

Genetic manipulation of vegetatively propagated crops like citrus is very tricky as the expression of transgenes over a long period during numerous cycles of graft propagation should be stable.

The first attempt to produce transgenic citrus was made in the 1980s wherein protoplast transformation was attempted but it was not successful. The first authentic report was published by Kaneyoshi et al. [74] who reported transforming NPT II and GUS genes into trifoliate orange through Agrobacterium. Epicotyls of the aforementioned citrus species were used to transform with the selectable marker gene as well as reporter gene and more than 25% transformation efficiency was achieved. Likewise, Yao et al. [72] reported the first successful transformation through gene gun. They transformed tangelo (*C. reticulata* × *C. paradisi*) embryogenic cells.

Since genetic transformation has successfully been performed in different species and hybrids including Carrizo citrange, Washington naval orange, *Poncirus trifoliata*, Sour orange, Mexican lime, sweet orange, *Citrus reticulata* [75], and a valuable rootstock, swingle citrumelo. Similarly, protocols have been optimized for the genetic transformation of different citrus species by using different explant tissues including seeds, embryogenic cells, epicotyls, embryogenic cells, callus, nodal stem segments, and protoplasts. The most responsive explant tissue has been epicotyl from the *in vitro* germinated seedlings and is preferably used for genetic transformation research. Duncan grapefruit was successfully transformed through *Agrobacterium* for the first time using epicotyl and confirmation of the transgene (NPTII and GUS) integration was carried in the resultant 25 transgenic plants by histochemical staining, PCR, and Southern blot hybridization. Transgenic grapefruit, sweet orange, and citrange plants were developed using epicotyls as target explant whereas selection was carried out on kanamycin [76]. Epicotyl has also been used for Agrobacterium-mediated transformation of citrange and sweet orange [77]. In addition, callus, as well as suspension cultures derived from different parts of flower and seed, have also been attempted to transform. The transformation efficiency attained, in this case, was lower than 0.5%. Genetic transformation has also been optimized in pomelo (*Citrus grandis*) and sour orange wherein internodal stem segments were used as explants and a promising transformation efficiency was achieved (91%) [78].

The biolistic transformation has also been performed successfully in tangelo (*C. paradisi* Macf. x *C. reticulata* Blanco) using nucellar embryogenic cells raised from the suspension culture and more than 15 stable whereas 600 transient transgenic lines were attained per bombardment. The transformed calli cells showed proficient growth on kanamycin selection medium and positive GUS activity but were not able to regenerate into plants. Calli treatment with 0.3 M osmoticum sorbitol and 0.3 M mannitol appeared to have positive effects for enhanced transformation efficiency for stable and transient transformation. Thin epicotyl segments of germinated seedlings were also targeted through the biolistic gun and more than %93 transformation efficiency was attained for transient transgene integration in *C. citrange*. The incubation of explant on culture medium before bombardment appeared to have profound effects on transformation efficiency which was further improved [79].

Since transgenic technology is the most reliable intervention having the massive potential to improve the citrus crop. The introduction of alien genes is only possible through this technology so any of the desired traits can be engineered. Recent research indicates that citrus growing farmers are facing severe problems due to biotic and abiotic stresses i.e. salinity, cold, drought, and diseases. Hence, the development of improved citrus varieties is direly needed to get a quality crop. Various citrus species have been engineered with alien genes to combat abiotic stresses including salinity and drought. Expression of HLA2 gene, isolated from yeast imparted salinity tolerance and resultant transgenic plants were able to tolerate a higher level of the salts as compared with non-transformants [80]. PsCBL and PsCIPK derived from *Pisum sativum* were transformed into *Citrus sinensis* and *Citrus reticulata* by targeting calli derived from mature seed. Bacterial strain LBA4404 was used to induce infection in the target calli cells. The putative transformants showed better performance as compared with control plants for salinity and drought tolerance when tested under *in vitro* conditions [81].

Citrus paradisi was transformed to improve carotenoid content by manipulating the genes involved in carotenoid biosynthesis i.e. phytoene synthase, lycopene-β-cyclase, and phytoene desaturase. The multigene transgenic citrus plants were aimed to supplement human nutrition with vitamin A along with antioxidants. Similarly, fruit juice quality has been attempted to improve in Valencia orange, a valuable variety that is majorly grown for its juice. Degradation of TSPME (thermostable pectin methylesterase) deteriorates the quality of the juice. This TSPME is encoded by the *CsPME4* gene. The protoplasts were isolated from embryogenic suspension cultures and transgene was introduced through the PEG mediated transformation method [82] aiming at down-regulation of the *CsPME4* resulting in the citrus with improved juice quality.

Citriculture is prone to be infected by a wide range of diseases that are controlled by chemicals, a drastic non-environmentally friendly strategy. Different types of viral, bacterial, and fungal pathogens infect these plants resulting in drastic losses to crop production and quality of the produce. A range of transgenic citrus lines have been developed varying from fully resistant to susceptible to the diseases. Coat protein (p25) from the CTV (Citrus Tristeza Virus) was expressed in Mexican lime and 33% of the transgenic plants were found to be resistant, neither symptoms appeared nor the viral load was detected. Accumulation of siRNA (small interfering RNA) in the transgenic lines resulted in resistant phenotype and plants were able to withstand viral infection [83]. Expression of viral coat protein (part of *p23* gene and the 3UTR), in the sense and antisense orientation also delayed viral infection in grapefruit [84].

Phytophthora is a noxious fungal pathogen that has been reported to infect a wide range of citrus species. Among these *Phytophthora parasitica* and *Phytophthora citrophthora* cause more severe damage in the citrus orchards and nurseries all over the world [85]. Expression of *bo* gene (bacterio-opsin) in Rangpur lime rootstock showed an elevated level of tolerance against *Phytophthora parasitica* infection. It was observed that expression of the aforementioned gene led to induce expression of defense-related proteins; chitinase, salicylic acid, and glucanase. Hence plants with an elevated level of transgene expression showed greater resistance against the oomycetic fungi including *Phytophthora parasitica*. Transgenic citrus plants expressing the tomato *PR5* gene showed an enhanced survival rate in the presence of pathogen (*P. citrophthora*). Transgenic grapefruit plants were able to better withstand citrus scab infection when transformed with the *attE* gene encoding for antimicrobial peptide [86].

Transgenic technology has also played a pivotal role to tackle another noxious disease in citrus i.e. Huanglongbing (HLB) which is supposed to be caused by phloem-restricted Gram-negative bacteria; *Candidatus Liberibacter* americanus, *Candidatus Liberibacter* asiaticus, and *Candidatus Liberibacter* africanus [87]. Various genetic strategies have been tested to develop HLB-resistant citrus lines with decreased susceptibility to the pathogen. These include the expression of anti-microbial peptides from a bacterial, fungal, plant, or animal origin and engineering host-pathogen interaction pathways. The expression of antimicrobial proteins under phloem-specific promoters has been an effective strategy to control this phloem-resident pathogen. Overexpression of synthetic cecropin B gene under phloem specific promoter resulted in reduced bacterial population after one year of inoculation and no disease symptoms appeared even after two years of inoculation [88]. Overexpression of modified methionine under double 35S promoter also appeared to have an inhibitory effect on bacterial growth and lowered down the CLas (*Candidatus Liberibacter* asiaticus) titer in the roots of transgenic Carrizo citrange (rootstock). Further, newly emerging leaves from the rough lemon, grafted on this transgenic rootstock, also had a non-detectable bacterial titer. Expression of AtNPR1 and chimeric proteins (ThioninLBP and Thionin-1D4E1) demonstrated elimination of CLas providing tolerance against HLB infection [89].

Another economically important disease, the citrus canker has also been addressed through transgenesis resulting in enhanced tolerance against *Xanthomonas citri.* Engineering sweet orange genome with *Xa21* gene showed a significant reduction in disease severity upon inoculation in three lines Hamlin, Pera, and Natal. Expression of the *Xa21* gene under its promoter appeared to be more effective in disease resistance when expressed in highly susceptible Anliucheng sweet orange [90]. Transgenic Carrizo citrange and sweet orange plants were developed by introducing RpfF from *X. fastidiosa* which encodes for a quorum-sensing factor and can disrupt bacterial communication by reducing activation of virulence factors, thus enhancing the ability to tolerate pathogen infection. Similarly, the expression of AMP sarcotoxin from flesh fly also enhanced tolerance against *X. citri* [91].

4.2 Genome editing

Genome editing through CRISPR-Cas9 has emerged as a breakthrough technology for the precise modification and manipulation of targeted genomic DNA. It has extensively been exploited by several research groups [92] and certain successes have been achieved. Three major sequence-specific engineered nucleases that have so far been used for genome editing are CRISPR-Cas (clustered regularly interspaced short palindromic repeats), TALENs (transcription activators such as effector nucleases), and ZFNs (zinc finger nucleases). Among these, the CRISPR-Cas9 editing system has been established in many plant species through gene activation, repression, mutation, and epigenome editing in wheat, rice, maize, tomato, potato, carrot, apple, grape, and citrus. Even a few of the genome-edited crops have been approved for commercial cultivation. Through this technology, field crops as well fruit crops can not only be manipulated for improved agronomic traits but can also be manipulated for improved nutritional value [93].

For citrus, genome editing research is at infancy, yet few successes have been achieved by editing its genome for enhanced resistance against diseases. The CRISPR-Cas9 system was firstly used to target the *CsPDS* gene in Duncan grapefruit and sweet orange. The target gene was successfully modified through a transient expression method, Xcc-facilitated agroinfiltration [94]. The modified *CsPDS* sequence was not detectable in the leaves of sweet orange indicating that CRISPR/Cas9 has induced the desired mutation successfully.

Most of the research studies were carried out to target the *LOB1* (LATERAL ORGAN BOUNDARIES 1) gene which has been characterized as a citrus susceptibility gene for *Xanthomonas citri.* The said gene has been explored to be upregulated by TAL (transcription activator-like) effector PthA4 which binds to the EBE (effector binding elements) in the promoter region of LOB1 thus activates expression of this canker-susceptibility gene [95]. Mutation in the single allele of effector binding elements of the *LOB1* gene resulted in minor alleviation of canker symptoms in Duncan grapefruit. Anyhow, a mutation in effector binding elements of both of the alleles of LOB1 promoters alleviated canker symptoms to great extent thus showing a high degree of resistance in Wanjincheng orange [96]. Another research group explored that editing the coding region of LOB1 in Duncan grapefruit, through the CRISPR-Cas9 system also provides resistance against *X. citri* infection.

A marker gene for pathogen triggered immunity (CsWRKY22) was knocked out in Wanjincheng orange and the resultant mutant plant showed a decreased level of susceptibility to the canker disease [97]. In addition to the CRISPR-Cas9 system, another improved genome editing system (CRISPR/Cas12a (Cpf1) has also been used to edit the *CsPDS* gene in the Duncan grapefruit gene. It appeared to be a more efficient editing system with lower off-target effects thus will prove a great milestone in citrus genome editing [98]. These studies indicate that CRISPR-mediated genome editing can be a promising pathway to generate disease-resistant citrus cultivars [99].

5. Multi-Omics technology: An integrated approach and useful strategy for the improvement of the citrus crop

MultiOMICS including genomics, transcriptomics, proteomics, metabolomics, interactomics, and phenomics approaches have massive potential for citrus improvement just like other crops and fruits. In all disciplines of OMICS, various techniques can be utilized for genome analyses, transcripts, proteins, metabolites, and interactions between different molecules to indicate the molecules which may result in crop improvement.

5.1 Genomics

The field of genomics is a highly applicable part of Omics technologies. It is based on sequencing technologies and the analysis of subsequent genome sequences. Many advanced techniques in genomics for example sequence determination DNA, marker-assisted selection, and transition from marker-assisted selection to genomic selection assist in quick varietal development. Genome sequencing technologies have brought about a revolution in the field of biology. It has also transformed the citrus breeding that helps to understand a relationship between the genetic makeup and response towards various abiotic and biotic stresses like Alternaria brown spot [100].

A specific pathotype of *Alternaria alternata* (Fr.) Keisel is a disease with heavy losses [101]. It causes necrosis and resultant lesions on the surface of fruits and young leaves. It leads towards defoliation and fruit drop [102]. Thus, exploitation of innate genetic resistance appears to be the most applicable and effective strategy of disease control. Currently, the control is primarily based on the application of 4–10 sprays of toxic and environmentally injurious fungicide per year [103]. Such limitations are compelling farmers to find alternative cultivars with resistant ones [104].

Usually, the female parent transmits the $2n$ gamete in 2x × 2x citrus crosses [105, 106, 107, 108]. Cuenca et al. [109] recognized ABS resistance locus containing genomic region by using BSA-genome scan combined with HTA based. Trait segregation in crosses between two heterozygous ABS-susceptible or between heterozygous ABS-susceptible parents and resistance was used to confirm the recessive inheritance of the ABS resistance in triploid populations. ABSr locus was first located at 10 cM from a centromere based on segregation of 368 SDR $2n$ gametes. A genomic region containing several markers with a high probability (> %99) of association with phenotype variation was identified on chromosome III by performing BSA over 93 triploid hybrids from a Fortune × Willow leaf population. This identified region contains 25 significant SNP markers within an interval of 13.1 cM. The size of the genomic region among these two markers is 15 Mb. Linkage genetic mapping was performed on identified genomic regions by developing new SNP and SSR markers. A -268diploid mapping population was performed by Cuenca et al. [110] from a heterozygous-susceptible × resistant hybridization. Fine mapping was performed to confirm the location of *ABSr* locus in a region of 1.1 cM between the markers SNP05/SNP06/SNP07/AT21 (at 0.7 cM) and SNP08 (at 0.4 cM). Another region containing eight genes with NBS-LRR repeats was identified by the SNP08 marker and considered ABS resistance genes.

In citrus plant, molecular markers are linked to some agronomic traits, e.g. SSR markers are linked to Citrus tristeza virus resistance from *Poncirus trifoliate*, PCR assay for the anthocyanin content of pulp [111], AFLP markers are associated with polyembryony [112] and RAPD markers are associated to dwarfism and fruit acidity [113]. Some other characteristics such as salinity tolerance and nematode resistance are linked to QTLs [114]. The selection of resistant genotype at the early growth stage was improved by the newly developed SNP08 marker. This marker was mapped at 0.4 cM from the ABD resistance gene and it has role in avoiding the selection of susceptible varieties. On the other side of the gene, some new markers were also identified at 0.7 cM from the *ABSr* locus. Combining these new markers with SNP08, the probability of selection of resistant genotype was increased by %0.0028. This marker appeared to be very helpful in the selection of resistant and susceptible genotypes and for analyzing the resistant germplasm to configure the ABS genes. So, it is a very valuable tool for the selection of susceptible heterozygous cultivars which may be used as breeding parents allowing manipulation of genetic diversity in citrus and prevents susceptible homozygous genotypes.

About 40 mandarin genotypes (susceptible and resistant) were tested by the SNP08 marker and were used as breeding parents. An ultimate association was observed between response to *Alternaria* infections and SNP08 marker. Recently SNP08 is used in breeding programs of citrus performed at CIRAD and IVIA for the selection of ABS-resistant citrus genotypes. About 2187 resistant hybrids were selected from 4517 total hybrids rising from 10 different parental combinations by using the SNP08 marker since its development. This analysis was very helpful to prevent the growth of more than 2000 susceptible lines which were removed at the early growth stage after selection so, a lot of time, cost, personnel, and resources were saved.

5.2 Proteomics and metabolomics approaches

Proteomics is the comprehensive analysis of all the proteins found in a cell. It includes the identification of proteins, their location in the cell, their interactions with other proteins and other biological components in the cell, and most importantly post-translational modifications that a protein undergoes in the cell [115]. Metabolites are referred to as the last product of any biological activity in a cell and are found in very small quantities [116]. Metabolites are small molecules including intermediates of various metabolic reactions, signaling molecules, hormones, and other regulatory products found in a cell. Hence, metabolomics is defined as the study of metabolites of a cell [117, 118]. It is estimated that around five thousand metabolites are found in any cell depending upon the physical and chemical complexity of that cell [119].

Huanglongbing (HLB) is considered one of the most devastating citrus diseases that affect not only the production but also the quality of citrus fruit and its juice. Using a combination of proteomics and metabolomics approaches it was found that in symptomatic fruit, the expression of proteins found in the cytoplasm for glycolysis, in mitochondria involved in the tricarboxylic acid (TCA) cycle, and in chloroplasts for the synthesis of amino-acids was downregulated. Similar downregulation was observed for genes involved in terpenoid metabolism for example valencene, limonene, 3-carene, linalool, myrcene, and aterpineol in fruit found on infected trees. Similar phenomena were observed for sucrose and glucose. Hence, the off-flavor found in symptomatic fruits was linked to the downregulation of the above genes and a decrease in the levels of the abovementioned secondary metabolites [120].

In another study, comparative iTRAQ proteomic profiling was carried out using the fruits of sweet orange which was grafted on sensitive and tolerant rootstocks infected by CaLas. The results showed that symptomatic fruit on sensitive rootstock exhibited a greater number of differentially expressed (DE) proteins as compared to the healthy fruit on a similar rootstock. It was also found that the expression level of various defense-related proteins was reduced in symptomatic fruit on sensitive rootstock, particularly the proteins related to the jasmonate biosynthesis, is signaling, protein hydrolysis, and vesicle trafficking. Hence, it was concluded that the down-regulation of these proteins is likely to be linked with the sensitivity of citrus to the CaLas pathogen [121].

5.3 Interactomics and metabolomics and phenomics

Interactomics bears a broad scope as it may cover a complete set of interactions in a cell [122]. It covers every type of interaction among interacting molecules including proteins and other molecules. It is a well-known fact that the Protein–protein interactions are major of all cellular processes [123].

To designate the complete phenotype of a plant, the term phenome is used. Similarly, a phenotype encloses a group of traits that are liable to be distinguished either by utilizing modern science analytical techniques or by a naked eye evaluation. These traits can also be attributed to being an interaction between external factors (environment) and Genotype. David Houle also termed phenomics as the collection of data from varying backgrounds and dimensions in a single entity [124]. Phenomics involves both "extreme phenotyping," referring to a comprehensive selection of a wide range of valid and correct phenotypes, and "phenome analysis" indicating towards an analysis of specimen and correlation between syndication of genotype and phenotype.

Plant phenomics utilizes screening of large populations to analyze genetic mutations found in the population for a specific trait (drought, salinity, or high-temperature stress tolerance). Various types of imaging techniques are employed in the phenotyping of plants for various growth and developmental processes. The techniques include visible-light imaging [125], Thermographic imaging [126], Hyperspectral imaging, Chlorophyll fluorescence, X-ray, MRI, PET [127].

Using the phenomics approach and tools, we can study the traits regarding plant growth, leaf growth, root growth, and architecture of soil/root interaction, etc. This extensive use of phenomics and its integration into OMICS is the need of the hour to combat food security issues and overcome adverse effects of climate change on crop production.

6. Conclusions

Conventional research has played a pivotal role in the improvement of citrus. Enhanced heterozygosity has helped in the development of genetically diverse germplasm in most of the citrus species and numerous varieties have been released for commercial cultivation. However, with the advent of modern biotechnological tools, the period involved in crop improvement through indirect mutagenesis and polyploidization could be further reduced and enhancing cost-effectiveness. Transgenic technology and OMICS have great potential to improve this fruit crop. MultiOMICS, integrative-OMICS, or panOMICS technologies may result in better crops having better agronomic traits, enhanced yield potential, and less prone to insect pests. It will ultimately lead towards food security and poverty alleviation. Various OMICS technologies have been used for crop improvement, yet their integrated use will further strengthen the application of this robust technology. Still, there are many challenges associated with tolerant varieties which need to be fine-tuned. Moreover, three thousand reports of enhanced drought and salinity tolerance in wheat, sorghum, canola and rice are present but none of them is in use by farmers. A fundamental reason for this is that salinity and drought are complex multigenic traits. So, to induce tolerance in plants every gene needs to be fine-tuned precisely. However, their evaluation in the field is a long way, and distribution at the commercial level is also a hurdle in their production.

References

1. Swingle WT. A new taxonomic arrangement of the orange subfamily, Aurantioideae. Journal of Washington Academy of Sciences 1938;28(1) 530-533

2. Food and Agriculture Organization. Crops country data. http://faostat.fao.org/site/339/default.aspx (assessed 20 May 2021)

3. endonca LBP, Badel JL, Zambolim L. Bacterial citrus diseases: Major threats and recent progress. Bacteriol Mycol Open Access. 2017;5(4):340-350

4. Swingle WT., Reece PC. The botany of Citrus and its wild relatives. In: Reuther W., Weber HJ., Batchelor LD. (eds.) The Citrus Industry Volume 1, History, World Distribution, Botany and Varieties. Berkeley: University of California Press; 1967. p 190-243

5. Garcia-Lor A, Curk F, Snoussi-Trifa H, Morillon R, Ancillo G, Luro F, Navarro L, Ollitrault P. A nuclear phylogenetic analysis: SNPs, indels and SSRs deliver new insights into the relationships in the 'true citrus fruit trees' group (Citrinae, Rutaceae) and the origin of cultivated species. Annals of Botany 2013;111(1) 1-9

6. Wu GA, Terol J, Ibanez V, López-García A, Pérez-Román E, Borredá C, Domingo C, Tadeo FR, Carbonell-Caballero J, Alonso R, Curk F, Du D, Ollitrault P, Roose ML, Dopazo J, Gmitter FG, Rokhsar DS, Talon M. Genomics of the origin and evolution of Citrus. Nature 2018;554(1) 311-316

7. Oueslati A, Salhi-Hannachi A, Luro F, Vignes H, Mournet P, Ollitrault P. Genotyping by sequencing reveals the interspecific C. maxima/C. reticulata admixture along the genomes of modern citrus varieties of mandarins, tangors, tangelos, orangelos and grapefruits. PloS One 2017;12(10) e0185618

8. Krueger R, Navarro L. Citrus germplasm resources. In: Khan IA. (ed.) Citrus Genetics, Breeding, and Biotechnology. Wallingford: CAB International; 2007. pp. 45-140

9. Grosser JW, Gmitter FG. Use of in vitro technique to facilitate seedless Citrus cultivars development. Proceedings of the XXIII International Horticultural Congress, 27 August-1 September 1990. Firenze, Italy: 1990a

10. Rana MA, Usman M, Fatima B, Fatima A, Rana IA, Rehman W, Shoukat D. Prospects of mutation breeding in grapefruit (Citrus paradisi Macf.). Journal of Horticultural Science & Technology 2020;3(2) 31-35

11. Ahloowalia BS., Maluszynski M, Nichterlein K. Global impact of mutation derived varieties. Euphytica 2004;135(1) 187-204

12. Dobres MS. Barriers to genetically engineered ornamentals: An industry perspective. In: da Silva JAT. (ed.) Floriculture, Ornamental and Plant Biotechnology; Advances and Topical Issues. Isleworth: Global Science Books; 2008. p. 1-14

13. Kharkwal MC. A brief history of plant mutagenesis. In: Shu QY, Forster BP, Nakagawa H. (eds.) Plant Mutation Breeding and Biotechnology. Wallingford: CAB International; 2012. p 21-30

14. Jankowicz-Cieslak J., Mba C, Till BJ. Mutagenesis for crop breeding and functional genomics. In: Jankowicz-Cieslak J, Tai T, Kumlehn J, Till B. (eds.) Biotechnologies for Plant Mutation Breeding. Switzerland: Springer Cham; 2017. p 3-18

15. Mutant variety database. MVD 2021. http://mvd.iaea.org/ (Accessed: 22 May 2021)

16. Lonnig WE. Mutation breeding, evolution, and the law of recurrent variation. Recent Research and Development in Genetics and Breeding 2005;2(1) 45-70

17. Ollitrault P, Navarro L. Citrus. In: Fruit breeding 2012 (pp. 623-662). Springer, Boston, MA

18. Mba C., Afza R, Shu QY. Mutagenic radiations: X-rays, ionizing particles and ultraviolet. In: Shu QY, Forster BP, Nakagawa H. (eds.) Plant Mutation Breeding and Biotechnology. Wallingford: CAB International; 2012. p 83-90

19. Goldenberg L, Yaniv Y, Porat R, Carmi N. Effects of gamma-irradiation mutagenesis for induction of seedlessness, on the quality of mandarin fruit. Food and Nutrition Sciences. 2014;5 943-952

20. Lamo K, Bhat DJ, Kour K, Solanki SPS. Mutation studies in fruit crops: A review. International Journal of Current Microbiology and Applied Sciences 2017;6(12) 3620-3633

21. Cimen B, Yesiloglu T, Incesu M, Yilmaz B. Studies on mutation breeding in citrus: Improving seedless types of 'Kozan' common orange by gamma irradiation. Scientia Horticulturae. 2021;278:109857

22. Amano E. Use of induced mutants in rice breeding in Japan. Plant Mutation Rep 2006;1(1) 21-24

23. Matsumura A, Nomitzu T, Furukani N, Hayashi K, Minamiyama Y, Hase Y. Ray florets colour and shape mutants induced by $^{12}C^{5+}$ ion beam irradiation in chrysanthemum. Scientia Horticulturae 2010;123(4) 558-561

24. Usman M., Fatima B. Mandarin (Citrus reticulata Blanco) breeding. In: Al-Khayri JM, Jain SM, Johnson DV. (eds.) Advances in Plant Breeding Strategies: Fruits. Switzerland: Springer Cham; 2018. p 465-533

25. Usman M., Khan MM, Al-Yahyai R, Fatima B. Lime breeding: A way forward. In: Yahia, E.D. (ed.). Achieving Sustainable Cultivation of Tropical Fruits. Cambridge: Burleigh Dodds Science Publishing; 2019b. p 1-45

26. Predieri S. Mutation induction and tissue culture in improving fruits. Plant Cell Tissue and Organ Culture 2001;64(2) 185-210

27. Campeanu G, Neaţa G, Darjanschi G, Stan R. Tree and fruit characteristics of various apple genotypes obtained through mutagenesis. Notulae Botanicae Horti Agrobotanici Cluj-Napoca 2010;38(1) 248-251

28. Saini HK, Gill MIS. Induction of mutation in rough lemon (Citrus jambhiri Lush.) using gamma rays. Journal of Horticultural Sciences 2009;4(1) 41-44

29. Williams T, Roose ML. Daisy SL, Fairchild SL and Kinnow SL-Three New, Very Low-Seeded, Mid Season Irradiated Selection of W. Murcott Mandarin from the University of California Riverside. In: International Citrus Congress 2008

30. Zhang YJ, Wang XJ, Wu JX, Chen SY, Chen H, Chai LJ, Yi HL. Comparative transcriptome analyses between a spontaneous late-ripening sweet orange mutant and its wild type suggest the functions of ABA, sucrose and JA during citrus fruit ripening. PLoS One 2014;9(12) e116056

31. Altaf S, Khan MM, Jaskani MJ, Khan IA, Usman M, Sadia B, Awan FS, Ali A, Khan AI. Morphogenetic characterization of seeded and seedless varieties of Kinnow Mandarin (Citrus reticulata Blanco). Australian Journal of Crop Science 2014;8(11) 1542-1549

32. Maluszynski M, Nichterlein K, Van-Zanten L, Aloowalia BS. Officially released mutant varieties – The FAO/IAEA database. Mutation Breeding Review 2000;12(1):1-84

33. Singh A, Naqvi SAMH, Singh S. Citrus Germplasm. Cultivars and Rootstocks. Ludhiana: Kalyani Publishers; 2002

34. Parry MA, Madgwick PJ, Bayon C, Tearall K, Hernandez-Lopez A, Baudo M, Rakszegi M, Hamada W, Al-Yassin A, Ouabbou H, Labhilili M. Mutation discovery for crop improvement. Journal of Experimental Botany 2009;60(10) 2817-2825

35. Luo M, Gilbert B, Ayliffe M. Applications of CRISPR/Cas9 technology for targeted mutagenesis, gene replacement and stacking of genes in higher plants. Plant Cell Reports 2016;35(7) 1439-1450

36. Oustric J, Morillon R, Ollitrault P, Herbette S, Luro F, Froelicher Y, Tur I, Dambier D, Giannettini J, Berti L, érémie Santini J. Somatic hybridization between diploid Poncirus and Citrus improves natural chilling and light stress tolerances compared with equivalent doubled-diploid genotypes. Trees. 2018;32(3) 883-895

37. Murata MM, Omar AA, Mou Z, Chase CD, Grosser JW, Graham JH. Novel plastid-nuclear genome combinations enhance resistance to citrus canker in cybrid grapefruit. Frontiers in plant science. 2019;7(9) 1858

38. Khalid MF, Hussain S, Anjum MA, Ali MA, Ahmad S, Ejaz S, Ali S, Usman M, Ehsan ul Haq, Morillon R. Efficient compartmentalization and translocation of toxic minerals lead tolerance in volkamer lemon tetraploids more than diploids under moderate and high salt stress. Fruits 2020; 75(5) 204-215

39. Soltis DE, Soltis PS. Polyploidy: Origins of species and genome evolution. Trends in Ecology and Evolution 1999;9(1):348-352

40. Usman M, Saeed T, Khan MM, Fatima B. Occurrence of spontaneous polyploids in Citrus. Horticultural Science (Prague) 2006;33(3) 124-129

41. Shafieizargar A, Awang Y, Juraimz AS, Othman R. Comparative studies between diploid and tetraploid Dez Orange (Citrus sinensis (L.) Osb.) under salinity stress. Australian Journal of Crop Science 2013;7(10) 1436-1441

42. Fatima B, Usman M, Khan IA, Khan MS, Khan MM. Identification of citrus polyploids using chromosome count, morphological and SSR markers. Pakistan Journal of Agricultural Sciences 2015;52(1) 107-114

43. Usman M, Fatima B, Usman M, Rana IA. Morphological and stomatal diversity in colchiploid germplasm of grapefruit. Pakistan Journal of Agricultural Sciences 2021;58(2) 555-560

44. Rouiss H, Cuenca J, Navarro L, Ollitrault P, Aleza P. Unreduced megagametophyte production in lemon occurs via three meiotic mechanisms, predominantly second-division restitution. Frontiers in Plant Science 2017;12(8) 1211

45. Xie KD, Xia QM, Peng J, Wu XM, Xie ZZ, Chen CL, Guo WW. Mechanism underlying 2n male and female gamete formation in lemon via cytological and molecular marker analysis. Plant Biotechnology Reports 2019;13(2) 141-149

46. Xia Q, Wang W, Xie K, Wu X, Deng X, Grosser JW, Guo W. Unreduced megagametophyte formation via second division restitution contributes to tetraploid production in interploidy crosses with 'Orah' mandarin (Citrus reticulata). Frontiers of Agricultural Science and Engineering 2017; https://doi.org/10.15302/J-FASE-2021385

47. Usman M, Fatima B, Gillani KA, Khan MS, Khan MM. Exploitation of potential target tissues to develop polyploids in citrus. Pakistan Journal of Botany 2008;40(4) 1755-1766

48. Yanagimoto Y, Kaneyoshi J, Yamasaki A, Kitajima A. Estimation of polyploidy of embryos and gametes based on polyploidy of endosperm remaining in the citrus seed. Horticultural Research (Japan). 2018;17(3) 293-302

49. Froelicher Y, Ollitrault P. Effects of the hormonal balance on Clausena excavata androgenesis. Acta Horticulturae 2000;535 139-146

50. Barrett HC, Hutchinson DJ. Spontaneous tetraploidy in apomictic seedlings of Citrus. Economic Botany 1978;32 27-45

51. Bachi O. Observaceous Citologicalem Citrus. I. Numero de cromosomas de algunas especies y variedades. Journal of Agronomy 1940;3(1) 249-258

52. Esen A, Soost RK. Tetraploid progenies from 2x x 4x crosses of Citrus and their origin. Journal of the American Society for Horticultural Science 1972;97 410-414

53. Froelicher Y, Luro F, Ollitrault P. Analysis of meiotic behavior of the tetraploid Clausea excavate species by Molecular Marker Segregation Studies. In: 9th ISC Congress. South Africa; 2000. p116

54. Jaskani MJ, Khan IA, Khan MM. Fruit set, seed development and embryo germination in interploid crosses of citrus. Scientia Horticulturae 2005;107(1) 51-57

55. Aleza P, Juárez J, Cuenca J, Ollitrault P, Navarro L. Extensive citrus triploid hybrid production by 2x× 4x sexual hybridizations and parent-effect on the length of the juvenile phase. Plant Cell Reports 2012;31(9) 1723-1735

56. Xie KD, Wang XP, Biswas MK, Liang WJ, Xu Q, Grosser JW, Guo WW. 2n megagametophyte formed via SDR contributes to tetraploidization in polyembryonic 'Nadorcott' tangor crossed by citrus allotetraploids. Plant Cell Reports 2014;33(10) 1641-1650

57. Xia Q, Wang W, Xie K, Wu X, Deng X, Grosser JW, Guo W. Unreduced megagametophyte formation via second division restitution contributes to tetraploid production in interploidy crosses. https://doi.org/10.15302/J-FASE-2021385

58. Aleza P, Cuenca J, Juárez J, Navarro L, Ollitrault P. Inheritance in doubled-diploid clementine and comparative study with SDR unreduced gametes of diploid clementine. Plant Cell Reports 2016;35(8) 1573-1586

59. Viloria Z, Grosser JW. Acid citrus fruit improvement via interploid hybridization using allotetraploid somatic hybrid and autotetraploid breeding parents. Journal of the American Society for Horticultural Science. 2005 May 1;130(3) 392-402

60. Grosser JW, Jiang J, Mourăo Filho FAA, Louzada ES, Baergen K, Chandler JL, Gmitter FG. Somatic hybridization, an integral component of Citrus cultivar improvement: I. Scion improvement. HortScience 1998;33(1) 1057-1059

61. Ohgawara T, Kobayashi S, Ohgawara E, Uchimaya H, Ishii S. Somatic hybrid plants obtained by protoplasm fusion between Citrus sinensis and Poncirus trifoliata. Theoretical and Applied Genetics 1985;71 1-4

62. Grosser JW, Mourăo Filho FAA, Gmitter FG, Louzada ES, Jiang J, Baergen K, Quiros A, Cabasson C, Schell JL, Chandler JL. Allotetraploid hybrids between Citrus and seven related genera produced by somatic hybridization. Theoretical and Applied Genetics 1996;92 577-582

63. Miranda M, Motomura T, Ikeda F, Ohgawara T, Saito W, Endo T, Omura M, Moriguchi T. Somatic hybrids obtained by fusion between Poncirus trifoliata (2x) and Fortunella hindsii (4x) protoplasts. Plant Cell Reports 1997;16(6) 401-405

64. Guo WW, Deng XX. Intertribal hexaploid somatic hybrid plants regeneration from electrofusion between diploids of Citrus sinensis and its sexually incompatible relative, Clausena lansium. Theoretical and applied genetics. 1999 1;98(3-4) 581-585

65. Grosser JW, Ollitrault P, Olivares-Fuster O. Somatic hybridization in citrus: an effective tool to facilitate variety improvement. In Vitro Cellular and Developmental Biology-Plant. 2000 1;36(6) 434-449

66. Cabasson CM, Luro F, Ollitrault P, Grosser J. Non-random inheritance of mitochondrial genomes in Citrus hybrids produced by protoplast fusion. Plant Cell Reports. 2001 20(7) 604-609

67. Khan MS, Ali S, Iqbal J. Developmental and photosynthetic regulation of Bacillus thuringiensis δ-endotoxin reveals that engineered sugarcane conferring resistance to 'dead heart' contains no toxins in cane juice. Molecular Biology Reports. 2011;38: 2359-2369

68. Mustafa G, Khan MS. Transmission of engineered plastids in sugarcane, a C4 monocotyledonous plant, reveals that sorting of preprogrammed progenitor cells produce heteroplasmy. Plants. 2021;10,26

69. Bock, R. and Khan, M.S. (2004). Taming plastids for a green future. Trends in Biotechnology 22: 311-318

70. Sun L, Ke F, Nie Z, Wang P, Xu J. Citrus genetic engineering for disease resistance: Past, present and future. International journal of molecular sciences 2019;20(21):5256

71. Dominguez A, Guerri J, Cambra M, Navarro L, Moreno P, Pena L. Efficient production of transgenic citrus plants expressing the coat protein gene of citrus tristeza virus. Plant cell reports. 2000; 19(4):427-433

72. Yao JL, Wu JH, Gleave AP, Morris BA. Transformation of citrus embryogenic cells using particle bombardment and production of transgenic embryos. Plant Science 1996;113(2):175-183

73. Fleming GH, Olivares-Fuster O, Del-Bosco SF, Grosser JW. An alternative method for the genetic transformation of sweet organce. In Vitro Cellular and Developmental Biology-Plant 2000;36(6):450-455

74. Kaneyoshi J, Kobayashi S, Nakamura Y, Shigemoto N, Doi Y. A simple and efficient gene transfer system of trifoliate orange (Poncirus trifoliata Raf.). Plant Cell Reports 1994;13(10):541-545

75. Singh S, Rajam MV. Citrus biotechnology: Achievements, limitations and future directions. Physiology and Molecular Biology of Plants 2009;15(1):3-22

76. Yang ZN, Ingelbrecht IL, Louzada E, Skaria M, Mirkov TE. Agrobacterium-mediated transformation of the commercially important grapefruit cultivar Rio Red (Citrus paradisi Macf.). Plant Cell Reports 2000;19(12):1203-1211

77. Yu C, Huang S, Chen C, Deng Z, Ling P, Gmitter FG. Factors affecting Agrobacterium-mediated transformation and regeneration of sweet orange and citrange. Plant Cell, Tissue and Organ Culture 2002;71(2):147-155

78. Xiao-hong Y, Zhong-hail S and Rui-jian T Optimizing culture system of Ri T-DNA transformed roots for Citrus grandis cv. Changshou Shatian You. Agric. Sci. China 2006; 5: 90-97

79. Bespalhok Filho JC, Kobayashi AK, Pereira LF, Galvão RM, Vieira LG. Transient gene expression of beta-glucuronidase in citrus thin epicotyl transversal sections using particle bombardment. Brazilian Archives of Biology and Technology 2003;46(1):1-6

80. Cervera M, Ortega C, Navarro A, Navarro L, Pena L. Generation of transgenic citrus plants with the tolerance-to-salinity gene HAL2 from yeast. The Journal of Horticultural Science and Biotechnology 2000;75(1):26-30

81. Hasan N, Kamruzzaman M, Islam S, Hoque H, Bhuiyan FH and Prodhan SH. Development of partial abiotic stress tolerant Citrus reticulata Blanco and Citrus sinensis (L.) Osbeck through Agrobacterium mediated transformation method. Journal of Genetic Engineering and Biotechnology 17:14

82. Guo W, Duan Y, Olivares-Fuster O, Wu Z, Arias CR, Burns JK, Grosser JW. Protoplast transformation and regeneration of transgenic Valencia sweet orange plants containing a juice quality-related pectin methylesterase gene. Plant Cell Reports 2005;24(8):482-486

83. Fagoaga C, López C, de Mendoza AH, Moreno P, Navarro L, Flores R, Peña L. Post-transcriptional gene silencing of the p23 silencing suppressor of Citrus tristeza virus confers resistance to the virus in transgenic Mexican lime. Plant molecular biology 2006;60(2):153-165

84. Febres VJ, Lee RF, Moore GA. Transgenic resistance to Citrus tristeza virus in grapefruit. Plant cell reports 2008;27(1):93-104

85. Boava LP, Cristofani-Yaly M, Mafra VS, Kubo K, Kishi LT, Takita MA, Ribeiro-Alves M, Machado MA. Global gene expression of Poncirus trifoliata, Citrus sunki and their hybrids under infection of Phytophthora parasitica. BMC genomics 2011;12(1):1-3

86. Mondal SN, Dutt M, Grosser JW, Dewdney MM. Transgenic citrus expressing the antimicrobial gene Attacin E (attE) reduces the susceptibility of 'Duncan' grapefruit to the citrus scab caused by Elsinoë fawcettii. European journal of plant pathology 2012;133(2):391-404

87. Caserta R, Teixeira-Silva NS, Granato LM, Dorta SO, Rodrigues CM, Mitre LK, Yochikawa JT, Fischer ER, Nascimento CA, Souza-Neto RR, Takita MA. Citrus biotechnology: What has been done to improve disease resistance in such an important crop?. Biotechnology Research and Innovation. 2020;3:95-109

88. Zou X, Jiang X, Xu L, Lei T, Peng A, He Y, Yao L, Chen S. Transgenic citrus expressing synthesized cecropin B genes in the phloem exhibits decreased susceptibility to Huanglongbing. Plant molecular biology 2017;93(4-5):341-353

89. Robertson CJ, Zhang X, Gowda S, Orbović V, Dawson WO, Mou Z. Overexpression of the Arabidopsis NPR1 protein in citrus confers tolerance to Huanglongbing. Journal of Citrus Pathology. 2018;5(1):1-8

90. Omar AA, Murata MM, El-Shamy HA, Graham JH, Grosser JW. Enhanced resistance to citrus canker in transgenic mandarin expressing Xa21 from rice. Transgenic research 2018;27(2):179-191

91. Kobayashi AK, Vieira LG, Bespalhok Filho JC, Leite RP, Pereira LF, Molinari HB, Marques VV. Enhanced resistance to citrus canker in transgenic sweet orange expressing the sarcotoxin IA gene. European Journal of Plant Pathology 2017;149(4):865-873

92. LeBlanc C, Zhang F, Mendez J, Lozano Y, Chatpar K, Irish VF, Jacob Y. Increased efficiency of targeted mutagenesis by CRISPR/Cas9 in plants using heat stress. The Plant Journal 2018;93(2):377-386

93. Knott GJ, Doudna JA. CRISPR-Cas guides the future of genetic engineering. Science 2018;361(6405):866-869

94. Jia H, Wang N. Xcc-facilitated agroinfiltration of citrus leaves: A tool for rapid functional analysis of transgenes in citrus leaves. Plant cell reports 2014;33(12):1993-2001

95. Hu Y, Zhang J, Jia H, Sosso D, Li T, Frommer WB, Yang B, White FF, Wang N, Jones JB. Lateral organ boundaries 1 is a disease susceptibility gene for citrus bacterial canker disease. Proceedings of the National Academy of Sciences 2014;111(4):E521-E529

96. Peng A, Chen S, Lei T, Xu L, He Y, Wu L, Yao L, Zou X. Engineering canker-resistant plants through CRISPR/Cas9-targeted editing of the susceptibility gene Cs LOB 1 promoter in citrus. Plant biotechnology journal 2017;15(12):1509-1519

97. Wang L, Chen S, Peng A, Xie Z, He Y, Zou X. CRISPR/Cas9-mediated editing of CsWRKY22 reduces susceptibility to Xanthomonas citri subsp. citri in Wanjincheng orange (Citrus sinensis (L.) Osbeck). Plant Biotechnology Reports 2019;13(5):501-510

98. Jia H, Orbović V, Wang N. CRISPR-LbCas12a-mediated modification of citrus. Plant biotechnology journal 2019;17(10):1928-1937

99. Razzaq A, Saleem F, Kanwal M, Mustafa, G, Yousaf S, Arshad, HMI, Hameed MK, Khan MS and Joyia FA (2019). Modern trends in plant genome editing: An inclusive review of the CRISPR/Cas9 toolbox. International Journal of Molecular Sciences. 20: 4045; doi:10.3390/ijms20164045

100. Hamblin MT, Buckler ES, Jannink JL. Population genetics of genomics-based crop improvement methods. Trends in Genetics 2011;27(3):98-106

101. Ajiro N, Miyamoto Y, Masunaka A, Tsuge T, Yamamoto M, Ohtani K, Fukumoto T, Gomi K, Peever TL, Izumi Y, Tada Y. Role of the host-selective ACT-toxin synthesis gene ACTTS2 encoding an enoyl-reductase in pathogenicity of the tangerine pathotype of Alternaria alternata. Phytopathology 2010;100(2):120-126

102. Peres NA, Timmer LW. Evaluation of the Alter-Rater model for spray timing for control of Alternaria brown spot on Murcott tangor in Brazil. Crop Protection 2006;25(5):454-460

103. Vicent A, Armengol J, García-Jiménez J. Protectant activity of reduced concentration copper sprays against Alternaria brown spot on 'Fortune' mandarin fruit in Spain. Crop Protection 2009;28(1):1-6

104. Vicent A, Armengol J, García-Jiménez J. Rain fastness and persistence of fungicides for control of Alternaria brown spot of citrus. Plant Disease 2007;91(4):393-399

105. Luro F, Maddy F, Jacquemond C, Froelicher Y, Morillon R, Rist D, et al. (Eds.), 2004. Identification and evaluation of diplogyny in clementine (Citrus clementina) for use in breeding. In International Society for Horticultural Science (ISHS), Leuven, Belgium

106. Cuenca J, Froelicher Y, Aleza P, Juárez J, Navarro L, Ollitrault P. Multilocus half-tetrad analysis and centromere mapping in citrus: Evidence of SDR mechanism for 2 n megagametophyte production and partial chiasma interference in mandarin cv 'Fortune'. Heredity. 2011;107(5):462-470

107. Cuenca J, Aleza P, Juarez J, García-Lor A, Froelicher Y, Navarro L, Ollitrault P. Maximum-likelihood method identifies meiotic restitution mechanism from heterozygosity transmission of centromeric loci: Application in citrus. Scientific Reports 2015;5(1):1-11

108. Rouiss H, Cuenca J, Navarro L, Ollitrault P, Aleza P. Tetraploid citrus progenies arising from FDR and SDR unreduced pollen in 4x X 2x hybridizations. Tree Genetics and Genomes 2017;13(1):10

109. Cuenca J, Aleza P, Vicent A, Brunel D, Ollitrault P, Navarro L. Genetically based location from triploid populations and gene ontology of a 3.3-Mb genome region linked to Alternaria brown spot resistance in citrus reveal clusters of resistance genes. PloS one 2013;8(10):e76755

110. Cuenca J, Aleza P, Garcia-Lor A, Ollitrault P, Navarro L. Fine mapping for identification of citrus alternaria brown spot candidate resistance genes and development of new SNP markers for marker-assisted selection. Frontiers in plant science 2016; 7:1948

111. Butelli E, Licciardello C, Zhang Y, Liu J, Mackay S, Bailey P, Reforgiato-Recupero G, Martin C. Retrotransposons control fruit-specific, cold-dependent accumulation of anthocyanins in blood oranges. The Plant Cell 2012;24(3):1242-1255

112. Wang X, Xu Y, Zhang S, Cao L, Huang Y, Cheng J, Wu G, Tian S, Chen C, Liu Y, Yu H. Genomic analyses of primitive, wild and cultivated citrus provide insights into asexual reproduction. Nature Genetics 2017;49(5):765-772

113. Fang DQ, Federici CT, Roose ML. Development of molecular markers linked to a gene controlling fruit acidity in citrus. Genome 1997;40(6):841-849

114. Ben-Hayyim G, Moore GA. Recent advances in breeding citrus for drought and saline stress tolerance. Advances in molecular breeding toward drought and salt tolerant crops. 2007:627-642

115. Ricroch AE, Bergé JB, Kuntz M. Evaluation of genetically engineered crops using transcriptomic, proteomic, and metabolomic profiling techniques. Plant physiology 2011;155(4):1752-1761

116. Randhawa HS, Asif M, Pozniak C, Clarke JM, Graf RJ, Fox SL, Humphreys DG, Knox RE, DePauw RM, Singh AK, Cuthbert RD. Application of molecular markers to wheat breeding in C anada. Plant Breeding 2013;132(5):458-471

117. Wishart TM, Rooney TM, Lamont DJ, Wright AK, Morton AJ, Jackson M, Freeman MR, Gillingwater TH. Combining comparative proteomics and molecular genetics uncovers regulators of synaptic and axonal stability and degeneration in vivo. PLoS Genetics 2012;8(8):e1002936

118. Aslam B, Basit M, Nisar MA, Khurshid M, Rasool MH. Proteomics: Technologies and their applications. Journal of chromatographic science. 2017 Feb 1;55(2):182-196

119. Gregersen N, Hansen J, Palmfeldt J. Mitochondrial proteomics—A tool for the study of metabolic disorders. Journal of inherited metabolic disease 2012;35(4):715-726

120. Yao L, Yu Q, Huang M, Hung W, Grosser J, Chen S, Wang Y, Gmitter FG. Proteomic and metabolomic analyses provide insight into the off-flavour of fruits from citrus trees infected with 'Candidatus Liberibacter asiaticus'. Horticulture research 2019;6(1):1-3

121. Yao L, Yu Q, Huang M, Song Z, Grosser J, Chen S, Wang Y, Gmitter Jr FG. Comparative iTRAQ proteomic profiling of sweet orange fruit on sensitive and tolerant rootstocks infected by 'Candidatus Liberibacter asiaticus'. PloS one 2020;15(2):e0228876

122. Graves PR, Haystead TA. Molecular biologist's guide to proteomics. Microbiology and molecular biology reviews. 2002 Mar 1;66(1):39-63

123. Singh J, Pandey P, James D, Chandrasekhar K, Achary VM, Kaul T, Tripathy BC, Reddy MK. Enhancing C3 photosynthesis: An outlook on feasible interventions for crop improvement. Plant Biotechnology Journal. 2014;12(9):1217-1230

124. Weckwerth W. Metabolomics in systems biology. Annual review of plant biology 2003;54(1):669-689

125. Oliver SG, Winson MK, Kell DB, Baganz F. Systematic functional analysis of the yeast genome. Trends in biotechnology 1998;16(9):373-378

126. Hansen J, Palmfeldt J, Vang S, Corydon TJ, Gregersen N, Bross P. Quantitative proteomics reveals cellular targets of celastrol. PLoS One 2011;6(10):e26634

127. Griffin JL, Vidal-Puig A. Current challenges in metabolomics for diabetes research: A vital functional genomic tool or just a ploy for gaining funding?. Physiological Genomics 2008;34(1):1-5

SECTION-2

CITRUS BIOTIC AND ABIOTIC STRESS MANAGEMENTS

Chapter-4

Integrated Management Approach to Citrus Fungal Diseases by Optimizing Cocoa-Based Agroforests Structural Characteristics

1. Introduction

Conventional and intensive agriculture has enabled a considerable increase in agricultural production since the 1950s. However, the resulting heavy ecological balance sheet discredits this unsustainable model of agriculture [1]. Thus, although the preferred strategies of intensive agriculture have shown undeniable benefits, their use is becoming increasingly worrying both for agriculture itself and for the environment and human health. This is partly due to the excessive use of pesticides and other chemicals [2, 3, 4, 5, 6]. The improvement of intensive production systems towards new models of sustainable agriculture, favoring the development of effective means of combating diseases that are sustainable and environmentally friendly, is becoming a matter of urgency. Tropical agroforestry systems, thanks to their high biodiversity and structural diversity, represent a privileged way out of the agroecological transition. Several studies have demonstrated the contribution of the structural characteristics of these systems in the integrated management of pests and diseases of perennial crops, particularly citrus [7, 8, 9].

Citrus represent a fruit crop of prime importance in socio-ecological terms in Cameroon [10, 11, 12]. Their significance lies in the fact of their high-quality nutritional value and their contribution to the diversification of producers' incomes in rural areas. Citrus are also important in the local pharmacopeia [13, 14, 15]. They are also known for their role in restoring ecological balances after deforestation [16]. However, despite the favorable agroecological conditions throughout the country, the number of production basins identified and even the density of trees in farms, production remains poor [12, 17, 18]. A diversity of diseases affecting citrus in the country humid zones is the main constraint to their production [8, 11, 19, 20].

Pseudocercospora leaf and fruit spot disease (PLFSD) caused by *Pseudocercospora angolensis*, citrus scab disease caused by *Elsinoe* spp.; and citrus diseases caused by Phytophthora (CDP) caused by *Phytophthora* spp., are the main soil diseases on citrus in Cameroon (Figure 1) [7, 21, 22, 23, 24, 25]. Damages caused by these diseases result in heavy crop losses [26]. Sorting deviations of up to 100% of the production can be recorded in the case of PLFSD, if no treatment measures are taken [27]. Concerning citrus scab, severe attacks on young *C. volcameriana* plants, for example in the nursery, result in their death [20, 28]. CDP significantly limits citrus production in the plots where it is present [29, 30, 31, 32, 33, 34]. In addition to reduced yields from the beginning of infection, the economic viability of orchards is reduced following tree death [7, 32, 35]. This strong and constantly changing diseases pressure, which not only causes enormous economic losses, but also leads (in the case of PLFSD) to quarantine and a ban on the export of citrus products to other production areas [25].

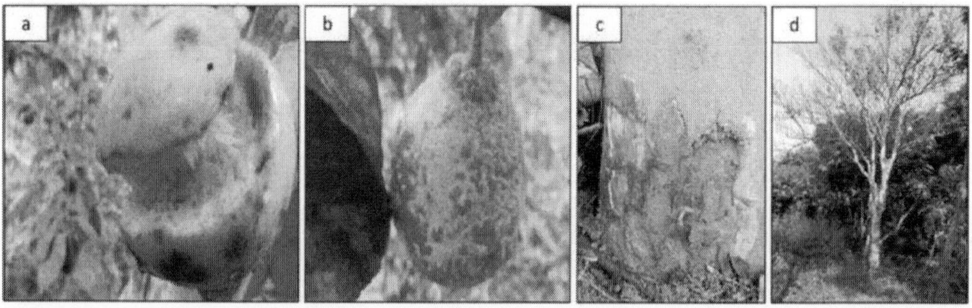

Figure 1. Symptoms of citrus diseases in Cameroon. C. paradisi fruit torn following a severe attack by Pseudocercospora angolensis (a), C. volcameriana fruit covered with scab spots due to Elsinoe spp. (b), lesions resulting from crown attacks due to Phytophthora spp. on C. sinensis tree(c) and dieback due to various diseases of an C. sinensis (d).

A variety of strategies are used to control citrus diseases. These include the practice of sanitation measures, the use of resistant cultivars and varieties, grafting, organic or mineral soil amendments, the use of plant extracts, biological control in a systemic approach and, above all, chemical control through fungicides [32, 36, 37, 38, 39, 40]. However, the high costs of these methods, development of resistance to chemical inputs, emergence of new diseases and growing concerns about environmental and soil health make these methods inadequate [41, 42, 43, 44, 45]. In addition, most of these techniques are unsuitable for the socio-cultural and even technological context of small-scale producers in inter-tropical regions. The development of targeted control protocols, taking into account the local socio-ecological context and existing production systems is therefore imperative. This chapter highlights the effect of the structural characteristics of the cocoa-based agroforestry system on the intensity of the three main citrus diseases in the humid forest zones of Cameroon.

2. Complex cocoa-based agroforest systems and structural characteristics

In Cameroon, citrus are mainly grown in cocoa-based agroforestry systems (CBAFS) [7, 46]. These are complex, highly biodiverse, natural forest-like cropping systems (Figure 2) [30, 47, 48]. In this system, several interactions of different nature and intensity can take place depending on the species present, their sizes and their positions [9, 49, 50]. One of these interactions is the action of diseases. The structural characteristics of CBAFS can contribute to control of these [20, 30, 51, 52]. Studies in these cropping systems and on various pathosystems have shown that spatial structure of species is important in reducing diseases development [49, 52]. Indeed, spatial structure has a twofold effect on the pathogen: firstly, the high plant biodiversity into CBAFS makes it possible to dilute the pathogens resource and thus reduce their presence and damage [53, 54, 55]. Secondly, multi-species agroecosystems are recognized for the high diversity of vertical and horizontal structures that can be adopted by the plant population [56]. This diversity of plant spatial structure affects diseases mainly through the microclimatic weathering mechanism [51, 52].

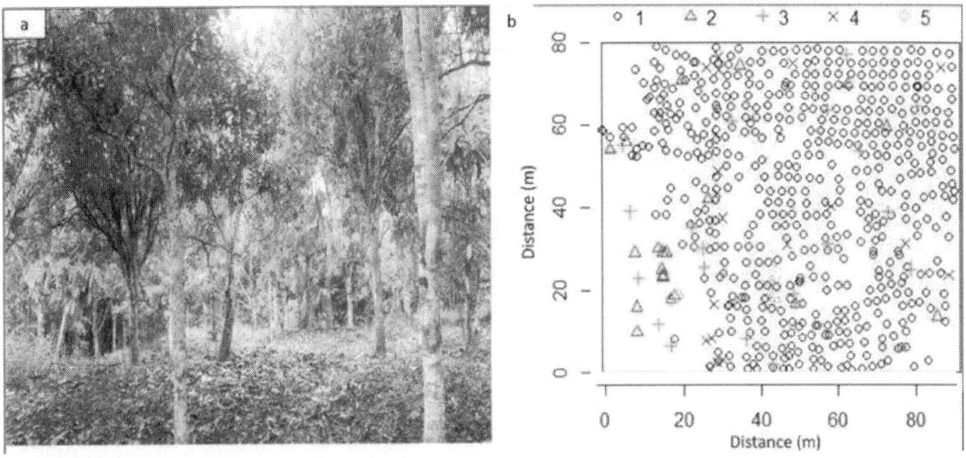

Figure 2. Illustration of a cocoa-based agroforestry system planted with citrus trees (a) and the horizontal structure of its plant population (b). 1 = cocoa trees, 2 = various forest tree species, 3 = various other fruit tree species, 4 = citrus and 5 = palm trees.

Previous works have supported interactions between individuals of a host population of pathogen and associated plants within intercropping systems [7, 8]. This type of interaction is likely to influence the presence of diseases. The action of shade trees on the understory microclimate decreases with decreasing distance between trees and pathogen transmission decreases as the distance between host individuals increases [51, 52, 57, 58, 59].

3. Effect of spatial structure of citrus into cocoa based agroforest systems on citrus diseases

The spatial structure of a plant community is the vertical and horizontal arrangement of constituent elements [51, 60, 61]. It reflects therefore the local environment around each individual [55, 62]. Within agroforest, non-host plants mainly perform a physical barrier effect on diseases [27, 51, 54]. The effect of the citrus spatial structure within CBAFS on the three main diseases affecting them in Cameroon has been assessed through various studies. A network of 27 plots in the three study sites was set up in Obala, Muyuka and Bokito sites. These are located in three ecological conditions into humid forest zone of Cameroon. CBAFS with at least 12 citrus trees in the plot area were selected. Each plot area was a square of at least 2500 m^2 (50 X 50 m). Plots were chosen in villages among the most productive areas and also representative of the study zone in terms of system diversity and variability of citrus species produced.

The analysis of the spatial structure of the citrus sub-population was done by the Ripley method [62]. Following the method illustrated in Ngo Bieng [55], a typology of spatial structure was build based on the spatial structure of the citrus trees in the study plots. In a first step, the horizontal spatial structure was characterized on the citrus trees in each plot, using the L(r) modified Ripley function [20, 61, 63] . The L (r) function is based on the calculation of the expected number of neighbor trees (Figure 2), within a distance ≤ to r of any point of the study pattern. This method enables to distinguish three types of tree spatial patterns: regular when L (r) is <0, aggregated when L (r) is >0, and random when L (r) =0. This function characterizes the neighborhood structure around a point. It is used for a simple, homogeneous and isotropic point process of density λ [64].

$$Kr=\lambda-1ErLr=Kr\pi-rE1$$

Subsequently, a hierarchical cluster analysis based on the Euclidean distance between the values of the L(r) function of the citrus trees in the different plots was made. It resulted in clusters of plots with a similar spatial structure, based on their trend to regular, random or aggregated spatial structure. This analysis was done with *ads* and *ade4* package R 3.2.2 software (Figure 3). Symptoms of CDP, PLFSD, and citrus scab were assessed by the visual recognition method. The intensity was assessed using a scale from 1 to 4.

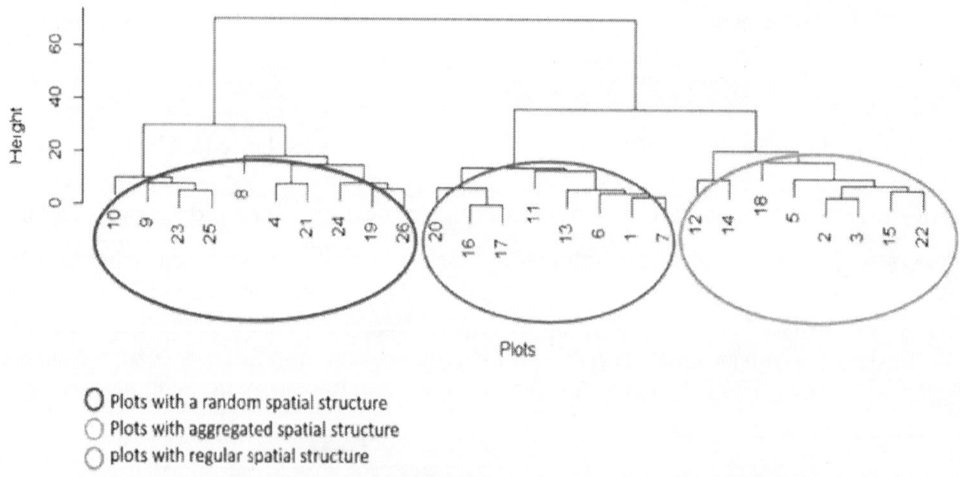

Figure 3. Hierarchical classification of plots according to the spatial structure of citrus trees in the experimental plots.

From this study it emerges that, the spatial structure has a significant influence on the intensity of the diseases observed. The analysis of variance and the mean comparison test reveal that plots in which citrus have an aggregated spatial structure, have a high intensity of citrus scab disease and PLFSD. On the other hand, plots in which citrus fruits have a regular spatial structure show a significantly low intensity of these same diseases (Table 1).

Table 1. Effect of the spatial structure of citrus trees in cocoa based agroforests systems on diseases.

In the same column, values with same letter are not significantly different (Tukey HSD test P < 0.05). *** indicates highly significant.

Maladies fongiques	Aggregated spatial structure	Random Spatial structure	Regular Spatial structure	Anova/ Tukey test	Df	F value	Pr(>F)
Pseudocercospora leaf and fruit spot disease	1.55 ± 0.36^{b}	1.43 ± 0.28^{a}	1.41 ± 0.26^{a}		2	10.08	5.14e-7***
Citrus scab disease	1.22 ± 0.56^{b}	1.11 ± 0.38^{ab}	1.08 ± 0.34^{a}		2	5.102	0.00638**
Phytophthora foot rot disease	1.59 ± 0.93^{a}	1.97 ± 0.94^{b}	1.78 ± 0.91^{ab}		2	6.297	0.002**

It is thus demonstrated that the aggregate spatial structure of citrus in CBAFS has a negative effect on diseases observed. These results are similar to those obtained by Ndo *et al.* [7, 8, 30] in the particular case of PLFSD. In addition to that, the involvement of spatial structure in the spread of various diseases has been shown [8, 53, 54]. Indeed, the aggregation of host populations favors the dispersion of diseases, while regularity would reduce it [65]. In addition, it is recognized that transmission of the pathogen decreases as the distance between host individuals increases [52, 58]. On the other hand, it has been shown that aggregation of host populations can reduce the incidence of pathogens [66]. Because the transmission of the pathogen between aggregates decreases with increasing distance between aggregates. These aspects would therefore explain the low intensity of CDP observed in plots where citrus fruits have an aggregated spatial structure.

4. Effect of shade intensity management on CDP and PLFSD

Depending on the situation of citrus in the CBAFS and in relation to the upper stratum, three levels of shading (dense, moderate and no shading) were defined. Shade trees play various roles in tropical agroforests. They can improve adverse weather conditions by modulating temperature variations [51, 52, 57, 59]. Shading has been recognized as one of the factors that can influence PLFSD dissemination [54, 67, 68]. Given that shading favors climatic conditions for the development of certain citrus pathogenic fungi such as *P. angolensis* or many *Phytophthora* spp. (high relative humidity (>%60) and cool temperature conditions (<°25C)) [18, 28, 69]. It is assumed that within a plot, trees under shade would have a higher incidence of the disease than those in full sunlight. On the other hand, given the role of shade trees in improving climatic and nutritional conditions, the growth of trees under these conditions can be improved, as well as their vigor and response to disease. In addition, shade trees can act as a barrier against wind and rain (the main factors in the spread of conidia) and slow the progress of the epidemic.

An experiment carried out in a fruit trees orchard in Foumbot in the Western region of the country demonstrated the effect of shade trees on the PLFSD epidemic. The trial was set up in the Institute of Agricultural Research for Development (IRAD) experimental orchard. This orchard comprises collection plots of mango (*Manguifera indica*), avocado (*Persea americana*) and various citrus trees separated from each other by fallow plots often reserved for annual crops. This experimental plot enabled to compare tree shading situations ie dense shade (under mango trees), light shade (under avocado trees) and full sun light (fallow plot). One-year old pomelo seedlings have been placed in three levels of shade i.e.: under mango trees (dense shade), under avocado trees (light shade) and on fallow land (no shade).

The results of this experiment showed that the higher the shade index (under mango trees), the lower the PLFSD severity. When the shade intensity is lower (under avocado trees), disease severity is also lower, however the differences are not significant (Figure 4). These results suggest that shading should be sufficient to significantly reduce PLFSD incidence. Otherwise too little shading will not have significant effect on disease severity. But in the meantime, the plant must receive sufficient sun radiation for good growth. So, it is necessary to determine an optimum shading that allows a good compromise between plant growth and reduces PLFSD incidence. This optimum can vary according to the climatic and sanitary conditions of the plantations under consideration [18, 20, 52].

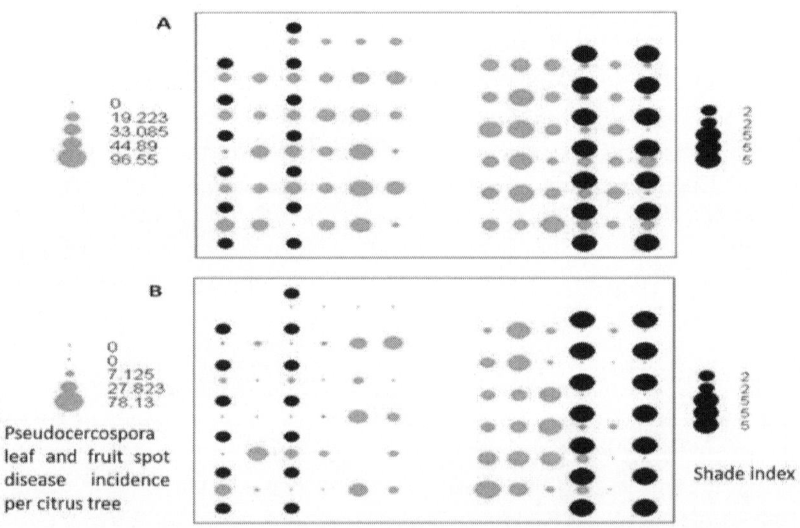

Figure 4. Graphical representation of the percentage of diseased leaves for each pomelo plant and the shade indices of the different shade trees during the first observation date (A) and the second date (B) in the Foumbot plot.

Another experiment conducted in 26 cocoa-based agroforestry systems showed the effect of shading on the spread of CDP and citrus scab disease. In this study, a total set of 476 citrus were observed under three shading conditions. Depending on the tree diversity and population, the various scenarios of shade density have been coded as follows: (1) "*dense shade*" when the citrus tree were placed under a direct and thick shading of the upper stratum; (2) "*light shade*" when the citrus tree received a mean shading of a higher stratum and finally (3) "*full sun*" when the citrus fruit did not receive a shade of a top stratum.

Results showed that, there is a variation in the citrus diseases intensity depending on whether they are located under dense shade, light shade or in full sunlight. The mean comparison test thus revealed significant differences in the intensities of CDP, according to the three citrus trees situations depending on shade. Citrus trees under dense shade are significantly less affected by CDP compared to those under light shade and those in full sunlight (Table 2).

Table 2. Effect of shade on citrus diseases.

Citrus diseases	Shade conditions			Anova/ Tukey test	Df	F value	Pr (>F)
	Full sunlight	Light shade	Dense shade				
Citrus scab disease	1.08 ± 0.34[a]	1.11 ± 0.41[a]	1.18 ± 0.50[a]		2	2.257	0.106
Phytophthora foot rot disease	1.92 ± 0.91[b]	1.98 ± 1.00[b]	1.56 ± 0.86[a]		2	11.18	$1.8 e^{-6***}$

In the same column, values with same letter are not significantly different (Tukey HSD test $P < 0.05$). *** indicates highly significant.

Results therefore showed that, the shade had a significant effect on diseases. This shading effect was positive on the intensity of CDP. In general, in CBAFS, shade reduces diseases intensity [52, 70]. Indeed, for pathogens, spore dispersal and germination are the two main phases of their life cycle. Shading promotes spore germination while the sensitivity of the dispersion of pathogen spores to microclimate depends on how it is dispersed [7, 59].

5. Mode of dispersal of citrus infectious pathogens and CBAFS structural characteristics

The majority of fungi require moisture for infection and production of conidia [9, 71, 72]. These conidia may be disseminated by wind or soil water runoff for telluric fungi like *Phytophthora*. Local dispersal is primarily favored by rain-splash as well as some insects moving on trees [24]. This mode of dissemination determines the spatial distribution of each disease.

In the case of PLFSD, the analysis of its spatial distribution indicates that the disease is distributed in clusters, and that above 12 m there is no spatial dependency. In fact, the disease spreads from one tree to its closest neighbors depending on wind speed and/or rainfall intensity. Infection will depend on the presence and quality of the host. If neighbors are susceptible hosts, infection continues and the epidemiological cycle continues. Otherwise, the course of the disease can be circumscribed. This may explain the aggregated spatial structure of diseases that usually have this mode of spread. However, Brown and Bolker [65] pointed out that the aggregation of host populations favors the dispersal of the diseases while its regulation reduces it. That is, the further away the trees are from each other, the slower the transmission of the disease [22].

With regard to CDP, the spread of *Phytophthora* in the field is primarily ensured by the use of infected plant material [32]. However, mechanical means of dispersal of *Phytophthora* have been illustrated. Cases of transmission from an infected root to a healthy root following their respective growing zones have been reported. The inoculum can also be spread by run-off water. Splashes can promote the spread of the inoculum from the soil to the aerial parts of the plant. This mode of disease spread may be promoted by aggregation of the host species of this pathogen. This hypothesis was confirmed by Akoutou *et al.* [7, 30]. These studies showed that citrus with an aggregated spatial structure were more attacked by CDP in contrast to those with a regular spatial structure. In fact, root diversity in the rhizosphere could limit contact between roots of the same species. Host trees planted at wide spacings and having non-host trees between them are less likely to come into contact and this would help to limit the spread of the inoculum. This effect of dilution of the pathogen's resource can also be applied to the mode of transmission through diseased fruits and contact with parts of the plant close to the soil.

In addition, it was shown that environmental factors play a critical role in the development, severity, dispersal and conservation of inoculum in the epidemiology of *Phytophthora* disease. The increase in temperature favors population growth of species such as *P. parasitica* and *P. palmivora*. Hot, dry climates are favorable to *P. parasitica*. These observations corroborate the conclusions drawn on the effect of shading on CDP development. Indeed, the cooler environmental conditions in the understory created by shade trees would make the habitat unfavorable for the pathogen. Citrus planted under dense shade would therefore be less exposed to the inoculum, which is therefore more intense in plot areas of the plot where there is no shade.

6. Conclusion

This study highlighted the effect of shading trees on citrus in agroforestry plots. In such plots citrus trees are mixed with plants belonging to different tree. Spatial structure has a significant influence on the observed diseases intensity. Plots in which citrus have an aggregated spatial structure have a high intensity of studied diseases, while plots in which citrus have a regular spatial structure are significantly less attacked by these diseases. Optimizing the structural characteristics of CBAFS could lead to the development of integrated control strategies against fungal diseases. These management strategies will be adapted to local agroecological contexts, respectful of the environment, and applicable by smallholders.

References

1. Tilman, D., Cassman, K. G., Matson, P. A., Naylor, R. and Polasky, S. (2002), 'Agricultural sustainability and intensive production practices', Nature 418(6898), 671-677
2. Dornbush, M.E., von Haden, A.C., 2017. Intensified agroecosystems and their effects on soil biodiversity and soil functions. In: Al-Kaisi, Mahdi M. (Ed.), Soil Health and Intensification of Agroecosytems. Iowa State University, Ames, IA, United States, pp. 173-193. https://doi.org/10.1016/B978-0-12-805317-1.00008-7
3. Eddleston, M., Karalliedde, L., Buckley, N., Fernando, R., Hutchinson, G., Isbister, G., Konradsen, F., Murray, D., Piola, J. C., Senanyake, N., Sheriff, R., Singh, S., Siwach, S. B. and Smit, L., 2002. Pesticide poisoning in the developing world - a minimum pesticides list', The Lancet 360 (9340), 1163-1167
4. Leibee, G. L. and Capinera, J. L., 1995. Pesticide Resistance in Florida Insects Limits Management Options', Florida Entomologist 78(3), 386-399

5. Martin, E.A., Feit, B., Requier, F., Friberg, H., Jonsson, M., 2019. Assessing the resilience of biodiversity-driven functions in agroecosystems under environmental change. In: Bohan, D.A., Dumbrell, A.J. (Eds.), In Resilience in Complex Socio-Ecological Systems, vol. 60, pp. 59-123. https://doi.org/10.1016/bs.aecr.2019.02.003

6. Schreinemachers, P. and Tipraqsa, P. (2012), 'Agricultural pesticides and land use intensification in high, middle- and low-income countries', Food policy 37 (6), 616-626

7. Akoutou Mvondo, E., Ndo, E.G.D, Tsouga Manga, M.L., Abaane, C.L., Abondo Bitoumou, J., Bella Manga, Bidzanga Nomo, L., Ambang, Z., Cilas, C., 2019. Effects of complex cocoa-based agroforests on citrus tree decline. Crop Protection Crop Protection 130 (2020) 105051. DOI: https://doi.org/10.1016/j.cropro.2019.105051

8. Ndo E.G.D., Akoutou M.E., Ambang Z., Manga B., Cilas C., Nomo L.B., Gidoin C., Ngo Bieng M.A., 2019a. Spatial organization influences citrus Pseudocercospora leaf and fruit spot disease severity in cocoa-based agroforestry systems. Am. J. Plant Sci. 10, 221-235. https://doi.org/10.4236/ajps.2019.101017

9. Winans, K.S., Tardif, A.S., Lteif, A.E., Whalen, J.K., 2015. Carbon sequestration potential and cost-benefit analysis of hybrid poplar, grain corn and hay cultivation in southern Quebec. Canada. Agroforest. Syst. 89, 421-433. https://doi.org/10.1007/s10457-014-9776-4

10. Economos, C., and Clay, W.D., 1998. Nutritional and health benefits of citrus fruits. Paper presented at the Twelth Session of the IntergovernmentalGroup on Citrus Fruit

11. Kuate, J., Bella-Manga, Damesse, F., Kouodiekong, L., Ndindeng, S.A., David, O. and Parrot, L. (2006) Fruit Trees Cultivated in Family Farms in the Humid Zone of Cameroon: A Survey. Fruits, 61, 373-387. https://doi.org/10.1051/fruits:2006037

12. Temple L., 2001. Quantification des productions et des échanges de fruits et légumes au Cameroun. Ca. Agri,10: 87-94

13. Bayaga, H.N., Guedje, N.M., Biye, E.H., 2017. Ethnopharmacological and ethnobotanical approach of medicinal plants used in the traditional treatment of Buruli ulcer in Akonolinga (Cameroon). IJBCS 11 (4), 1523-1541. https://www.ajol.info/index.php/ijbcs/article/download/164026/153514

14. Boeing, H.; Bechthold, A.; Bub, A.; Ellinger, S.; Haller, D.; Kroke, A.; Leschik-Bonnet, E.; Mueller, M.J.; Oberritter, H.; Schulze, M.; Stehle, P.; Watzl, B. Critical review: Vegetables and fruit in the prevention of chronic diseases. Eur. J. Nutr. 2012, 51, 637-663

15. Radhika, G.; Sudha, V.; Sathya, R.M.; Ganesan, A.; Mohan, V., 2008, Association of fruit and vegetable intake with cardiovascular risk factors in urban South Indians. Brit. J. Nutr. 99, 398-405

16. Westphal, E., Embrechts, J., Ferweda, J.D., van-Gils-Meeus, H.A.E., Mutsaers, H.W., & Wesphal-Stevels, J.M.C., 1985. Cultures vivrières tropicales avec référence spéciale au Cameroun. Wageningen, Netherlands

17. Ndo E. G. D. 2007. Analyse du risque épidémiologique des populations d'agrumes vis à vis de la cercosporiose, du scab et de la gommose dans les zones humides du Cameroun. Thèse de MSc, Faculté d'Agronomie et des Sciences Agricoles Université de Dschang, Cameroun, 85 p

18. Ndo E.G.D., 2011. Evaluation des facteurs de risque épidémiologique de la Pseudocercosporiose des agrumes dans les zones humides du Cameroun. Thèse de doctorat, Sup Agro Montpelier, 202 pp. https://agritrop.cirad.fr/562661/1/do cument_562661.pdf
19. Mariau, D., 1999. Les Maladies des Cultures Pérennes Tropicales (CIRAD ed.)
20. Ndo, E.G.D., Kuate, J., Sidjeu Wonfa, C.S., Tchio, F., Ndzana Abanda, F.X., Mbieji Kemayou, C., Akoutou Mvondo E., Amele Ndjoumoui, C., Amang A Mbang, J., 2019. Tolerance of citrus genotypes towards Pseudocercospora leaf and fruit spot disease in western highlands zone of Cameroon. Crop Protection 124. https://doi.org/10.1016/j.cropro.2019.05.022
21. Agostini, J.P., Bushong, P. M., Bahia, A., & Timmer, L. W., 2003. Influence of environmental factors on severity of citrus scab and melanose. Plant Disease,87, 1102-1106
22. Boccas, B., and Laville, E. (Eds.). (1978). Les maladies à Phytophthora (IRFA Paris ed.)
23. Graham, J.H., and Timmer, L.W. (2003). Phytophthora diseases of citrus
24. Kuate, J. (1998). Cercosporiose des agrumes causés par Phaeoramularia angolensis. Cahier Agriculture,7(2), 121-129
25. Ndo, E.G.D., Bella-Manga, Ndoumbe-Nkeng, M., and Cilas C., 2019. Distribution of Pseudocercospora fruit and leaf spot, Phytophthora foot rot and scab diseases and their effect on Citrus tree decline prevalence in the humid zones of Cameroon. Fruits 74(5), 249-256; https://doi.org/10.17660/th2019/74.5.5
26. Yesuf, M., 2013. Pseudocercospora leaf and fruit spot disease of citrus: achievements and challenges in the citrus industry: a review. Agric. Sci. 4, 324-328
27. De Vallavieille-Pope C., Fraj M. B., Mille B., et Meynard J.-M., 2004. Les associations de variétés: accroître la biodiversité pour mieux maîtriser les maladies. Doss de l'environnement de. Dossiers de l'environnement de l'INRA, 30: 101-109
28. Kuate, J., Fouré, E., Foko, J., Ducelier, D., & Tchio, F., 2002. La phaeoramulariose des agrumes au Cameroun due à Phaeoramularia angolensis:expression parasitaire à différentes altitudes. Fruits,54(04), 207-218
29. Adaskaveg, J.E., Hao, W., and Forster, H., 2015. Postharvest Strategies for Managing Phytophthora Brown Rot of Citrus using Potassium Phosphite in Combination with Heat Treatments. Plant Dis. 99:1477-1482. https://doi.org/10.1094/PDIS-01-15-0040-RE
30. Akoutou Mvondo, E., Ndo, E. G. D., Ngo Bieng, M.A., Ambang, Z., Bella Manga, Cilas, C., Tsouga Manga, M. L., Bidzanga Nomo L., 2017. Assessment of the interaction between the spatial organization of citrus trees populations in cocoa agroforests and Phytophthora foot rot disease of citrus severity. AgroForestry Systems, 10 p. http://dx.doi.org/10.1007/s10457-017-0140-3
31. Chaudhary, S., David, A. Laughlin, M.S.J.V., da Graça, M.K., O.J. Alabi, K.M., Crosby, K.L. Ong, and Veronica A., 2020. Incidence, Severity, and Characterization of Phytophthora Foot Rot of Citrus in Texas and Implications for Disease Management. APS, Published Online: 30 Jun 2020 https://doi.org/10.1094/PDIS-07-19-1493-RE

32. Graham, J.H., Timmer, L.W., 2008. Florida Citrus pest management guide: Phytophthora foot rot and root rot. Retrieved 15.02.2013, from University of Florida IFAS extension. http://edis.ifas.ufl.edu/cg009

33. Mekonen, M, Ayalew, A, Weldetsadik, K. and Seid, A., 2015. Assessing and Measuring of Citrus gummosis (Phytophthora spp.) in Major Citrus Growing Areas of Ethiopia. J Horticulture 2: 154. doi:10.4172/2376-0354.1000154

34. Sakupwanya, M.N., Labuschagne, N., loots, T., and Apostolides, Z., 2018. Towards develoloping a metabolic-marker based predictive model for Phytophthora nicotianae tolerance in citrus rootstocks. Journal of Plant Pathology, 100(2), 269-277. Doi:1007/s42161-018-0080-4

35. Graham, J., and Feichtenberger, E., 2015. Citrus phytophthora diseases: Management challenges and successes. J. Cit. Pathol. iocv_journalcitruspathology_27203

36. Cacciola, S.O. and Di San Lio, G.M., 2008. Management of citrus diseases caused by Phytophthora spp. In: Ciancio A, Mukerji KG (eds) Integrated management of diseases caused by fungi, bacteria and Phytoplasma. Springer, Netherlands, pp. 61-84

37. Gade and Lad, 2019. Strategies for management of Phytophthora diseases in citrus in India. Bio-intensive approaches: Application and effectiveness in plant diseases management (2019): 435-451

38. Ippolito, A., Schena L., Nigro, F., Soleti Ligorio, V., & Yaseen, T., 2006. Real-time detection of Phytophthora nicotianae and P. citrophthora citrus roots and soil. European Journal of Plant Pathology, 110, 833-843

39. Ippolito, A., Schena, L., & Nigro F., 2002. Detection of Phytophthora nicotianae and P. citrophthora in citrus roots and soils by nested PCR. European Journal of Plant Pathology, 108, 855-868

40. Jazet-Dongmo P. M., Kuate J., Boyom F. F., Ducelier, D., Damesse, F. et Zollo, P.H. A., 2008. Composition chimique et activité antifongique in vitro des huiles essentielles de citrus sur la croissance mycélienne de Phaeoramularia angolensis. Fru. 57(2): 95-104

41. Eddleston, M., 2020. Poisoning by pesticides. Medicine 48 (3), 214-217. https://doi.org/10.1016/j.mpmed.2019.12.019

42. Jia, Z.-Q., Zhang, Y.-C., Huang, Q.-T., Jones, A.K., Han, Z.-J., Zhao, C.-Q., 2020. Acute toxicity, bioconcentration, elimination, action mode and detoxification metabolism of broflanilide in zebrafish, Danio rerio. Journal of Hazardous Materials 394. https://doi.org/10.1016/j.jhazmat.2020.122521 122521

43. Mohamed, A. Hassaan, Ahmed El Nemr, 2020. Pesticides pollution: Classifications, human health impact, extraction and treatment techniques. Egyptian Journal of Aquatic Research 46 (2020) 207-220. https://doi.org/10.1016/j.ejar.2020.08.007

44. Panth, S.C. Hassler and F. Baysal-Gurel., 2020. Methods for Management of Soilborne Diseases in Crop Production. Agriculture 2020, 10, 16; doi: 10.3390/agriculture10010016 www.mdpi.com/journal/agriculture

45. Qiu, Y.-W., Zeng, E.Y., Qiu, H., Yu, K., Cai, S., 2017. Bioconcentration of polybrominated diphenyl ethers and organochlorine pesticides in algae is an important contaminant

route to higher trophic levels. Science of The Total Environment 579, 1885-1893. https://doi.org/10.1016/j.scitotenv.2016.11.192

46. Ndo E. G. D., Bella-Manga F., Atanga Ndindeng S., Ndoumbe-Nkeng M., Ajong Dominic Fontem A. D., et Cilas C., 2010. Altitude, tree species and soil type are the main factors influencing the severity of Phaeoramularia leaf and fruit spot disease of citrus in the humid zones of Cameroon. Eur J Plant Pathol., 128: 385-397

47. Laird, S. A., Leke-Awung, G., & Lysinge, R. J., 2007. Cocoa farms in the Mount Cameroon region: biological and cultural diversity in local livelihoods. Biodiversity and Conservation,16, 2401-2427

48. Sonwa DJ, Nkongmeneck BA, Weise SF, Tchatat M, Adesina AA, Janssens MJJ (2007) Diversity of plants in cocoa agroforests in humid forest areas of Southern Cameroon. Biodiv Conserv 16:2385-2400

49. Cerda R., Allinne C., Gary C., Tixier P., Harvey C.A., Krolczyk L., Mathiot C. Clément E., Auberto J.N., Avelino J., 2017. Effects of shade, altitude and management on multiple ecosystem services in coffee agroecosystems. Europ. J. Agronomy. 82(2017) 308-319. http://dx.doi.org/10.1016/j.eja.2016.09.019

50. Rao M. R., Nair P. K. R., et Ong, C. K., 1998. Biophysical interactions in tropical agroforestry systems. Agrofor. Syst., 38: 3-50

51. Gidoin C., 2013. Relations entre structure du peuplement végétal et bioagresseurs dans les agroforêts à cacaoyers. Application à trois bioagresseurs du cacaoyer: la moniliose au Costa Rica, la pourriture brune et les mirides au Cameroun. Thèse de doctorat, Centre International d'Etudes Supérieures en Sciences Agronomiques de Montpellier. 189 p

52. Schroth, G., Krauss, U., Gasparotto, L., Aguilar, J. A. D. and Vohland, K., 2000. Pests and diseases in agroforestry systems of the humid tropics', Agroforestry Systems 50(3), 199-241

53. Gidoin, C., Avelino, J., Deheuvels, O., Cilas, C. and Bieng, M.A.N., 2014. Shade Tree Spatial Structure and Pod Production Explain Frosty Pod Rot Intensity in Ca-cao Agroforests, Costa Rica. Phytopathology, 104, 275-281. https://doi.org/10.1094/PHYTO-07-13-0216-R

54. Ngo Bieng M. A., Laudine A. Chloé C. and Philippe T., 2017. Tree spacing impacts the individual incidence of Moniliophthora roreri disease in cacao agroforests. Pest Manag Sci (2017). DOI 10.1002/ps.4635. wileyonlinelibrary.com/journal/ps

55. Ngo Bieng, M. A., Gidoin, C., Avelino, J., Cilas, C., Deheuvels, O. and Wery, J. (2013), 'Diversity and spatial clustering of shade trees affect cacao yield and pathogen pressure in Costa Rican agroforests', Basic and Applied Ecology 14(4), 329-336

56. Malezieux, E., Crozat, Y., Dupraz, C., Laurans, M., Makowski, D., Ozier-Lafontaine, H., Rapidel, B., de Tourdonnet, S. and Valantin-Morison, M., 2009. Mixing plant species in cropping systems: concepts, tools and models. A review', Agronomy for Sustainable Development 29(1), 43-62

57. Guenat, S., Kaartinen, R., Jonsson, M., 2019. Shade trees decrease pest abundances on brassica crops in Kenya. Agrofor. Syst. 93, 641-652. https://doi.org/10.1007/s10457-017-0159-5

58. Ostfeld, R. S., Glass, G. E. and Keesing, F., 2005. Spatial epidemiology: an emerging (or re-emerging) discipline', Trends in Ecology & Evolution 20(6), 328-336

59. Somarriba, E. (2005), '¿Cómo evaluar y mejorar el dosel de sombra en cacaotales ?', Agroforestería en las Américas (41-42)

60. Akoutou, M.E., 2015. Effect of the spatial structure of citrus on the severity of three fungal diseases (Pseudocercospora angolensis, Phytophthora disease and Elsinoe) in cocoa-based agroforestry systems in wetlands in Cameroon. Msc degree, Université de Yaoundé 1, Faculté des Sciences. Yaoundé, 71p

61. Goreaud F., 2000. Apports de l'analyse de la structure spatiale en forêt tempérée à l'étude et à la modélisation des peuplements complexes., PhD thesis, L'ENGREF, centre de Nancy, France, 526 p

62. Ngo Bieng, M. A., Ginisty, C. and Goreaud, F. (2011), 'Point process models for mixed sessile forest stands', Annals of Forest Science 68(2), 267-274

63. Ripley, B. D. (1977), 'Modeling Spatial Patterns', Journal of the Royal Statistical Society Series B-Methodological 39 (2), 172-212

64. Walter J.M.N., 2001. La méthode de ripley pour l'analyse des structures spatiales ponctuelles en écologie. DEA de Géographie Physique et Aménagement, Université Louis Pasteur, Strasbourg, 2000-2001

65. Brown, D. H. and Bolker, B. M., 2004. The effects of disease dispersal and host clustering on the epidemic threshold in plants. Bulletin of Mathematical Biology 66 (2), 341-371

66. Watve, M. G. and Jog, M. M., 1997. Epidemic diseases and host clustering: An optimum cluster size ensures maximum survival. Journal of Theoretical Biology 184(2), 165-169

67. Abdulai, I., Jassogne, L., Graefe, S., Asare, R., Van Asten, P., La Ederach, P., 2018. Characterization of cocoa production, income diversification and shade tree management along a climate gradient in Ghana. PLoS One 13 (4), e0195777. https://doi.org/10.1371/journal.pone.0195777

68. Barrios, E., Valencia, V., Jonsson, M., Brauman, A., Hairiah, K., Mortimer, P.E., Okubo, S., 2018. Contribution of trees to the conservation of biodiversity and ecosystem services in agricultural landscapes. IJBESM 14 (1), 1-16. https://doi.org/10.1080/21513732.2017.1399167

69. Seif A.A., Hillocks R.J., 1998. Some factors affecting infection of citrus by Phaeoramularia angolensis. Journal of phytopathology 146 (8-9), 385-391

70. Babin, R., Anikwe, J. C., Dibog, L. and Lumaret, J. P., 2011. Effects of cocoa tree phenology and canopy microclimate on the performance of the mirid bug Sahlbergella singularis. Entomologia Experimentalis Et Applicata 141(1), 25-34

71. Diallo, M.T.S., 2003. Vers une lutte contre la cercosporiose des agrumes en Guinée. Fruits,58(6), 329-344

72. Pretorius, M. C., Crous, P. W., Groenewald, J. Z., & Braun, U., 2003. Phylogeny of some cercosporoid fungi from Citrus. Sydowia,55, 286-305

Chapter-5

A Current Overview of Two Viroids Prevailing in Citrus Orchards: Citrus Exocortis Viroid and Hop Stunt Viroid

1. Introduction

Viroids are circular, highly structured, single-stranded RNA (ssRNA) phytopathogens. Although they do not code for any peptide, these enigmatic pathogens have evolved the capacity to replicate within cellular organella, the nucleus and chloroplast for *Pospiviroidae* and *Avsunviroidae*, respectively [1, 2, 3, 4]. Viroid replication is ensured through an RNA-based rolling-circle mechanism [1]. Intriguingly, viroids can induce severe diseases in susceptible host plants similar to those caused by numerous plant viruses [5, 6, 7]. From the seven known citrus viroids only, two, namely, CEVd and HSVd, have been reported to be associated with citrus diseases that can pose significant economic risks to global citrus production. These diseases are exocortis and cachexia, respectively [8]. Since their original description in 1948 and 1950, respectively, both diseases have been reported to be present in almost all citrus-growing areas of the world, as well as in early citrus budwood registration programs [9, 10]. Given the importance of and rapid research progress in citrus virology in recent years, this review emphasizes recent findings related to CEVd and HSVd, the most serious viroids associated with citrus. It comprises reviews and research articles covering broad research areas on the characterization of both viroids and their symptoms, the development of reliable and rapid diagnosis methods, and management strategies. A brief snapshot of the present situation of CEVd and HSVd in the Mediterranean region, with an emphasis on their spread in citrus-growing areas of Morocco, is included.

2. Citrus exocortis viroid

2.1 Taxonomy

Citrus exocortis is a destructive disease infecting citrus species [9, 11]. The agent of this disease, citrus exocortis viroid [*Pospiviroidae*; *Pospiviroid*; CEVd], is a small, covalently closed ssRNA of about 371 nucleotides (nts) [12, 13, 14]. CEVd molecules can exist as either linear or circular [9]. As all viroids allocating to the *Pospiviroid* genus, CEVd lacks RNA self-cleavage activity and has a central conserved region (CCR), composed of two sets of conserved nucleotides in the upper and lower strands of its rod-like secondary structure, and a terminal conserved region (TCR) [15]. The rod-like secondary structure of CEVd takes the form of a model of five structural-functional domains. The latter is the Central (C), the Pathogenic (P), the Variable (V), the Terminal Left (TL), and the Terminal Right (TR) domains [16]. Based on their biological properties, CEVd sequences have been classified into two groups by using tomato as an experimental host: severe "Class A" and mild "Class B". Both classes of sequences differ by a minimum of 26 nts. These mutations affect two genomic regions, designated P_L and P_R, located respectively within the P and the V domains [17, 18, 19]. It is important to emphasize that CEVd strains of both classes cause distinct symptoms in gynura (*Gynura aurantiaca* (Blume) DC.) [20]. However, they induce only subtle differences in trifoliate orange (*Poncirus trifoliate* L. Raf.) used as a rootstock and a similar overall performance of the infected trees [21]. Sequencing of additional CEVd isolates revealed that further strains different from those of "Class A" and "Class B" existed. Furthermore, it seems that the sequence/pathogenicity relationship was more complex than originally anticipated [22]. Infectivity assays carried out with chimeric cDNA clones suggested that P_L is the pathogenicity-modulating domain. Although it remains to be explored how this domain modulates pathogenicity (i.e. stunting and epinasty). The role of the P_R domain is not known. However, infectivity assays suggested that it may influence the efficiency of viroid infection or replication in the plant [18]. Further infectivity assays of CEVd chimeras and another viroid of the *Pospiviroid* genus, tomato apical stunt viroid [TASVd], have been done to identify the role of individual structural domains. Firstly, it has been demonstrated that symptom severity is modulated by the TL and the P domains. Secondly, it has been shown that the V and the TR domains are involved in regulating viroid replication and/or accumulation [23].

2.2 Symptoms and economical impact

Citrus exocortis could affect a various part of the tree including the rootstock (at bark and wood levels), scion, leaves, and fruits, thus causing different types of damages such as bark scaling and cracking, bumps, severe stunting, low fruit-bearing, the poor appearance of the canopy [21, 24, 25, 26, 27], and poor tree performance [28]. CEVd-infected trees in the orchard show typical symptoms. The most characteristic one is bark scaling on trifoliate orange rootstock, yellow stem blotch on trifoliate orange and its hybrids and Rangpur lime (*Citrus limonia* Osb.), and stunting on trifoliate orange or its hybrid rootstocks [9, 11]. It is important to note that classic exocortis symptoms are not always closely associated with all CEVd isolates. For instance, only transient flaking (Washington Navel orange L) or a fine reticulum of surface cracks (Washington Navel orange 3536) on the trifoliate orange rootstock have been observed on two CEVd-infected trees in Australia [28]. Additionally, no bark scaling symptoms have been observed in CEVd-infected Washington Navel orange trees grafted on Carrizo citrange although these trees presented lesions and blisters in the roots [26]. It is important to emphasize that bark scaling symptoms could be caused by viroids other than CEVd. Indeed, it has been proved that, in certain viroid combinations, synergistic effects occur and cause exocortis scaling symptoms in the absence of CEVd [24]. Furthermore, in Australia, a large number of trees showing exocortis-like symptoms including dwarfing and/or bud union abnormalities produced only mild epinasty when grafted on the indicator plant Etrog citron. The presence of viroids other than CEVd has been highlighted [28]. Bark scaling symptoms could be also the consequence of tree exposure to abiotic stresses such as sunburn [29].

A reduction in vegetative growth has been observed on commune clementine (*Citrus clementine* Hort. ex Tanaka) trees infected by CEVd as it has been determined by the height and rootstock and scion circumferences [21]. Similar but milder symptoms have been reported in CEVd-infected Washington Navel orange trees grafted on Carrizo citrange [26].

The major susceptible citrus rootstocks, which show exocortis bark scaling symptoms, are trifoliate orange and its hybrids, Palestine sweet lime (*Citrus limettioedes* Tan.) or Rangpur lime [11]. Trees grown on trifoliate orange are the most severely affected, with symptoms of bark scaling and severe stunting usually developing when the trees are around 4 years old [29, 30]. Cracking and peeling of the bark below the bud union appear when bark scaling occurs on these rootstocks [30]. Symptoms of exocortis have been also reported on citrange and Swingle citrumelo rootstocks. However, unlike trifoliate orange, bark scaling symptom does not always occur on trees grown on citrange rootstocks. Trees grafted on these rootstocks exhibit symptoms somewhat late and the level of tree stunting is usually less severe than that on trifoliate orange. On another susceptible rootstock, CEVd-infected trees showed symptoms of stunting, yellowing of the canopy, and general tree decline, and occasional flaking of the rootstock bark. On these trees, fruit quality is not affected. However, tree yield is severely reduced since the viroid causes tree stunting [30]. The time required for disease expression by citron scions is believed to be directly associated with the inherent vigor of the rootstock, the environmental temperature, and cultural practices [31]. CEVd does not induce any symptoms in most sweet orange (*Citrus sinensis* (L.) Osbeck), mandarin (*Citrus reticulate* Blanco), and grapefruit (*Citrus paradisi* Macfad) scion cultivars. However, when CEVd-infected budwoods are grafted on one of the previous susceptible rootstocks, distinct symptoms may appear [11].

The type and severity of symptoms induced by citrus exocortis disease depend not just on the selected rootstock as described above, but also on the amount of viroid present in the scion and the infection with other citrus viroids. High temperatures can also accelerate the development of symptoms [30]. The results of a long-term field trial carried out with clementine trees grafted on the trifoliate orange rootstock revealed that CEVd-induced effects might be both reduced or increased when CEVd-infected trees were exposed to mixed viroid infections. In other words, several interactions among viroids including CEVd have been revealed through comparative assays between symptoms developed on trees infected with CEVd alone or co-infected with other viroids. The most clear-cut interaction occurs between CEVd and Citrus viroid IV [*Pospiviroidae*; *Cocadviroid*; CVd-IV]. This interaction is manifested by the attenuation of bark-scaling or bark-cracking symptoms as a result of the occurrence of antagonism between both viroids. CVd-IV limits the negative effects of CEVd on tree performance. The reduction of tree size and fruit yield occurs mainly in trees infected with combinations containing CEVd or CVd-III and, to a lesser extent, those containing Citrus bent leaf viroid [*Pospiviroidae*; *Apscaviroid*; CBLVd] [24].

Numerous field trials have been conducted on different citrus species, varieties, and rootstocks under three different agroecosystems, to evaluate the effect of CEVd on vegetative growth and yield (Table 1). The first field trial has been conducted to assess the effect of CEVd infection on commune clementine trees grafted on Pomeroy trifoliate orange. CEVd-infected trees have been periodically monitored for a period of 12 years (from 1990 to 2002) for symptom expression, growth, and fruit yield. CEVd-infected trees showed a significant reduction of growth and yield, which became increasingly apparent over time with infection. Cumulative yield varied from 291,1 to 570,3 kg in 2001 for CEVd and the control, respectively. This equated to 50% cumulative yield lost. This yield attenuation was associated mainly with the loss of large fruit production. Indeed, it has been shown that CEVd reduced fruit production significantly for calibers 2 to 5. Cumulative weights were smaller than the control for caliber 0–1 and small calibers 6 and 7–8, with some significant difference [21]. The quality of fruits from CEVd-infected orange trees (Washington Navel) grafted on Carrizo citrange rootstock has been evaluated from 2004 to 2007. The results of this experiment showed that the quality of the fruit was not affected by CEVd infection [26].

Table 1. Results summary of the known field trials carried out in three citrus-growing countries of the Mediterranean area to evaluate the effect of CEVd and HSVd on vegetative growth and yield of different citrus scion and rootstock combinations.

Combination		Effect on			References
Scion	Rootstock	Height	Canopy	Cumulative yield	
CEVd					
Clementine	Orange Pomeroy trifoliate	**SE*** (Reduction of almost 14% for CEVd-117)	NSE	**SE** (Reduction of almost 50%)	[21]
Oranger Washington Navel	Carrizo citrangec	**SE** (Reduction of almost 10% and 15% for CEVd-129 and CEVd-117, respectively)	NSE (Reduction of almost 17% and 27% for CEVd-129 and CEVd-117, respectively)	NSE (Low reduction of about 2% and 10% for CEVd-129 and CEVd-117, respectively)	[26]
Orange Maltaise demi sanguine	Soor orange (*Citrus aurantium* L.)a	NSE	NSE	NSE (Reduction of almost 20%)	[46]
	Alemow *Citrus macrophylla* Webster b	NSE	NSE	NSE	
	Carrizo citrangec	NSE	NSE	NSE	
	Citrus volkameriana Ten. And Pasq.c	NSE	**SE** (Reduction of almost 33%)	NSE (Reduction of almost 20%)	
	Cleopatra mandarin (*Citrus reshni* Hort. ex Tan.)c	NSE	NSE (Reduction of almost 25%)	NSE (Reduction of almost 20%)	
	Swingle citrumelo (Citru)c	NSE	NSE	NSE	
	Rangpur lime (*Citrus limonia* Osb.)c	NSE	NSE (Reduction of almost 28%)	NSE (Reduction of almost 20%)	
	Trifoliate orangec	**SE** (Reduction of almost 25%)	**SE** (Reduction of almost 49%)	NSE	
HSVd					

Combination		Effect on			References
Scion	**Rootstock**	**Height**	**Canopy**	**Cumulative yield**	
Clementine	Orange Pomeroy trifoliatec	Little or no real impact	NSE	**SE*** (Reduction of almost 34% for CVd-IIc)	[21]
Orange Washington Navel	Carrizo citrangec	NSE (Reduction of almost 15% for CVd-IIc)	NSE (Reduction of almost 8% for CVd-IIb)	No effect	[26]
Orange Maltaise demi sanguine	Soor orange (*C. aurantium*)[a]	NSE	NSE	NSE (Reduction of almost 20%)	[46]
	Alemow *C. macrophylla* Webster b	**SE** (Reduction of almost 33%)	**SE** (Reduction of almost 77%)	**SE** (Reduction of almost 76%)	
	Carrizo citrangec	NSE	NSE	NSE	
	C. volkameriana Ten. And Pasq.c	NSE	**SE** (Reduction of almost 30%)	NSE (Reduction of almost 20%)	
	Cleopatra mandarin (*C. reshni* Hort. ex Tan.)c	NSE	NSE	NSE (Reduction of almost 20%)	
	Swingle citrumelo (Citru)c	NSE	NSE	**SE** (Reduction of almost 36%)	
	Rangpur lime (*C. limonia* Osb.) c	NSE	NSE	NSE (Reduction of almost 20%)	
	Trifoliate orangec	**SE** (Reduction of almost 26%)	**SE** (Reduction of almost 45%)	**SE** (Reduction of almost 66%)	

[a] Susceptible to citrus tristeza virus [*Closteroviridae*; *Closterovirus*; CTV] but viroids tolerant.

[b] Susceptible to CTV stem-pitting and cachexia.

[c] CTV tolerant.

[*] A function of the used viroid isolates.

SE: Significant effect. NSE: No significant effect.

2.3 Transmission and epidemiology

All citrus viroids are distributed primarily by the introduction and propagation of infected budwoods and subsequently by mechanical transmission, and CEVd is no exception [11]. Mechanical transmission of CEVd has been already reported. It took place on secateurs, tools, knives, and hedging equipment [9, 11, 27, 29] especially from lemon (*Citrus lemon* Bum. f.) to lemon [11]. Further, it has been shown that CEVd can survive for 8 days on steel knife blades. CEVd infectivity was not affected over a wide range of time intervals between knife contamination and transfer to citron or by 2 sequential transfers by this method. CEVd spread to susceptible hosts by contaminated tools was accomplished from numerous tested citrus species of great economic importance such as lemon, sweet orange, grapefruit, tangerine, and a trifoliate hybrid [31]. Another transmission assay carried out under greenhouse conditions showed that CEVd can be mechanically transmitted from citron to healthy citron [32, 33] and gynura herbaceous plant [33] by a single slash with a knife blade [32, 33]. CEVd-retransmission from infected gynura back to citron was successful [33]. Natural grafts of citrus roots seem to be associated with CEVd propagation. That is the case for example for a budwood multiplication block of an Australian nursery where the propagation of CEVd by natural grafts of roots induced the infection of some healthy lemon mother trees on which neither hedging nor pruning operations took place before their removal. This may be mainly linked to the fact that citrus trees were planted close to each other (within 2 m). The role of root grafting in CEVd transmission was assessed by excavating root systems [29]. CEVd root transmission, though possible, would be overshadowed by mechanical transmission. CEVd is not known to be a vector- or seed-transmitted [9, 11]. The role of gots as possible vectors of viroids, including CEVd, has been investigated. The experiment was carried out by rubbing healthy citrus plants with goat horns previously rubbed for 24 h on infected Etrog stems. Results highlighted the detection of CEVd in the tested plants. Therefore, transmission through gots could have facilitated the long-range spread of CEVd among both cultivated and wild plants and *vice versa* and also among graft-incompatible plants [34].

3. Hop stunt viroid

3.1 Taxonomy

Cachexia is a destructive disease infecting citrus species [11]. The agent of this disease, hop stunt viroid [*Pospiviroidae*; *Hostuviroid*; HSVd], is a small covalently closed ssRNA of about 300 nts [3, 35]. HSVd is a single member of the genus *Hostuviroid* [36]. HSVd molecules can exist as either circular or linear [35]. HSVd isolates are divided into five groups: three major and two minor groups. The first groups, composed of "plum-type", "hop-type" and "citrus-type", are composed of isolates from a limited number of isolation hosts. As to the second group, it has been suggested that they are the results of the occurrence of recombination events between members of the main groups [37]. Like CEVd, HSVd takes the form of a model of five structural-functional domains within the rod-like secondary structure: C, P, V, TR, and TL [15]. However, HSVd has a genus-specific CCR and a terminal conserved hairpin (TCH) and lacks a TCR [10]. It is worth mentioning that two HSVd-related Group II citrus viroids that differ by a "cachexia expression motif" have been described. It includes a cachexia disease non-pathogenic variant (CVd-IIa) and two pathogenic variants (CVd-IIb and CVd-IIc) [38, 39, 40]. Electrophoretic profiles obtained with single-stranded polymorphism (SSCP) allowed deciphering the variability among and within cachexia-inducing sources of citrus isolates of HSVd. SSCP allowed discrimination between non-cachexia and cachexia sources of HSVd. Sequence analysis showed that the V domain was extremely conserved among all the cachexia variants. Indeed, 5 nts differences, affecting both the upper (3 nts) and the lower (2 nts) strands of the V domain, were identified as the most characteristic motif allowing the discrimination between cachexia and non-cachexia sequences. It has been suggested, therefore, that the 5 nts affect the organization of a short helical region and two flanking loops of the V domain, thus modifying the three-dimensional geometry of the molecule [41]. Subsequently, it has been shown that only a single change in HSVd modulates citrus cachexia symptoms [38].

3.2 Symptoms and economical impact

Cachexia could affect a various part of the tree including the trunk, bark, twigs, branches, leaves, and fruits, thus causing different types of damages such as bark and trunk gumming with a rough and rugose appearance, bark-cracking, moderate and severe tree stunting, chlorosis, decline and death of severely affected trees, brown stipple spotting on the underside of the leaves, and the appearance of small pits on the wood [9, 21, 24]. Cachexia disease mainly affects some mandarins and their hybrids such as tangelos, and *Citrus macrophylla* Wester. Most other citrus species seem to be symptomless unless grafted on susceptible rootstocks [10]. Cachexia-inducing variants were proven to cause gummy bark disease of sweet orange [42, 43] and split bark disorder of sweet lime (*Citrus limetta* Risso) [44]. HSVd variants have been reported to induce yellow corky vein disease of Kagzi lime (*Citrus aurantifolia* (Christm.) Swingle) [45] and sweet orange [44] in India and Iran, respectively. It was subsequently shown that cachexia and a similar disorder previously described in Palestine sweet lime, known as xyloporosis, are caused by the same type of HSVd variants [40].

As mentioned before, for CEVd, the type and severity of citrus cachexia symptoms depend also on the presence of other citrus viroids in the tree. The results of a long-term field trial carried out with clementine trees grafted on the trifoliate orange rootstock revealed that HSVd-induced effects might be both reduced or increased when HSVd-infected trees were exposed to mixed viroid infections. The most clear-cut interaction occurs between HSVd and CVd-IV. This interaction is manifested by a slight increase in fruit yield and reduction of scion circumferences [24].

The same field trials described before to evaluate the effect of CEVd on vegetative growth and yield (Table 1) were used for the same purpose for HSVd. Little or no effect in vegetative growth has been observed on commune clementine trees infected by HSVd as it has been determined by the measure of height and rootstock and scion circumferences [21]. Cumulative yield varied from 377,6 to 570,3 kg in 2001 for HSVd (CVd-IIc isolate) and the control, respectively. This equated to 34% cumulative yield lost [21]. The negative impact of HSVd infection on cumulative yield has been reported in another study carried out on Orange Maltaise demi sanguine grafted on Alemow (*C. macrophylla*). HSVd-infected trees have been periodically monitored for a period of 12 years (from 2005 to 2017) for growth and fruit yield. HSVd-infected trees showed a significant reduction of yield of about 76% compared to healthy control [46]. As for CEVd, the effect of HSVd infection on the quality of fruit from Washington Navel orange trees grafted on Carrizo citrange rootstock has been evaluated from 2004 to 2007. The results of this experiment showed that the quality of the fruit was not affected by HSVd infection. However, a reduction occurred in the diameter of the harvested fruits [26].

3.3 Transmission and epidemiology

As pointed out before, for CEVd, propagation of infected budwoods and mechanical inoculations with contaminating tools were reported as the principal causes for the omnipresence of multiple viroid species, including HSVd, among citrus orchards [34]. Mechanical transmission of HSVd has been already reported. Indeed, the results of a transmission assay carried out under greenhouse conditions revealed that all HSVd strains are mechanically transmitted from citron to healthy citron by a single slash with a knife blade [32]. As for CEVd, the potential involvement of gots in HSVd spread has been shown under controlled conditions [34]. Top working, a common practice in Mediterranean countries, seems to have largely contributed to HSVd spread in Mediterranean citrus orchards [9]. HSVd is not known to be seed-borne [47] in citrus or to have natural vectors [11, 48].

4. Signaling pathways in citrus exocortis and cachexia pathogenesis

It is usually accepted that although the mechanisms through which viroids interact with their hosts are beginning to be dissected, the key triggering events and molecular mechanisms underlying viroid pathogenesis remain unclear [49, 50], and CEVd and HSVd are no exception. As demonstrated by various types of citrus pathogens [51, 52], further investigation of the molecular basis of viroid-host interactions is crucial to better understand the pathogenesis of viroids, and thus help to develop effective strategies to combat viroid diseases [50, 53]. Important changes occur in the chloroplast, cell wall, peroxidase, and symporter activities upon infection of Etrog citron with CEVd [54]. The CEVd-infected citron system has been subsequently used for studying the feedback regulation mechanism using transcriptomic analysis. The analysis of the woody host response to CEVd revealed the activation of basic defense and RNA-silencing mechanisms following CEVd infection. In other words, a large number of genes (about 1530) encoding key proteins involved in the RNA silencing pathway, and proteins related to basic defense responses are expressed following CEVd infection [53]. Furthermore, a recent study elucidates the role of phytohormone pathways, particularly those linked to ethylene, in disease development and ribosomal stress caused by CEVd infection by using tomato as an experimental host [55]. For HSVd, a small RNA-mediated gene silencing response has been highlighted upon the infection of lemon by HSVd. The large amounts of HSVd-small interfering RNA (siRNA) from both central and variant domains have been suggested to be involved in interference with host gene and symptom development [56].

5. Detection methods

5.1 Biological indexing and cross protection

Generally, biological assays based on indicator host plants expressing typical symptoms of infection and able to withstand higher levels of viroid replication played an essential role in both viroid detection and characterization [57]. CEVd and HSVd are viroid diseases present in citrus orchards around the world [11]. The biological diagnosis through indexing method is considered as an efficient tool to test the health status of a plant, regarding a disease by inoculation with the grafting of the budwood or any other infected tissue in indicator plants that allow viroid replication, symptoms expression [11, 58], and the enhancement of viroid concentration [59]. However, bioassays for CEVd and HSVd detection and identification may require a panel of indicator host plants [60]. Certain considerations need to be respected for the proper indexing of citrus viroids. These include the use of excellent plants, the work under warm temperatures, and the use of citron index plants grown one per container. As mentioned previously, citrus viroids are highly mechanically transmissible and tools must be disinfected to avoid their spread [9].

The citron test is a very sensitive and diagnostic index for determining the presence of CEVd [9]. However, indexing, *in vivo* for CEVd diagnosis is time-consuming, labor-intensive, and requires technician greenhouses [59]. It can take 90 days after inoculation or grafting onto indicator plants [61, 62, 63]. Symptoms of slight to severe epinasty leaves wrinkled and twisted to the reverse with light to dark brown cracks in petiole and branches, blisters in the petiole, corking of the midrib, and reduced growth are the main symptoms observed on Etrog citron Arizona 861-S indicator plants graft-inoculated with CEVd-infected budwoods [28, 29, 59, 63]. An *in vitro* indexing procedure has been developed to minimize the risks of epidemics caused by viroids including CEVd. It has been proved that the *in vitro* indexing of CEVd is efficient as well as the *in vivo* diagnosis, and requires between 20 and 40 days less to reach the maximum incidence after inoculation. Epinasty, growth reduction, and rugged leaves with dry tips, and reduced size are the main symptoms observed on the sprouts planted *in vitro* and grafted with CEVd-infected callus [59]. The same symptoms have been reported for sprouts grafted with CEVd infected cortex [62]. Cuban Shaddock (*Citrus maxima* (Burm.) Merr.) has been proved to be the best rootstock, compared to rough lemon (*Citrus jambhiri* Lush.) or Volkamer lemon (*Citrus volkameriana* Ten.), for symptom expression on Arizona 861 S1- citron indicator plants for indexing exocortis [64]. Gynura is also considered as an excellent indicator for CEVd. This latter reacts strongly in this host plant [9].

As to cachexia, Parson's special mandarin budded on vigorous root-stock such as rough lemon or Volkamer lemon is reported as an excellent indicator for the disease [9, 65]. The biological indexing may take up to one year before symptoms are seen. The reaction of Parson's Special mandarin may differ depending on HSVd isolates. In other words, some isolates are very mild reacting, whereas others are quite severe in their reaction to the indicator plant. Indeed, a mild strain reaction consists of just a slight browning at the bud union or cut back region of the Parson's Special mandarin while a severe reaction consists of the appearance of gum in the wood that may extend via the entire plant [9]. Cuban Shaddock has been proved to be the best rootstock, compared to rough lemon or Volkamer lemon, for symptom expression on Clemeline 11–20 indicator plants for indexing cachexia. Furthermore, the application of 0,5% foliar urea sprays, alone or in combination with 20 ppm gibberellic acid showed to produce more intense expression of cachexia symptoms in the indicator Clemeline 11–20 than the unsprayed control [64].

Cross protection is a biological assay, in which the infection of a plant with a viroid strain ensures protection from infection with another strain of the same viroid. This bioassay can be used for indirect viroid biological indexing. It has been applied in the diagnosis of several viroids including CEVd and HSVd. Typically, the principle of this method is based on the infection of the plant with a mild strain of a viroid, followed by its inoculation with inoculum from a plant suspected to be infected with a severe strain of the same viroid. Positive indexing of the viroid is revealed by the non-expression of symptoms in the tested plants [60]. It has been shown in a cross-protection assay, performed with CEVd-129 as a "protecting" strain against the severe type strain of CEVd that a mild strain of CEVd could lead to apparent "protection" against challenge inoculation with the severe strain. However, it is important to highlight that variability has been shown in the induced protection effect. The latter varied from only a brief delay to almost total impairment of symptom expression. The level of protection depends on the length of the interval between the inoculations with the mild and severe strains [66].

5.2 Nucleic acid-based methods

Since viroids lack a protein capsid, serological techniques used routinely in plant viruses' detection are not applicable [67]. Nucleic acid-based methods, including polyacrylamide gel electrophoresis (PAGE), hybridization (dot- and northern-blots and micro–/ macroarrays), amplification (reverse transcription-polymerase chain reaction (RT-PCR) and reverse transcription loop-mediated isothermal amplification (RT-LAMP)) and sequencing (next-generation sequencing and Sanger sequencing), offer rapid cost-effective, and reliable diagnosis of viroids [60].

PAGE is considered as the first molecular technique used for the rapid (2–3 day period) identification of viroid infected plants. This technique continues to play a crucial role in the identification of new viroids since it is the only diagnostic method that is sequence-independent. PAGE analysis under denaturing conditions showed that many *Citrus* species may harbor numerous viroids including CEVd and HSVd [57]. PAGE and ethidium bromide or silver staining is considered as the first molecular technique applied for CEVd detection [22, 68]. However, it seems that the sensitivity of this technique requires an adequate viroid accumulation level [22]. In other terms, the PAGE procedure was used successfully to directly detect CEVd from field-grown sweet orange and grapefruit trees. The key was reported to be the use of large (50 g) samples of succulent, expanding-flush tissue collected during the summer season. However, samples collected from field-grown trees in January and February did not give consistent detection in trees known to be CEVd-infected, presumably because lack of new growth and low temperatures do not favor CEVd replication [69]. PAGE analysis can routinely resolve as many as four different viroids in the same sample. For instance, it has been shown that this technique can resolve two HSVd variants differing in length by only four nucleotides i.e. 303 nts vs. 299 nts [57].

Since the beginning of their use in the 1980s, dot blot hybridization and hybridization of tissue imprints began to replace PAGE for routine viroid detection. This is mainly because these methods allow the processing of a large number of samples [57]. A northern hybridization protocol, which relied on the analysis of preparations from bark tissues, was proved to be more sensitive than PAGE to detect CEVd and HSVd from field-grown plants of different citrus species and cultivars [70]. A citrus viroids-multiprobe composed of full-length clones of HSVd, CEVd, and two other citrus viroids has been constructed for the simultaneous detection of viroids associated with citrus trees. All the tested viroids were effectively detected with this multiprobe when tested by both northern hybridization and dot blot methods. It is important to highlight that this multiprobe does not allow the identification of the viroid type species resulting in a positive signal [71].

Due to the small size of viroids, numerous RT-PCR approaches can be applied for both their detection and subsequent characterization. In the case of CEVd and HSVd, numerous RT-PCR and real-time RT-PCR approaches have been developed and proved to allow the detection of CEVd and HSVd in both singleplex or multiplex assays [63, 72, 73, 74, 75, 76]. A list of some RT-PCR and related tests developed to detect CEVd and HSVd are presented in Table 2. Some of these tests allow also the discrimination between mild and severe CEVd strains and the identification of HSVd isolates associated with cachexia symptoms [77]. The multiplex one-step RT-PCR assay developed by Wang et al. [75] is considered a good tool streamlining the simultaneous detection of up to five citrus viroids, including CEVd and HSVd. This enables to reduce time and labor without affecting sensitivity and specificity. Indeed, serial dilution experiments showed that the singleplex RT-PCR sensitivity was similar to that of multiplex RT-PCR for all the tested viroids [75]. This type of assay could be used in high throughput screenings of viroids associated with citrus in field surveys, germplasm banks, nurseries, as well as in other viroid disease management programs [74]. Similarly, the multiplex RT-TaqMan PCR assay developed by Papayiannis [76] enables accurate discrimination between CEVd and HSVd with a diagnostic sensitivity and specificity of 100%. It is important to emphasize that in conventional RT-PCR tests, the overall sensitivity and specificity were lower and varied between 97 and 98% for HSVd, and 94 and 95% for CEVd. Therefore, this essay presented 1000-fold more analytical sensitivity [76]. The specificity of the tests described previously was confirmed by including healthy controls and/or plant tissue infected with other citrus graft transmissible virus and bacteria pathogens and non-targeted citrus viroids. Both singleplex and multiplex assays did not cross-react with any non-inoculated negative controls or other citrus pathogens [63, 74, 76]. To date, PCR-based approaches have been proven efficacy on viroid direct detection. However, false positives and negatives due to amplicon contamination and failure to generate cDNA of suitable size during reverse transcription, respectively, are not uncommon and therefore preclude the application of RT-PCR for large scale indexing [70].

Table 2. Primer sequences and their annealing temperature (Tm), primer/probe location, and expected size of PCR products for each primer pair when used to amplify CEVd and HSVd by RT-PCR and related tests (this is not a full or exclusive list).

Name of RT-PCR test Primer/Probe Name	Sequence 5′-3′	Tm (°C)	Genomic coordinates	Size of the expected product	References
RT-PCR					
Singleplex					
CEVd-R	GGGGATCCCTGAAGGACTT	60	80-98a	371 bp	[72, 85]
CEVd-F	GGAAACCTGGAGGAAGTCG		99-117a		
HSVd-F 27-mer VP-20	CGCCCGGGGCAACTCTTCTCAGAATCC	60	78–102	251 bp	[37, 73]
HSVd-R 26-mer VP-19	GCCCCGGGGCTCCTTTCTCAGGTAAG		60–85		
HSVd VP-98	CTCCAGAGCACCGCGGCCCTC	DN	120–140	140 bp	[37]
HSVd VP-99	CTGGGGAATTCTCGAGTTGCCGC		1–23		
Multiplex					
CEVd-F194	TTTCGCTGCTGGCTCCACA	58	194–212	196 bp	[63]
CEVd-R18	ACCTCAAGAAAGATCCCGA		371–18		
HSVd-F1	GGGGCAACTCTTCTCAGAATCC		81–102	302 bp	
HSVd-R1	GGGGCTCCTTTCTCAGGTAAGTC		58–80		

Name of RT-PCR test / Primer/Probe Name	Sequence 5′-3′	Tm (°C)	Genomic coordinates	Size of the expected product	References
CEV-R	CCGGGGATCCCTGAAGGACTT	58	78-98a	371 bp	[75]
CEV-F	GGAAACCTGGAGGAAGTCGAG		99-119a		
HSVd-R	CCGGGGCTCCTTTCTCAGGTAAGT		59-82b	302 bp	
HSVd-F	GGCAACTCTTCTCAGAATCCAGC		83-105b		
Real time-RT-PCR					
Singleplex					
CEVd –161 F	GTCCAGCGGAGAAACAGGAG	60	181-200c	105 bp	[74]*
CEVd –258 R	AGAGAAGCTCCGGGCGA		270-286c		
CEVd –187 P FAM	TCCTTCCTTTCGCTGCT		212-228c		
HSVd-208 F	GAGACGCGACCGGTGG	60	216-231d	88 bp	
HSVd-295 R	GCTCAAGAGAGGATCCGCG		286-304d		
HSVd-226 P TET	TCACCTCTCGGTTCGTC		234-250d		
Multiplex					
CEVd-RTR_F	GTCGCCGCGGATCACT	60	142–159	64 bp	[63]
CEVd-RTR_R	CCAGCAGCGAAAGGAAGGA		187–205		
HSVd-RTR_F	GGAATTCTCGAGTTGCCGCA		5–24	127 bp	
HSVd-RTR_R	CCGCGGCCCTCTCT		118–131		
CEVd-RTR_P	CCAGCGGAGAAACAG		163–177	—	
HSVd-RTR_P	CAACTCTTCTCAGAATCC		85–102	—	
CEVdF	GCGTCCAGCGGAGAAACA	60	158–175e	68 bp	[76]
CEVdR	CAGCGACGATCGGATGTG		226–208e		
CEVdTAQ	{FAM}-TCGTCTCCTTCCTTTCGCTGCTGG-{BHQ1}		181–204e		
HSVdF	GCCTTCGAAACACCATCGA		159–177f	71 bp	
HSVdR	CACCGGTCGCGTCTCATC		230–213f		
HSVdTAQ	{HEX}-CGTC-CCTTCTTCTTTACCTTCTCCTG-GCTC-{BHQ2}		179–208f		

*The same primers have been also tested in multiplex Real time-RT-PCR test.

[a]GenBank Accession no. NC-001464.

[b]GenBank Accession no. NC-001351.

[c]GenBank Accession no. CEVd-HQ284019.

[d]GenBank Accession no. HSVd-KJ810553.

[e]GenBank Accession no. U21126.

[f]GenBank Accession no. GQ249348. R: antisense primer. F: sense primer.

Next-generation sequencing (NGS) technologies are currently becoming routinely applied in different fields of virus and viroids studies. These advanced technologies have therefore contributed to a revolution in both the detection and discovery of plant viruses and viroids [78, 79, 80, 81]. NGS has also provided an alternative method to identify viroids in the citrus cultivars. In other words, transcriptome sequencing has shown efficacy in citrus viroid diagnostics. Indeed, this method enabled the simultaneous identification of numerous viroids from various citrus samples, including CEVd and HSVd [82]. A deep sequencing approach, combined with bioinformatics analysis, is already being implemented for HSVd detection in *C. lemon* in China. This finding suggests that HSVd could infect this host and potentially be a pathogen that causes disease on *C. lemon* trees [56].

6. Control strategies

In vitro somatic embryogenesis, from both style and stigma cultures, has been proved to be a highly effective sanitation method leading to the complete elimination of the main virus and virus-like diseases associated with citrus. Furthermore, it has been shown efficacy to eliminate diseases induced by viroids, and the production of healthy citrus plants [83, 84, 85]. This method was applied to eliminate CEVd and HSVd from some *Citrus* species [83]. For example, somatic embryogenesis has been tested on 13 genotypes, belonging to the Algerian germplasm collection of two different *Citrus* species, lemon and sweet orange, infected by at least one graft-transmissible agent, including CEVd and HSVd. This method has shown efficacy to eliminate CEVd from 13/12 tested genotypes. However, HSVd was proved to be the most infectious viroid since it has been eradicated only from 13/5 tested genotypes. It is important to emphasize that no somaclonal variability has been highlighted in lemon regenerated plants. However, a genetic instability has been observed in some of the regenerated orange plants Washington navel 251 [83]. Sanitation by *in vitro* shoot-tip grafting has also been proved to be a very effective method for citrus graft-transmissible diseases eradication including citrus viroids (success rate of about 88 ,87 ,86] (%100]. For instance, it has been reported that CEVd and HSVd can be routinely eliminated from citrus by shoot-tip grafting. Since citrus viroids are extremely tolerant of heat, the use of thermotherapy as a sanitary method is not effective in eliminating viroids from citrus budwoods [9].

No naturally occurring durable resistance has been observed in most species, despite non-hosts for viroids exist. Therefore, the effective control methods for viroid diseases consist mainly of detection and eradication, and cultural controls [50].

7. Viroids situation in the mediterranean region: focus on Morocco

Exocortis and cachexia are widespread diseases in the Mediterranean region. CEVd and HSVd have been reported in most Mediterranean countries and are among the most prevalent citrus viroids in the region [9]. The development of reliable diagnostic methods facilitated extensive surveys for CEVd and HSVd in different parts of the region. Both viroids were successively identified in many countries, including Morocco [89, 90, 91, 92], Cyprus [33], Spain [26], Egypt [43, 93], Italy [61, 94], Tunisia [46, 95], France [21], Syria [96], and Turkey [42].

In Morocco, exocortis and cachexia are among the major citrus viroid diseases [90, 91]. These diseases are prevalent in citrus orchards and can be found in all *Citrus* species and varieties [89, 90, 91, 92, 97]. Mechanical transmission of citrus viroids, including CEVd and HSVd, via working tools seems to be behind the widespread of these phytopathogens and their detection in both old and young plantings in all surveyed citrus orchards [92]. Research, recently completed from 2008 to 2018, to monitor CEVd and HSVd prevalence, in the main citrus-growing areas of Morocco (Gharb, Haouz, Loukkos, Moulouya, Souss, and Tadla), showed that CEVd and HSVd are omnipresent in almost all citrus-growing areas of the country with relatively high prevalence. That is the case for example for the Gharb area where CEVd and HSVd were detected at a prevalence of 85% [89] and 21% [92], respectively. Concerning genetic analysis, a first sequence comparison among six Moroccan HSVd isolates collected in the six main citrus-growing areas of Morocco has been recently reported by Afechtal et al. [92]. Phylogenetic analysis showed that the six HSVd isolates are clustered into one group within the "citrus-type". Furthermore, it seems that sequence variability is neither a function of host plant nor a function of the symptoms [92].

8. Conclusions

Citrus viroids, including CEVd and HSVd, are distributed mainly by the introduction and propagation of infected budwoods, by top working, and by mechanical transmission [9, 11]. Both viroids are known for their ability to infect a large number of host plants [36]. CEVd and HSVd are destructive to certain citrus varieties and, can cause yield losses that may be as high as 34 to 76 percent depending on the combination viroid-rootstock-scion [21, 46]. The mechanisms through which CEVd and HSVd interact with their hosts and induce pathogenesis are beginning to be deciphered. In other words, the involvement of RNA-silencing and basic defense mechanisms following CEVd and HSVd infection has been highlighted [54, 56].

Once introduced and established in a country, both viroids can spread relatively rapidly because of their ability to be transmitted via mechanical means [9, 11]. Since CEVd and HSVd have a high resistance to heat, the chemical treatment appears to be the best method to disinfect CEVd- and HSVd-contaminated tools. For instance, a 0,25 to 0,5 and a 1 percent solution of sodium hypochlorite appears to be the best option to eliminate CEVd and HSVd, respectively, from contaminated hedging and budwood cutting tools [11, 98]. Like all citrus viroids, CEVd and HSVd seem to be successively eliminated from propagative material by shoot-tip grafting or by the deployment of nucellar budlines. Being extremely tolerant of heat, CEVd and HSVd have not been successfully eliminated from budwood by applying thermotherapy [9]. Certification programs must include measures to control viroid spread in nurseries [32]. The majority of rootstocks that are tolerant to the citrus tristeza virus [*Closteroviridae*; *Closterovirus*; CTV] are susceptible to citrus viroids. Therefore, in the absence of a certification program, exocortis disease usually follows upon the replantation of these rootstocks [9]. Since no useful sources of natural resistance to viroid disease are known, diagnostic tests continue to play a key role in efforts to control viroid diseases [67]. Nowadays, several nucleic acid-based methods for detecting CEVd and HSVd exist, including PAGE, hybridization, amplification, and sequencing [60]. Although biological assay has several disadvantages, it will always play a pivotal role in viroid research. Indeed, Cuban Shaddock has been proved to be the best rootstock for symptom expression on Arizona 861 S-1 citron and Clemeline 11–20 indicator plants for indexing exocortis and cachexia, respectively [64]. Besides, gynura seems to be an excellent indicator for CEVd [9]. A combination of both molecular and biological assays should lead to the most effective means for viroid identification and characterization [60].

Complicated interactions, including antagonism and synergy, occur between viroids coinfecting the same citrus host. These interactions may lead to different symptoms, canopy volumes, fruit yields, and commercial performance. Although no obvious physiological changes in citrus hosts have been described in mixed infections of CEVd and HSVd and both viroids do not induce severe symptoms in citrus [24, 99], their interaction was intriguing because they are commonly found simultaneously infecting different citrus cultivars and they have identical biological properties within the same host. The relationship between the two viroids has been investigated over 3 years (from 2011 to 2013). Results showed a positive correlation between CEVd and HSVd in specific tissues of two citrus cultivars (blood orange and Murcott mandarin). This result has been supported by three findings: titer enhancement, localization similarity, and lack of symptom aggravation under mixed-infection conditions. Compared to their concentrations under single-infection conditions, a significant increase in the CEVd and HSVd population has been observed under mixed-infection during 6 and only 1 season of the 12 monitored seasons, respectively. This result is somewhat surprising because no competition phenomenon for host resources occurs between the two viroids although they have the same biological functions and share identical cellular and subcellular spaces [27]. This issue merits consideration in future research.

Regarding the current situation of CEVd and HSVd in Morocco, this chapter provides a general overview of their spread in the Moroccan citrus-growing areas. Preventing the introduction and the establishment of exocortis and cachexia diseases in the Moroccan citrus orchards can be set up through the use of viroid-free (certified) planting material, disinfection of pruning tools, regular monitoring of citrus orchards to ensure early detection of both diseases, and by avoiding top working practice. This review pointers to new research avenues in exocortis and cachexia diseases in Morocco or elsewhere. These research fields could include for instance the characterization of CEVd and HSVd isolates, searching for secondary hosts, and developing sustainable control strategies. Investigating the prevalence of CEVd and HSVd infection in numerous natural host plants, and the characterization of the viroid sequence variants is valuable especially that a cross-transmission phenomenon between different hosts seems to be possible for HSVd [100].

Studying functional genomics through transcriptomic analysis and/or proteomic approaches in citrus-CEVd/HSVd interaction would be an interesting approach to shed more light on the full mechanisms underlying the complex and varied events associated with such interactome, and thus contribute to the development of novel diagnostic methods and plant protection strategies. This further advanced research will expand our understanding of CEVd and HSVd epidemiology and the mechanisms behind their spread across the world in general and Morocco in particular, and could potentially help in devising innovative management strategies of both viroids.

References

1. Flores, R.; Gas, M.E.; Molina-Serrano, D.; Nohales, M.A.; Carbonell, A.; Gago, S.; De la Peña, M.; Daròs, J.A. Viroid replication: Rolling-circles, enzymes and ribozymes. Viruses2009, 1, 317-334, doi:10.3390/v1020317

2. Carbonell, A.; Martínez de Alba, Á.E.; Flores, R.; Gago, S. Double-stranded RNA interferes in a sequence-specific manner with the infection of representative members of the two viroid families. Virology2008, 371, 44-53, doi:10.1016/j.virol.2007.09.031

3. Marquez-Molins, J.; Gomez, G.; Pallas, V. Hop stunt viroid: A polyphagous pathogenic RNA that has shed light on viroid–host interactions. Mol. Plant Pathol.2020, 1-10, doi:10.1111/mpp.13022

4. Palukaitis, P. What has been happening with viroids? Virus Genes2014, 49, 175-184, doi:10.1007/s11262-014-1110-8

5. Adkar-Purushothama, C.R.; Perreault, J.P. Current overview on viroid–host interactions. Wiley Interdiscip. Rev. RNA2019, 11, 1-21, doi:doi:10.1002/wrna.1570

6. Ding, B.; Itaya, A. Viroid: A useful model for studying the basic principles of infection and RNA biology. Mol. Plant-Microbe Interact.2007, 20, 7-20, doi:10.1094/MPMI-20-0007

7. Bani Hashemian, S.M.; Serra, P.; Barbosa, C.J.; Juárez, J.; Aleza, P.; Corvera, J.M.; Lluch, A.; Pina, J.A.; Duran-Vila, N. Effect of a field-source mixture of citrus viroids on the performance of "nules" clementine and "navelina" sweet orange trees grafted on carrizo citrange. Plant Dis.2009, 93, 699-707, doi:10.1094/PDIS-93-7-0699

8. Semancik, J.S.; Roistacher, C.N.; Rivera-Bustamante, R.; Duran-Vila, N. citrus Cachexia Viroid, a New Viroid of Citrus: Relationship to Viroids of the Exocortis Disease Complex. J. Gen. Virol.1988, 69, 3059-3068, doi:10.1099/0022-1317-69-12-3059

9. Roistacher, C.N. Diagnosis and Management of Virus and Virus like Diseases of Citrus. In Diseases of Fruits and Vegetables; Naqvi, S.A.M.H., Ed.; 2004; Vol. I, pp. 109-189

10. Hataya, T.; Tsushima, T.; Sano, T. Hop Stunt Viroid. In Viroids and Satellites; Hadidi, A., Flores, R., Randles, J.W., Palukkaitis, P., Eds.; Sara Tenny, 2017; pp. 199-210

11. Roistacher, C.N. Graft-transmissible diseases of citrus, hand book for detection and diagnosis. Food Agric. Organ. United Nations1991, p.285

12. Gross, H.J.; Krupp, G.; Domdey, H.; Raba, M.; Jank, P.; Lossow, C.; Alberty, H.; Ramm, K.; Sanger, H.L. Nucleotide Sequence and Secondary Structure of Citrus Exocortis and Chrysanthemum Stunt Viroid. Eur. J. Biochem.1982, 121, 249-257, doi:10.1111/j.1432-1033.1982.tb05779.x

13. Visvader, J.E.; Gould, A.R.; Bruening, G.E.; Symons, R.H. Citrus exocortis viroid: nucleotide sequence and secondary structure of an Australian isolate. FEBS Lett.1982, 137, 288-292, doi:10.1016/0014-5793(82)80369-4

14. Duran-Vila, N.; Roistacher, C.N.; Rivera-Bustamante, R.; Semancik, J.S. A Definition of Citrus Viroid Groups and Their Relationship to the Exocortis Disease. J. Gen. Virol.1988, 69, 3069-3080, doi:https://doi.org/10.1099/0022-1317-69-12-3069

15. Flores, R.; Randles, J.W.; Ownes, R.A.; Bar-Joseph, M.; Diener, T.O. The Suviral Agents. In Virus Taxonomy, Eighth Report of the International Committee on Taxonomy of Viruses; Fauquet, C.M., Mayo, M.A., Maniloff, J., Desselberger, U., Ball, L.A., Eds.; 2005; pp. 1145-1161 ISBN 0122499514
16. Keese, P.; Symons, R.H. Domains in viroids: Evidence of intermolecular RNA rearrangements and their contribution to viroid evolution. Proc. Natl. Acad. Sci. U. S. A.1985, 82, 4582-4586, doi:10.1073/pnas.82.14.4582
17. Visvader, J.E.; Symons, R.H. Eleven new sequence variants of citrus exocortis viroid and the correlation of sequence with pathogenicity. Nucleic Acids Res.1985, 13, 2907-2920, doi:10.1093/nar/13.8.2907
18. Visvader, J.E.; Symons, R.H. Replication of in vitro constructed viroid mutants: location of the pathogenicity-modulating domain of citrus exocortis viroid. EMBO J.1986, 5, 2051-2055, doi:10.1002/j.1460-2075.1986.tb04465.x
19. Garcia-Arenal, F.; Pallas, V.; Flores, R. The sequence of a viroid from grapevine closely related to severe isolates of citrus exocortis viroid. Nucleic Acids Res.1987, 15, 4203-4210
20. Chaffai, M.; Serra, P.; Gandía, M.; Hernández, C.; Duran-Vila, N. Molecular characterization of CEVd strains that induce different phenotypes in Gynura aurantiaca: structure-pathogenicity relationships. Arch. Virol.2007, 152, 1283-1294, doi:10.1007/s00705-007-0958-5
21. Vernière, C.; Perrier, X.; Dubois, C.; Dubois, A.; Botella, L.; Chabrier, C.; Bové, J.M.; Duran Vila, N. Citrus viroids: Symptom expression and effect on vegetative growth and yield of clementine trees grafted on trifoliate orange. Plant Dis.2004, 88, 1189-1197, doi:10.1094/PDIS.2004.88.11.1189
22. Duran-Vila, N. Citrus Exocortis Viroid. In Viroids and Satellites; 2017; pp. 169-179
23. Sano, T.; Candresse, T.; Hammond, R.W.; Diener, T.O.; Owens, R.A. Identification of multiple structural domains regulating viroid pathogenicity. Proc. Natl. Acad. Sci. U. S. A.1992, 89, 10104-10108, doi:10.1073/pnas.89.21.10104
24. Vernière, C.; Perrier, X.; Dubois, C.; Dubois, A.; Botella, L.; Chabrier, C.; Bové, J.M.; Duran Vila, N. Interactions Between Citrus Viroids Affect Symptom Expression and Field Performance of Clementine Trees Grafted on Trifoliate Orange. Virology2006, 96, 356-368, doi:10.1094/PDIS.2004.88.11.1189
25. Vernière, C.; Botella, L.; Dubois, A.; Chabrier, C.; Duran-Vila, N. Properties of Citrus Viroids: Symptom Expression and Dwarfing. In Proceedings of the Fifteenth International Organization of Citrus Virologists Conference Proceedings-Viroids; 2002; pp. 240-248
26. Murcia, N.; Bani Hashemian, S.M.; Serra, P.; Pina, J.A.; Duran-Vila, N. Citrus viroids: Symptom expression and performance of washington navel sweet orange trees grafted on carrizo citrange. Plant Dis.2015, 99, 125-136, doi:10.1094/PDIS-05-14-0457-RE
27. Lin, C.Y.; Wu, M.L.; Shen, T.L.; Hung, T.H. A mutual titer-enhancing relationship and similar localization patterns between Citrus exocortis viroid and Hop stunt viroid co-infecting two citrus cultivars. Virol. J.2015, 12, 1-11, doi:10.1186/s12985-015-0357-6

28. Gillings, M.R.; Broadbent, P.; Gollnow, B.I. Viroids in Australian Citrus: Relationship to Exocortis, Cachexia and Citrus Dwarfing. Aust. J. Plant Physiol.1991, 18, 559-570, doi:10.1071/PP9910559

29. Broadbent, P.; Gollnow, B.I.; Gillings, M.R.; Bevington, K.B. Root Grafting and Mechanical Transmission of Citrus Exocortis Viroid Within a Citrus Budwood Multiplication Block. In Proceedings of the International Organization of Citrus Virologists Conference; 1988

30. Hardy, S.; Donovan, N.; Barkley, P.B. Citrus Exocortis. Primefact2008, 772, 1-3

31. Allen, R. Mechanical transmission of exocortis virus. Citograph1970, 55, 145-148

32. Barbosa, C.J.; Pina, J.A.; Pérez-Panadés, J.; Bernad, L.; Serra, P.; Navarro, L.; Duran-Vila, N. Mechanical Transmission of Citrus Viroids. Plant Dis.2005, 89, 749-754, doi:10.1094 /PD-89-0749

33. Kyriakou, A.P. Incidence in Cyprus of citrus exocortis viroid and its mechanical transmission. Plant Pathol.1992, 41, 20-24, doi:10.1111/j.1365-3059.1992.tb02310.x

34. Cohen, O.; Batuman, O.; Moskowits, Y.; Rozov, A.; Gootwine, E.; Mawassi, M.; Bar-Joseph, M. Goat horns: Platforms for viroid transmission to fruit trees? Phytoparasitica2005, 33, 141-148, doi:10.1007/BF03029972

35. Semancik, J.S.; Roistacher, C.N.; Duran-Vila, N. A New Viroid is the Causal Agent of the Citrus Cachexia Disease. In Proceedings of the Tenth International Organization of Citrus Virologists Conference; 1988; pp. 125-135

36. Eiras, M.; Targon, M.L.P.N.; Fajardo, T.V.M.; Flores, R.; Kitajima, E.W. Citrus exocortis viroid and Hop Stunt viroid doubly infecting grapevines in Brazil. Fitopatol. Bras.2006, 31, 440-446, doi:https://doi.org/10.1590/S0100-41582006000500002

37. Amari, K.; Gomez, G.; Myrta, A.; Di Terlizzi, B.; Pallás, V. The molecular characterization of 16 new sequence variants of Hop stunt viroid reveals the existence of invariable regions and a conserved hammerhead-like structure on the viroid molecule. J. Gen. Virol.2001, 82, 953-962, doi:10.1099/0022-1317-82-4-953

38. Serra, P.; Gago, S.; Duran-Vila, N. A single nucleotide change in Hop stunt viroid modulates citrus cachexia symptoms. Virus Res.2008, 138, 130-134, doi:10.1016/j.virusres.2008.08.003

39. 39.Reanwarakorn, K.; Semancik, J.S. Regulation of pathogenicity in hop stunt viroid-related group II citrus viroids. J. Gen. Virol.1998, 79, 3163-3171, doi:10.1099/0022-1317-79-12-3163

40. Reanwarakorn, K.; Semancik, J.S. Correlation of hop stunt viroid variants to cachexia and xyloporosis diseases of citrus. Phytopathology1999, 89, 568-574, doi:10.1094/PHYTO.1999.89.7.568

41. Palacio-Bielsa, A.; Romero-Durbán, J.; Duran-Vila, N. Characterization of citrus HSVd isolates. Arch. Virol.2004, 149, 537-552, doi:10.1007/s00705-003-0223-5

42. Önelge, N.; Cinar, A.; Szychowski, J.A.; Vidalakis, G.; Semancik, J.S. Citrus viroid II variants associated with "Gummy Bark" disease. Eur. J. Plant Pathol.2004, 110, 1047-1052, doi:10.1007/s10658-004-0815-2

43. Sofy, A.R.; Soliman, A.M.; Mousa, A.A.; Ghazal, S.A.; El-Dougdoug, K.A. First record of Citrus viroid II (CVd-II) associated with gummy bark disease in sweet orange (Citrus sinensis) in Egypt. New Dis. Reports2010, 21, 24, doi:10.5197/j.2044-0588.2010.021.024
44. Bagherian, S.A.A.; Izadpanah, K. Two novel variants of hop stunt viroid associated with yellow corky vein disease of sweet orange and split bark disorder of sweet lime. In Proceedings of the 21st International Conference on Virus and other Graft Transmissible Deseases of Fruit Crops; 2010; pp. 105-113
45. Roy, A.; Ramachandran, P. Occurrence of a Hop stunt viroid (HSVd) variant in yellow corky vein disease of citrus in India. Curr. Sci.2003, 85, 1608-1612
46. Najar, A.; Hamdi, I.; Varsani, A.; Duran-Vila, N. Citrus viroids in Tunisia: Prevalence and molecular characterization. J. Plant Pathol.2017, 99, 787-792, doi:10.4454/jpp.v99i3.3989
47. Bar-Joseph, M. A Contribution to the Natural History of Viroids. In Proceedings of the Thirteenth International Organization of Citrus Virologists Conference Proceedings (1957-2010); 2010; pp. 226-229
48. Fuchs, M.; Almeyda, C. V.; Al Rwahnih, M.; Atallah, S.S.; Cieniewicz, E.J.; Farrar, K.; Foote, W.R.; Golino, D.A.; Gómez, M.I.; Harper, S.J.; et al. Economic Studies Reinforce Efforts to Safeguard Specialty Crops in the United States. Plant Dis.2020, PDIS-05-20-1061, doi:10.1094/pdis-05-20-1061-fe
49. Thibaut, O.; Claude, B. Innate immunity activation and RNAi interplay in citrus exocortis viroid—tomato pathosystem. Viruses2018, 10, doi:10.3390/v10110587
50. Kovalskaya, N.; Hammond, R.W. Molecular biology of viroid-host interactions and disease control strategies. Plant Sci.2014, 228, 48-60, doi:10.1016/j.plantsci.2014.05.006
51. Marmisolle, F.E.; Arizmendi, A.; Ribone, A.; Rivarola, M.; García, M.L.; Reyes, C.A. Up-regulation of microRNA targets correlates with symptom severity in Citrus sinensis plants infected with two different isolates of citrus psorosis virus. Planta2020, 251, 1-11, doi:https://doi.org/10.1007/s00425-019-03294-0
52. Reyes, C.A.; Ocolotobiche, E.E.; Marmisollé, F.E.; Robles Luna, G.; Borniego, M.B.; Bazzini, A.A.; Asurmendi, S.; García, M.L. Citrus psorosis virus 24K protein interacts with citrus miRNA precursors, affects their processing and subsequent miRNA accumulation and target expression. Mol. Plant Pathol.2016, 17, 317-329, doi:10.1111/mpp.12282
53. Wang, Y.; Wu, J.; Qiu, Y.; Atta, S.; Zhou, C.; Cao, M. Global transcriptomic analysis reveals insights into the response of 'etrog' citron (Citrus medica L.) to Citrus Exocortis viroid infection. Viruses2019, 11, doi:10.3390/v11050453
54. Rizza, S.; Conesa, A.; Juarez, J.; Catara, A.; Navarro, L.; Duran-Vila, N.; Ancillo, G. Microarray analysis of etrog citron (Citrus medica l.) reveals changes in chloroplast, cell wall, peroxidase and symporter activities in response to viroid infection. Mol. Plant Pathol.2012, 13, 852-864, doi:10.1111/j.1364-3703.2012.00794.x
55. Vázquez Prol, F.; López-Gresa, M.P.; Rodrigo, I.; Bellés, J.M.; Lisón, P. Ethylene is involved in symptom development and ribosomal stress of tomato plants upon citrus exocortis viroid infection. Plants2020, 9, 1-15, doi:10.3390/plants9050582

56. Su, X.; Fu, S.; Qian, Y.; Xu, Y.; Zhou, X. Identification of Hop stunt viroid infecting Citrus limon in China using small RNAs deep sequencing approach. Virol. J.2015, 12, 1-5, doi:10.1186/s12985-015-0332-2

57. Owens, R.A.; Sano, T.; Duran-Vila, N. Methods in Molecular Biology. In Antiviral Resistance in Plants-Methods and Protocols; Watson, J.M., Wang, M.B., Eds.; 2012; pp. 253-271

58. Hajeri, S.; Ramadugu, C.; Manjunath, K.; Ng, J.; Lee, R.; Vidalakis, G. In vivo generated Citrus exocortis viroid progeny variants display a range of phenotypes with altered levels of replication, systemic accumulation and pathogenicity. Virology2011, 417, 400-409, doi:10.1016/j.virol.2011.06.013

59. Alcántara-Mendoza, S.; Alcántara-Mendoza, S.; García-Rubio, O.; Cambrón-San- doval, V.H.; Cambrón-San- doval, D.; Cambrón-San- doval, C. Characterization of Citrus exocortis viroid in different conditions of indexing. Rev. Mex. Fitopatol.2017, 35, 284-303

60. Nie, X.; Singh, R.P. Viroid Detection and Identification by Bioassay. In Viroids and Satellites; 2017; pp. 347-356

61. Malfitano, M.; Barone, M.; Duran-Vila, N.; Alioto, D. Indexing of viroids in citrus orchards of Campania, Southern Italy. J. Plant Pathol.2005, 87, 115-121

62. Kapari-Isaia, T.; Kyriakou, A.; Papayiannis, L.; Tsaltas, D.; Gregoriou, S.; Psaltis, I. Rapid in vitro microindexing of viroids in citrus. Plant Pathol.2008, 57, 348-353, doi:10.1111/j.1365-3059.2007.01774.x

63. Lin, C.Y.; Wu, M.L.; Shen, T.L.; Yeh, H.H.; Hung, T.H. Multiplex detection, distribution, and genetic diversity of Hop stunt viroid and citrus exocortis viroid infecting citrus in Taiwan. Virol. J.2015, 12, 1-11, doi:10.1186/s12985-015-0247-y

64. Pérez, R.; Correa, E.; del Valle, N.; Otero, O. Reduced Indexing Time for Cachexia and Exocortis Diseases in Citrus. In Proceedings of the Fourteenth International Organization of Citrus Virologists Conference; 2000; pp. 386-387

65. Ito, T.; Furuta, T.; Ito, T.; Isaka, M.; Ide, Y.; Kaneyoshi, J. Identification of cachexia-inducible Hop stunt viroid variants in citrus orchards in Japan using biological indexing and improved reverse transcription polymerase chain reaction. J. Gen. Plant Pathol.2006, 72, 378-382, doi:10.1007/s10327-006-0303-y

66. Duran-Vila, N.; Semancik, J.S. Variations in the "cross protection" effect between two strains of citrus exocortis viroid. Ann. Appl. Biol.1990, 117, 367-377

67. Flores, R.; Ownes, R.A. Encyclopedia of Virology. In Encyclopedia of Virology; Mahy, B., Regenmortel, M.H.V.V., Eds.; 2008; pp. 332-342

68. .Duran-Vila, N. Viroids as companions of a professional career. Viruses2019, 11, 1-10, doi:10.3390/v11030245

69. Baksh, N.; Lee, R.F.; Gasnsey, S.M. Detection of Citrus Exocortis Viroid by Polyacrylamide Gel Electrophoresis. In Proceedings of the Ninth International Organization of Citrus Virologists Conference Proceedings (1957-2010); 1984; pp. 343-352

70. Murcia, N.; Serra, P.; Olmos, A.; Duran-Vila, N. A novel hybridization approach for detection of citrus viroids. Mol. Cell. Probes2009, 23, 95-102, doi:10.1016/j.mcp.2008.12.007

71. Cohen, O.; Batuman, O.; Stanbekova, G.; Sano, T.; Mawassi, M.; Bar-Joseph, M. Construction of a multiprobe for the simultaneous detection of viroids infecting citrus trees. Virus Genes2006, 33, 287-292, doi:10.1007/s11262-006-0067-7

72. Semancik, J.S.; Szychowski, J.A.; Rakowski, A.G.; Symons, R.H. Isolates of citrus exocortis viroid recovered by host and tissue selection. J. Gen. Virol.1993, 74, 2427-2436, doi:10.1099/0022-1317-74-11-2427

73. Kofalvi, S.A.; Marcos, J.F.; Cañizares, M.C.; Pallás, V.; Candresse, T. Hop stunt viroid (HSVd) sequence variants from Prunes species: Evidence for recombination between HSVd isolates. J. Gen. Virol.1997, 78, 3177-3186, doi:10.1099/0022-1317-78-12-3177

74. Osman, F.; Dang, T.; Bodaghi, S.; Vidalakis, G. One-step multiplex RT-qPCR detects three citrus viroids from different genera in a wide range of hosts. J. Virol. Methods2017, 245, 40-52, doi:10.1016/j.jviromet.2017.03.007

75. Wang, X.; Zhou, C.; Tang, K.; Zhou, Y.; Li, Z. A rapid one-step multiplex RT-PCR assay for the simultaneous detection of five citrus viroids in China. Eur. J. Plant Pathol.2009, 124, 175-180, doi:10.1007/s10658-008-9386-y

76. Papayiannis, L.C. Diagnostic real-time RT-PCR for the simultaneous detection of Citrus exocortis viroid and Hop stunt viroid. J. Virol. Methods2014, 196, 93-99, doi:10.1016/j.jviromet.2013.11.001

77. Bernad, L.; Duran-Vila, N. A novel RT-PCR approach for detection and characterization of citrus viroids. Mol. Cell. Probes2006, 20, 105-113, doi:10.1016/j.mcp.2005.11.001

78. Pecman, A.; Kutnjak, D.; Gutiérrez-Aguirre, I.; Adams, I.; Fox, A.; Boonham, N.; Ravnikar, M. Next generation sequencing for detection and discovery of plant viruses and viroids: Comparison of two approaches. Front. Microbiol.2017, 8, 1-10, doi:10.3389/fmicb.2017.01998

79. 79.Wu, Q.; Ding, S.W.; Zhang, Y.; Zhu, S. Identification of Viruses and Viroids by Next-Generation Sequencing and Homology-Dependent and Homology-Independent Algorithms. Annu. Rev. Phytopathol.2015, 53, 425-444, doi:10.1146/annurev-phyto-080614-120030

80. Martinez, G.; Donaire, L.; Llave, C.; Pallas, V.; Gomez, G. High-throughput sequencing of Hop stunt viroid-derived small RNAs from cucumber leaves and phloem. Mol. Plant Pathol.2010, 11, 347-359, doi:10.1111/j.1364-3703.2009.00608.x

81. Adkar-Purushothama, C.R.; Perreault, J.P. Impact of nucleic acid sequencing on viroid biology. Int. J. Mol. Sci.2020, 21, 1-27, doi:10.3390/ijms21155532

82. Wang, Y.; Atta, S.; Wang, X.; Yang, F.; Zhou, C.; Cao, M. Transcriptome sequencing reveals novel Citrus bark cracking viroid (CBCVd) variants from citrus and their molecular characterization. PLoS One2018, 13, 1-12, doi:10.1371/journal.pone.0198022

83. Meziane, M.; Frasheri, D.; Carra, A.; Boudjeniba, M.; D'Onghia, A.M.; Mercati, F.; Djelouah, K.; Carimi, F. Attempts to eradicate graft-transmissible infections through

somatic embryogenesis in Citrus ssp. and analysis of genetic stability of regenerated plants. Eur. J. Plant Pathol.2016, 148, 85-95, doi:10.1007/s10658-016-1072-x

84. Meziane, M.; Boudjeniba, M.; Frasheri, D.; D'Onghia, A.M.; Carra, A.; Carimi, F.; Haddad, N.; Haddad, S.; Braneci, S. Regeneration of Algerian Citrus germplasm by stigma/style somatic embryogenesis. African J. Biotechnol.2012, 11, 6666-6672, doi:10.5897/ajb10.2485

85. Ben Mahmoud, K.; Najar, A.; Jedidi, E.; Hamdi, I.; Jemmali, A. Detection of two viroids in the Tunisian sweet orange (Citrus sinensis L) cv . Maltese and sanitation via somatic embryogenesis. J. Chem. Pharm. Res.2017, 9

86. Roistacher, C.N.; Navarro, L.; Murashige, T. Recovery of citrus selections free of several viruses, exocortis viroid, and Spiroplasma citri by shoot-tip grafting in vitro. In Proceedings of the Seventh International Organization of Citrus Virologists Conference Proceedings (1957-2010); 1976; pp. 186-193

87. Carimi, F.; De Pasquale, F.; Fiore, S.; D'Onghia, A.M. Sanitation of citrus germplasm by somatic embryogenesis and shoot-tip grafting. In Options Méditerranéennes, Série B n. 33; 2001; pp. 61-65

88. Kapari-Isaia, T.; Minas, G.J.; Polykarpou, D.; Iosephidou, E.; Arseni, S.; Kyriakou, A. Shoot-tip Grafting in vitro for Elimination of Viroids and Citrus psorosis virus in the Local Arakapas Mandarin in Cyprus. In Proceedings of the Fifteenth International Organization of Citrus Virologists Proceedings; 2002; pp. 417-419

89. Ouantar, M.; Bibi, I.; Chebli, B.; Ait Friha, A.T.; Afechtal, M. Prospection et première caractérisation moléculaire de l'exocortis (Citrus exocortis viroid, CEVd) dans la région du Gharb. Rev. Marocaine des Sci. Agron. Vétérinaires2018, 6, 14-18

90. Bibi, I.; Afechtal, M.; Chafik, Z.; Bamouh, A.; Benyazid, J.; Bousamid, A.; Kharmach, E. Occurrence and distribution of virus and virus-like Diseases of citrus in North-Est of Morocco - Moulouya perimeter. In Proceedings of the Onzième Congrès de l'Association Marocaine de Protection des plantes, 26-27 Mars 2019, Rabat, Maroc; 2019; pp. 119-134

91. Bibi, I.; Kharmach, E.; Chafik, Z.; Benyazid, J.; Bousamid, A. Incidence of Citrus exocortis viroid and Hop stund viroid in commercial citrus groves from Morocco. Moroccan J. Agric. Sci.2020, 1, 142-144

92. Afechtal, M.; Kharmach, E.; Bibi, I. Survey and molecular characterization of Hop stunt viroid (HSVd) sequence variants from citrus groves in Morocco. Moroccan J. Agric. Sci.2020, 1, 145-148

93. Abdel-Salam, A.M.; Abdou, Y.A.; Abou-Zeid, A.A.; Abou-Elfotoh, M.A. Studies on Citrus Exocortis Viroid (CEVd) in Egypt. In Options Méditerranéennes : Série B. Etudes et Recherches; n. 43; D'Onghia, A.M., Djelouah, K., Roistacher, C.N., Eds.; 2002; pp. 105-108

94. Ragozzino, E.; Faggioli, F.; Barba, M. Distribution of citrus exocortis viroid and hop stunt viroid in citrus orchards of central Italy as revealed by one-tube one-step RT-PCR. Phytopathol. Mediterr.2005, 44, 322-326

95. Najar, A.; Duran-Vila, N. Viroid Prevalence in Tunisian Citrus. Plant Dis.2004, 88, 1286, doi:10.1094/PDIS.2004.88.11.1286B

96. Abou Kubaa, R.; Saponari, M.; El-Khateeb, A.; Djelouah, K. First identification of citrus exocortis viroid (CEVd) and citrus dwarf viroid (CVd-III) in citrus orchards in Syria. J. Plant Pathol.2016, 98, 171-185, doi:http://dx.doi.org/10.4454/JPP.V98I1.045

97. Belabess, Z.; Afechtal, M.; Benyazid, J.; Sagouti, T.; Rhallabi, N. Prévalence des phytovirus associés aux agrumes dans la région de Berkane. African Mediterr. Agric. Journal-Al Awamia2020, 129, 144-160

98. Garnsey, S.M.; Weathers, L.G. Factors Affecting Mechanical Spread of Exocortis Virus Inoculation Procedures. 1967

99. Vidalakis, G.; Pagliaccia, D.; Bash, J.A.; Semancik, J.S. Effects of mixtures of citrus viroids as transmissible small nuclear RNA on tree dwarfing and commercial scion performance on Carrizo citrange rootstock. Ann. Appl. Biol.2010, 157, 415-423, doi:10.1111/j.1744-7348.2010.00430.x

100. Zhang, Z.; Zhou, Y.; Guo, R.; Mu, L.; Yang, Y.; Li, S.; Wang, H. Molecular characterization of Chinese Hop stunt viroid isolates reveals a new phylogenetic group and possible cross transmission between grapevine and stone fruits. Eur. J. Plant Pathol.2012, 134, 217-225, doi:10.1007/s10658-012-9983-7

Chapter-6

Indexing Virus and Virus-Like Diseases of Citrus

1. Introduction

Citrus belongs to family Rutaceae and holds an important position among fruits all around the globe. It is the most cultivated fruit in the world after grapes. Citrus is believed to be originated from southeastern Asian region [1]. Northern hemisphere accounts for about 70% of the total citrus production and approximately 80 citrus species are native to India and other tropical and sub-tropical areas of Asia [2]. Citrus being a perennial fruit tree is usually produced through vegetative propagation of scion on rootstock. Combination and compatibility of scion and rootstock can result in high yielding citrus plants. The United States, China, Brazil and the Mediterranean countries contribute two third of global citrus production and are regarded as major citrus producing countries [3]. Citrus products and by-products provide the basis for local agricultural industries, which generate employment and raise income, and in many cases, this industry constitutes an important source of foreign revenue for developed and developing countries such as Pakistan.

A number of factors and certain conditions are collectively responsible for fluctuations in citrus production. Selection of rootstock, agronomic practices and management in citrus nurseries and orchards, propagation methods and biotic and abiotic factors contribute their share to some extent in reduced citrus production. Like other commercial crops, number of diseases, insect pests and genetic problems affect the citrus production. Diseases are one of the major limiting factors for the low citrus production and gives a serious threat to citrus industry. These diseases are caused by fungi, prokaryotes, nematodes, viroids, viruses and virus-like pathogens. Among these, viruses and virus-like pathogens play a major role in citrus decline. These pathogens incur varying degree of damages to citrus plants and make their life span shorter, causing low yield and deterioration of quality and ultimately loss of economy which leads towards the citrus decline [4].

Citrus decline is the matrix of all above mentioned factors and conditions. The common diseases, playing an important role in citrus decline are citrus gummosis caused by *Phytophthora* sp. and *Fusarium* sp., citrus canker caused *by Xanthomonas* sp., Huánglóngbìng caused by *Candidatus laberibactor* sp., citrus stubborn caused by *Spiroplasma citri* and one of the most devastating citrus viruses *i.e.* citrus tristeza virus. Citrus viruses play a vital role in its decline by using the prevailing conditions and many other factors as these are bud/graft-transmissible and have systemic infections. A variety of symptoms has been observed regarding the infection of citrus viruses resulting in systemic infection. No viral diseases on citrus was under discussion or the hot issue before 1940 but during and after 60 years, thirty economically important viruses and virus-like diseases of citrus were recognized as a cause of citrus decline in different parts of the world [5, 6, 7]. Unfortunately, citrus orchards are short lived and decline within 15 years as against their potential of 50 years or more. This is mainly attributed, among other factors, to the prevalence of graft-transmissible virus and virus-like diseases, faulty nursery operations and poor orchard management. However, most of the problems originate from nurseries.

Therefore, it is the time when citrus nurseries should operate on highly technical and scientific lines and start providing disease-free and certified plants to the growers. In the first instance, nurseries should be registered and indiscriminate multiplication and sale of uncertified citrus plants must end. For this purpose, the most imperative points such as the prevalence and detection of citrus viral diseases, selection of material, production of disease-free material and streamlined screening procedures are highlighted in this bulletin. If the guidelines are properly followed and certified bud-wood becomes available for producing disease-free citrus plants, the problem of citrus decline can be minimized.

1.1 Citrus pathology

Citrus pathology is the study of citrus diseases caused by biotic (pathogens) and abiotic factors. It is now being considered as a major part in the field of plant pathology. Being a major fruit crop in the world, citrus production always remains important for the citrus industry. Physiology, morphology, biochemistry and behavior of the citrus tree towards the prevailing climatic conditions are the key areas to be kept in mind while investigating the citrus diseases. Etiology of citrus diseases and their detection methods help to manage these diseases. A plenty of information regarding the diseases of citrus and their control has been published around the world.

2. Virus and virus-like diseases of citrus

Virus, viroids and virus-like diseases, however, infecting different citrus species could not receive due attention because of the lack of laboratories with proper facilities for their proper identification. These diseases are also known as 'graft-transmissible diseases' (GTDs) and the term used for the casual agents is 'citrus graft-transmissible pathogens' (CGTPS) [8]. These are an emerging threat for citrus industry. Major viruses and virus-like pathogens include citrus tristeza virus (CTV), citrus yellow vein clearing virus (CYVCV), citrus variegation virus (CVV), concave gum, psorosis, cristacortis, ringspot, exocortis, *Cachexia-xyloprosis*, *Candidatus liberibacter asiaticus* and *Spiroplasma citri* [9, 10]. A brief description of these virus and virus-like pathogens is summarized below (Table 1).

Table 1. Major virus and virus-like diseases of citrus in Pakistan, their transmission and hosts*.

Sr. No.	Citrus disease	Pathogen	Transmission	Incidence %	Host	Importance
1.	Citrus greening disease (CGD) Huánglóngbìng	Bacterium-like organism	Psyllid: Diaphorina citri	20–90	Sweet orange, grapefruit, orange jessamine	Associated with citrus decline
2.	Citrus tristeza virus (CTV)	Closterovirus	Aphid species (Aphis gossypii, Toxoptera citricida)	7–18 Up to 48	Sweet orange, lime, mandarin	Economically important
3.	Gummy bark (GB)	Virus probable	Grafting, mechanical	20–30	Mandarin, sweet orange	-do-
4.	Bud union crease (BUC)	Virus probable	Grafting, mechanical	20–30	Mandarin, sweet orange	-do-
5.	Cristacortis	Virus probable	Not known	10	All citrus species	-do-
6.	Exocortis	Viroid	Mechanical	7–10 Up to 16	Sweet lime	-do-
7.	*Cachexia-xyloporosis*	Viroid	Mechanical	4–10	Mandarin	-do-
8.	Citrus stubborn disease (CSD)	Prokaryote	Leaf hopper (*Neoaliturus haemocops*)	2–7	Sweet orange, grapefruit, periwinkle	-do-
9.	Yellow vein clearing (YVC)	Virus	Grafting, vector not known	2	Lemon, sour orange	Minor importance
10.	Ring spot/ Variegation	Virus	Not known	2–3	Sweet orange	Minor importance

*The above summarized information is extracted from the work of [10, 11, 12].

Note: All diseases are graft-transmissible. No adequate information on vector transmission is available except their identity; viroids problems are favored by warm conditions.

Although plant pathologists have put their efforts for the identification and management of virus and virus-like diseases of citrus but there are some areas need to be investigated. A comprehensive book has been written by Roistacher in 1991 regarding the detection of virus and virus-like diseases of citrus. These diseases reduce the citrus yield and ultimately result in the loss of low foreign exchange. Diseases caused by viruses and virus-like pathogens are infectious, contagious and devastating due to their systemic nature. They are transmitted through different means in nature; through vegetative propagation, by insect vectors and horticultural tools used for the routine activities in citrus orchards and nurseries. These diseases have a considerable economic importance because of their involvement in the citrus decline [4]. Millions of citrus trees have been died due to CTV. The CGTPS usually have two types of effects either quick decline or long term losses.

These diseases are very difficult to control or manage unless or until by the application of integrated management practices. The appropriate diagnosis or indexing method plays an important role for the management of CGTPS [8].

The major symptoms due to virus and virus-like pathogens are vein clearing, bark cracking, yellowing of leaves, leaf dropping, gummosis, mosaic, rugosity, bark scaling, stem pitting, dwarfing, chlorosis and mottling [10, 13]. The virus and virus-like diseases, infecting different citrus species in Pakistan, have been neglected for a long time due to lack of proper facilitations in the research laboratories and skilled personnel for their detection and characterization. A brief description is presented in Table 2 regarding the citrus species and viruses and virus-like diseases in Pakistan. Indexing facilities are very important for the diagnosis of plant pathogens. Similarly, unlike other pathogens viruses and virus-like pathogens are very sensitive to their indexing through different techniques. Pathogen detection system always played an important role in management of virus and virus-like pathogens. Proper indexing facilities help in the characterization and differentiation of different viruses and their isolates. Management of viruses and virus-like pathogens is only possible when appropriate indexing procedures and facilities are available.

Table 2. Citrus species and presence of viruses and virus-like diseases.

Citrus species	Virus				Viroid		Prokaryote		Virus-like symptoms					PS	DE
	CTV	IVV	RS	YUC	Ex	CX	GR	ST	BS	GB	BU	BP	Misc.		
Sweet orange (Mosambi)	+		+		+	+	+	+		+	+	+	+	+	+
Sweet orange (Mosambi)	+		+		+	+	+	+		+	+	+	+	+	+
Mandarin	+					+	+		+	+	+	+	+	+	+
Sweet lime					+						+	+	+	+	+
Grapefruit								+		+	+	+	+	+	
Lemon	+	+	+	+											
Acid lime	+								+				+		+
Rough lemon		+	+				+							+	
Sour orange	+	+		+									+		
Orange jessamine					+		+			+					

Note: "+" is the indication of presence of infection on the citrus varieties.

CTV = Citrus Tristeza Virus, IVV = Infection variegation, RS = Ring spot; Ex = Exocortis, CX = Cachexia xyloporsis, GR = Greening disease, ST = Stubburn Disease, BS = Bark Scaling, GB = Gummy Bark, BU = Bud Union Disease, PS = Psorosis, DE = Decline. [10, 12].

3. Insects as vectors of virus and virus-like pathogens

Insect pests have always been key role players in the direct or indirect transmission of plant pathogens in agricultural and horticultural crops [14, 15, 16]. Citrus tristeza, cachexia-xyloporosis, greening or Huánglóngbìng, infectious variegation, vein enation, yellow vein clearing, exocortis and stubborn are the most conspicuous viral diseases of citrus all over the world including Pakistan [11, 17]. These diseases are usually graft-transmissible and phloem-restricted. Although these diseases along with other fungal, bacterial or mycoplasmic infections of citrus are usually spread through unhealthy mechanical intrusions and by the use of infected uncertified bud, scion or rootstock in plant propagation, many type of sap-feeding insect pests play important role in the transmission of these diseases such as leafhoppers, aphids, psyllids, whiteflies and thrips [17, 18, 19, 20].

Among the vector borne viral diseases of citrus, citrus tristeza (CTV) which is caused by a *Closterovirus* is the most dominant and widely studied viral diseases of citrus. It is transmitted by different aphid species primarily by black citrus aphid (*Toxoptera citricida* Kirk.) and cotton-melon aphid (*Aphis gossypii* Glov.) [17, 21]. Another emerging viral disease of citrus is the yellow vein clearing (CYVCV) caused by a *Mandarivirus*. It was first observed in Pakistan in 1988 in the orchards of sour orange (*C. aurantium* L.) and lemon (*C. limon* L.) [22], and later on it was reported in China, India, Iran and Turkey [23, 24, 25, 26]. This CYVCV is reported to be vectored by e transmitted by whiteflies and aphids (*Aphis craccivora* and *A. spiraecola*) [25, 27]. Although not virus borne, citrus stubborn is a destructive disease being caused by a bacterium *Spiroplasma citri*. It is usually transmitted by many species of leafhoppers, primarily by *Scaphytopius nitridus* and Circulifer tenellus in citrus-growing suburbs of California and Arizona and by *Circulifer haematoceps* in the Mediterranean zones [17].

4. Indexing strategies

Indexing is an indispensable procedure to produce and diagnose disease-free plants. Different techniques or combination of techniques have been applied in this regard and the effectiveness of each depends upon the facilities available. Generally indexing can be divided into two types.

Field indexing; also known as biological indexing including the mechanical inoculation through direct contact or vegetative propagation and/or through insect transmission.

Laboratory indexing; also known as quick indexing including serological, molecular and chemical assays.

Commonly used indexing methods are tissue grafting, budding, insect transmission for biological indexing and enzyme-linked immunosorbent assay (ELISA) and polymerase chain reaction (PCR) for quick indexing strategies. Although all viruses and virus-like pathogens can be detected through PCR and its derivatives, polyacrylamide gel electrophoresis (PAGE) is commonly used for the detection of viroids.

4.1 Biological indexing

Biological indexing is the inoculation or introduction of virus source (infected sample) into the indicator plants for detection and purification. It involves one of the common indexing methods such as vegetative propagation of infected scion (grafting/budding) to indicator plants, mechanical inoculation of indicator plants or transmission of virus through the insect vector (*e.g.* aphids for CTV, psyllids for greening and leaf hoppers for prokaryotic diseases). Biological indexing is usually time consuming, require glasshouse facilities and takes about 12–6 months for results. At least 4–3 plants are required per treatment. Biological indexing of graft-transmissible pathogens, indicator plants and symptoms in the indicator plants are summarized in Table 3.

Table 3. Biological indexing of citrus graft-transmissible pathogens.

Sr. No.	Disease	Appropriate number of test plants	Indicator plants (for inoculation)	Favorable temperature	Symptoms on the indicator (indexed) plant
1.	Citrus tristeza	5	Mexican lime (C. aurantifolia), sweet orange on sour orange root stock, Duncan grapefruit	65-68° F Cool-warm	Vein clearing, stem pitting, leaf cupping, decline on sour orange stem pitting on grapefruit.
2.	Yellow vein clearing	5	Lemon (C. limon), sour orange (*C. aurantium*)	Cool	Yellowing and clearing of veins
3.	Ringspot/ Infections variegation	4	Citron (C. medica), cowpea (Vigna unguiculata)	Cool	Ringspot, necrotic local lesions, distorted leaves.
4.	Psorosis	4	Sweet orange (C. sinensis)	Cool	Flecking on leaves
5.	Cachexia-xyloporosis	5	Mandarin (C. reticulata) Parsons special	Warm >95°F	Gum in bark, scion and at bud union
6.	Exocortis	5	Citron (*C. medica*)	Warm	Tip browning, leaf epinasty
7.	Cristacortis	4	Grapefruit, sweet orange	Cool- warm	Flecking
8.	Concave gum	4	Sweet orange	Cool	Oak leaf pattern, narrowing of leaves
9.	Citrus greening disease	5	Mandarin (C. *reticulata*), Grapefruit (*C. paradisi*), Murraya sp.	Cool- warm	Leaf blotches, chlorosis
10.	Citrus stubborn disease	4	Sweet orange, grapefruit, periwinkle (Catharanthus roseus)	Warm	Stunted shoot, smelling of leaves, Zn deficiency like signs.
11.	Yellows (probably Aster)	5	Grapefruit, Periwinkle (C. roseus)	Warm	Chlorosis
12.	Bud union crease	5	Sweet orange	Warm	Brown line at bud union
13.	Gummy bark	5	Mandarin	Warm	Gum in the bark

Note: The above information is extracted from the work of [10, 12] and personal communication of Dr. S.M. Mughal. (Dr. Mughal was also in the team of Dr. Catara during the surveys of citrus growing areas of Pakistan in early 1980's).

Detailed methodology for biological indexing has been described much in literature [28, 29, 30, 31]. Followings are the generalized and simplified steps to be kept in mind during the biological indexing on the basis of available literature.

- Sow the seeds of test plants (usually Mexican lime or acid lime) in the sand in germinating tray. Transplant the seedlings in pots having potting media (sand, soil and moss @ 1:1:1 ratio) after 17–28 days of germination, depending on the germinating conditions.
- Inoculate the seedlings at 4–6 leaf stage.
- Keep the indicator plants in insect-free chambers before and after inoculation.
- **Mechanical inoculation:** grind the virus infected samples in phosphate buffer (pH 7.2). Dust the carborundum powder on to the indicator plants, but not too much, before inoculation (excessive dusting of carborundum will cause the necrosis and misled results).
- Pass the crude virus extract from double layer muslin cloth and then apply to the indicator plants with the help of forefinger and leave for 3–4 min. Remove the excess sap from indicator plants under tap water.
- **Insect transmission:** Do rearing of vector insects in the laboratory or collect the vector insects directly from the field and keep them in laboratory for few days to be acclimatized with the rearing conditions. Provide them with the artificial or natural diet.
- Observe the fasting period according to the nature of transmission (non-persistent, persistent or semi-persistent) before allowing them to be fed on virus source to acquire the virus.
- For non-persistent transmission pre-acquisition fasting time is 30–90 min. Long fasting period enhances the chances of quick acquisition of virus from the infected source.
- Transfer the insects on to the young leaves with virus symptoms/virus infected plants with the help of camel brush and allow them to feed for few min.
- After few min, immediately transfer the viruliferous insects on to the indicator plants and keep the plants in insect-free chamber to avoid the contamination from other insects.
- Maintain the insect population on indicator plants for at least 24 hr. and eradicate them after that through insecticides.
- For persistent transmission and semi-persistent transmission pre-acquisition fast has no effect. In both cases long acquisition feeding period enhances the chance of transmission.
- Maintain the insect population for a week and eradicate them with insecticide in case of semi-persistent transmission while maintain the insect population through transferring them on new indicator plants till they are alive.
- **Vegetative propagation:** Take the infected bud-wood from the virus source and graft on to the rootstock/indicator plants and keep the indicator plants in insect-free zones.

- ❖ Temperature range between 65 and 95°F helps the appearance of symptoms on indicator plants for viruses and virus-like pathogens. Observation time also varies from 3 to 16 months for different viruses, virus-like pathogens and viroids.

Laboratory indexing/advanced detection methods

There are rapid methods, highly specific, routinely applicable and some of which test large number of samples. These methods are summarized in Table 4. ELISA is the main laboratory indexing method used for the detection of CTV, PAGE for viroids and PCR for all diseases. Mother plants (plants recovered by nucellar embroyony *in-vivo* or *in-vitro*, by thermotherapy or micro-grafted plants or by micro-budding may be indexed by any of the above methods. Although 'chromatography' is a useful in chemical indexing of certain virus and virus-like pathogens but it is less reliable than vegetative propagation indexing. Electron microscopy is also helpful for the detection of greening and stubborn diseases other than the viruses. Moreover, *S. citri* can be cultured on a specific medium. Now-a-days, commercial kits are available for the ELISA, PCR and other detection methods along with the instructions.

Table 4. Advanced methods for the detection of viral diseases of citrus.

Sr. No.	Methods	Tested for	Advantages and limitations	References
1.	Immunofluorescence, tissue staining Azure A, Light microscopic observations	Citrus tristeza virus, yellow vein virus, greening diseases	Simple, economical, limited number of samples, time consuming, non-specific	[29, 32]
2.	Gel immunodiffusion	Citrus tristeza virus	Economical, time consuming, require quality antiserum, where ELISA facilities are not available.	[6]
3.	Enzyme-linked Immunosorbant Assay (ELISA and its variants)	Citrus tristeza virus and some prokaryotes	Rapid, economical, specific, routinely applied for large number of samples, quantitative, sensitive	[33]
4.	Electron microscopy (EM)	Citrus tristeza viru and other viruses	Quick for elongated viruses (CTV, CYVV), requires proper facilities	[6]
5.	Immunosorbant electron microscopy (ISEM)- Decoraton Technique	Citrus tristeza virus	Quick, specific, require antiserum and proper facilities, limited sampling	[6]
6.	Polyacrylamide Gel Electrophoresis (PAGE)	Citrus viroids (Exocortis, Cachexia)	Excellent for viroid detection and characterization, requires purification of viroids and proper conditions and facilities	[34]
7.	Molecular hybridization (RNA/DNA Probes), Polymerase Chain Reaction (PCR)	Virus and virus like diseases, CTV.	Highly sensitive, routinely applicable, time consuming, require primers and equipment facilities	[6]

Note: Large scale screening of material is possible with any of the method(s) mentioned above. However, there are several limitations including time and availability of proper facilities and trained manpower.

4.2 Serological assays

Serology involves the quick indexing of plant viruses, based on the antibody–antigen reaction. Enzyme-linked immunosorbent assay (ELISA) is one of the widely used in detection of plant viruses. It is relatively cheap and can test large number of samples.

ELISA with its derivatives, direct (DAS-ELISA) and indirect (DAC-ELISA), is the main serological indexing tool used for most of the citrus viruses at large scale samples.

Followings are some general steps followed during the ELISA based detection or indexing [35].

- ❖ **For DAS-ELISA;** Coat the antibodies in the ELISA plate and keep at °4C for overnight or °37C for 4 hr. for incubation.
- ❖ Wash the plate with washing buffer for 3 times with the interval of 5 min.
- ❖ Add the antigen (virus sap extracted from infected samples) into the wells of ELISA plate and incubate as above.
- ❖ Repeat the washing and coat the ELISA plate with enzyme-labeled antibodies and incubate as above.
- ❖ Repeat the washing step and add the substrate followed by incubation for 30 to 90 min for visual observation of color change and read the micro-plate through ELISA reader/spectrophotometer for quantitative data.
- ❖ **For DAC-ELISA:** Add the antigen and incubate the plate as above.
- ❖ Wash the ELISA plate as in DAS-ELISA.
- ❖ Add the primary antibody and incubate.
- ❖ Add the secondary antibody and incubate.
- ❖ Add the enzyme-labeled antibodies and incubate.
- ❖ Add the substrate and then observe the color change after incubation and read the plate through ELISA reader for quantitative data.

Note: Repeat the washing step after every step before adding the substrate. Stop the reaction in both types of ELISA with the help of 1 N NaOH.

4.3 Molecular assays

Molecular detection of citrus viruses and virus-like diseases has revolutionized the subject and provided the platform to detect the early stages of infection to reduce the economic losses. The molecular hybridization techniques supplemented with nucleic acid amplification methods based on PCR, in which high-throughput sequencing approaches can be adopted to identify the strains in relation to evolutionary history or phylogenetic assemblages [36, 37]. Although, nucleic acid based methods are highly sensitive and discriminatory allowing specific strain typing, but it bears the problems in reproducibility [38, 39]. Progressive efforts have been made to decrease the troubleshoots and hurdles to improve the amplification systems by improving the sensitivity and specificity of detection by limiting the high contents of plant related enzyme inhibitors. In contest, nested and multiplex PCR provides high sensitivity and make the possible to detect several targets in single assay [40]. Moreover, highly sensitive technologies by conducting the amplification of nucleic acids in an isothermal reaction, nucleic acid sequence-based amplification (NASBA) and reverse transcription loop-mediated isothermal amplification (RT-LAMP) provides specific detection of viruses and virus-like diseases.

The addition of real-time PCR for high-throughput testing allows the automation of PCR by combing the fluorimeteric approaches to detect and quantify the targets simultaneously [41, 42]. The combination of different protocols including the serological techniques and molecular approaches will increase the accuracy and reliability of virus diagnostic. Furthermore, in future prospects, nucleic acid arrays and biosensors assisted by nanotechnology will open new corridors to revolutionize the detection of plant viruses and virus-like diseases.

Citrus tristeza virus (CTV) is the most dangerous citrus disease all over the world and is also known as quick decline disease reducing the population of citrus trees significantly [43, 44, 45]. However, the utilization of advanced diagnostic methods, such as, biological indexing, electron microscopy (EM), ELISA and PCR or reverse transcriptase PCR (RT-PCR) is providing promising detection of the virus particles and leading towards the management strategies of CTV [46]. The application of conventional PCR is sensitive and specific under optimized and controlled conditions. However, sometimes, it is not possible to judge the amount of pathogens in the samples. Therefore, researchers have to employ other subsequent techniques for complete detection and quantification. Meanwhile, with real-time PCR approach, users can monitor the reaction and also the quantification of the specific pathogen in the sample. While setting up the real-time reaction for virus detection, it is the basic requirement to adapt the specific conditions of the detection system and instrument, and the characteristics of the reaction reagents and cycling procedures in which the most important are primer design, reaction components and conditions. The real-time PCR works well with small amplicons (5–200 bp), while standard PCR allows amplification of several hundred bases without sensitivity and specificity. Moreover, concentrations of $MgCl_2$, primers, and dNTPs are usually higher than conventional PCR [47].

The new developing chemistries are setting up the protocols with different characteristics depending upon the target and assay requirements. In addition to the most widely working chemistries (SYBRGreen, TaqMan, Scorpion, Molecular Beacons), there are more novel chemicals or technologies such as Amplifluor; Locked Nucleic Acid (LNA) Probes, Sigma Proligo; Cycling Probe Technology (CPT), Takara; Light Upon eXtension (Lux) Fluorogenic Primers, Invitrogen Corporation; Plexor Technology, Promega [48, 49]. Real-time technology is being used also in multiplex formatting for the specific detection and strain identification for several viruses [50, 51, 52, 53, 54, 55]. Furthermore, real-time reaction in multiplex system is difficult to optimize due to different ratio between the targets and the reaction. The replacement of conventional PCR with real-time PCR is providing new horizons towards the multiple detection system of plant viruses especially of the citrus viruses and virus-like diseases.

5. Detection of citrus viroids

After the discovery of viroid group of pathogens as an infectious agent to the plants, new aspects in virology were come in front of researchers to be addressed. Viroids are the smallest pathogens which consist of 246 to 401 nucleotides. They are low molecular weight, circular and single stranded RNAs. Viroids exist as free RNA because they lack protein coat [56]. Since viroids do not code for protein and enzyme, they rely on host enzyme for protein synthesis system and replication. To date, 38 viroids have been identified and they are classified into 2 families *i.e. Pospiviroidae* and *Avsunviroidae* [57].

The major economic important viroids in different plants are coconut viroids (CCCVd), citrus viroids (Exocortis and cachexia and variants), Hop stunt viroid and Potato spindle tuber viroids [57]. The origin of viroids is still questionable as they do not have natural host relationship [58, 59].

Citrus production is also affected by viroids. These are the emerging threat to citrus industry. To date, seven citrus viroids have been detected so far in citrus *viz. Citrus Exocortis viroid* (CEVd), *Citrus Bent Leaf viroid* (CBLVd), *Hop Stunt viroid-citrus* (HPSVd-cit), *Citrus Dwarfing viroid* (CDVd), *Citrus Bark Cracking viroid* (CBCVd), *Citrus viroid V* (CVd V) and *Citrus viroid VI* (CVd VI-OS). These have been distributed in different geographical areas as shown in Table 5 [70]. Diseases caused by citrus viroids are citrus exocortis disease (CED), citrus cachexia disease (CCD), citrus leaf bending disease (CLBD), citrus bark cracking disease (CBCD) and citrus dwarfing disease (CDD). Among these, citrus exocortis and citrus cachexia-xyloporosis are the most devastating and widely distributed [57]. These diseases cause a reduction in yield, size of fruit and quality of production [8]. These are transmitted directly and through propagation [71]. Stunting, dwarfing, bark cracking, yellowing of leaves, backward leaf bent, pin holing, yield loss and ultimately tree decline are the common symptoms of citrus viroid diseases [63, 71, 72]. Citrus viroids alone or with other viruses or prokaryotes in the host contribute considerably in tree decline [73]. Exocortis and cachexia are the major viroids which are widely distributed in citrus orchards. Other citrus viroids have also been detected from citrus orchards in different parts of the world [73]. Unlike viruses, viroids do not have protein coat, therefore, these are very difficult to detect through serological methods. For this purpose, molecular techniques such as PCR, PAGE are available for the detection of citrus viroids. These are sensitive, sophisticated and rapid detection techniques. Molecular techniques not only help in the detection but also in the characterization of viroids.

5.1 Pathogen description and characterization

All citrus viroids are classified in different genus under *Pospiviroidae* as mentioned in Table 6.

Table 5. Geographical distribution of citrus viroids.

Viroid	Geographical distribution	References
Citrus excocortis viroid (CEVd)	Australia, Argentina, Brazil, Japan, Taiwan, Corsica, China, India, Israel, Spain, Pakistan, South Africa USA, Uruguay, Iran	[8, 57, 60, 61, 62]
Citrus bent leaf viroid (CBLVd)	Israel, Japan, Australia, China, Uruguay, Pakistan, UAE, Iran, Spain	[8, 60, 61, 63, 64, 65]

Viroid	Geographical distribution	References
Hopstunt viroid (CVd- II),	Israel, Brazil, Uruguay	[8, 60, 62]
Citrus dwarfing viroid (CDVd)	USA, Uruguay, Pakistan	[8, 60, 65, 66]
Citrus bark cracking viroid (CBCVd)	USA, Uruguay	[60, 67]
Citrus viroid V (CVd-V)	Spain, Iran	[61, 68]
Citrus viroid VI (CVd- VI)	Japan	[69]

Table 6. Classification of citrus viroids (king et al., 2011; Hammond & Owens, 2006).

Genus	Citrus viroid	Length (nucleotides)	Diseases
Pospiviroid	*Citrus excocortis viroid* (CEVd)	371	Citrus exocortis disease
Hostuviroid	*Hop stunt viroid-citrus* (CVd-IIa) *Citrus cachexia viroid* (CVd-IIb)	299–302	Citrus cachexia disease.
Cocadviroid	*Citrus viroid IV*	284–286	Citrus bark cracking disease
Apscaviroid	*Citrus bent leaf viroid* (CVd-I),	318	Citrus leaf bending disease
	Citrus dwarfing viroid (CVd-III)	294–297	Citrus dwarfing disease
	Citrus viroid V	295	
	Citrus viroid VI	330	

Note: The above information is extracted from the work of Hammond & Owens (2006) and King et al. (2011).

5.2 Diagnostic methods for citrus viroids

Biological indexing is done through graft inoculation in indicator plants. It is very suitable to check the symptoms produced by citrus viroids and their severity. The most important host for indexing CEVd is Etrog citron (*Citrus medica*, Arizona 861) because of its great sensibility and rapid symptom expression [74]. According to Nakahara et al. [75], bioassay on Etrog citron is the most sensitive technique in detection of viroids although it takes more time compared to other methods. **Molecular tools** are now widely being used in the detection of citrus viroids. Combinations of several molecular techniques are very useful for reducing the time and to allow large numbers of samples to be examined and to identify each citrus viroid species [75, 76]. **PAGE** is also used to separate variation based on molecular weight. PAGE is not suitable for indexing large number of samples because it is not cost-effective. It is used to test the circularity of viroid RNA by two-dimensional denaturing PAGE (2D-PAGE) [77, 78, 79]. Sequential-PAGE is also commonly used and capable of detecting all citrus viroids [71].

Reverse transcription- polymerase chain reaction (RT-PCR) is the most commonly used method to detect citrus viroids. It is also a reliable method for quick screening and detection of citrus viroids [80]. It is known for its high specificity and ability to detect unknown viroids or variants [57]. **Multiplex RT-PCR** is another approach to detect simultaneously more than a viroid by using several set of primers. For instance, CEVd, CBLVd, CVd -1LSS, CVd-II, CVd-III, CVd-IV and CVd-VI were successfully detected simultaneously via multiplex RT-PCR [69]. **Real-time RT-PCR** is also used to detect citrus viroid. It is a quantitative PCR technology basically same as RT-PCR but it measures and quantifies products generated during each cycle of PCR [81]. **Molecular hybridization** is based on the specific interaction between complementary purine and pyrimidine bases forming A-U and G-C base pairs. According to Targon et al. [82], imprint hybridization technique is fast, sensitive and economic methods to be used as a routine for citrus viroid indexing in the certification programs. However, dot-blot technique is required an appropriate amount of extracted nucleic acids [75], and it is not suitable for detection of new or unknown viroids. Another molecular approach to detect viroids is Northern blot hybridization. CEVd, CBLVd, CVd-II, CVd-III and CVd-VI were successfully detected by Northern blot hybridization using specific probes in inoculated Etrog citron [83].

5.2.1 RT-PCR for detection of citrus viroids

5.2.1.1 Samples collection

Collect the leave samples based on virus and viroids-like symptoms in the field. Bring the leaves samples to laboratory for processing and preservation until use as follows;

Collect the leave samples in sterile plastic bags and place in ice box.

In the laboratory, wash the samples first in 10% bleach followed by distilled water.

Dry the samples and put in the plastic bags.

Label the plastic bags and store them at –80°C until further use.

5.2.1.2 Nucleic acid extraction

Extract the nucleic acids from leave samples using the TESLP buffer [84] as follows;

- ❖ Grind the 2-3 g of leaves (about 10–12 leaves) using mortar and pestle with liquid nitrogen. The slurry needs to be transferred to 50 ml screw cap tubes.
- ❖ Add 10 ml of TESLP buffer [0.13 M Tris–HCl (pH 8.9), 0.017 M EDTA (pH 7.0), 1 M LiCl, 0.83%SDS, 5%PVP] into the tube.
- ❖ Add 16 µl of 2-mercapthoethanol into the mixture.
- ❖ Incubate the mixture for 30 min at room temperature in the rotary shaker.
- ❖ Centrifuge the mixture at 11000 rpm for 15 min.
- ❖ The supernatant needs to be transferred to a new 50 ml screw tube.
- ❖ Add phenol:chloroform:iso-amyl (PCA, 25:24:1) @ 3:2 and mix well using vortex followed by centrifugation for 15 min, 11,000 rpm at room temperature.
- ❖ Transfer the supernatant into a new 15 ml screw tube and add CA (24:1) @ 4:3. The mixture needs to be mixed well using vortex and repeat the step 7.
- ❖ The supernatant is obtained into a new 15 ml screw tube @ 1 volume of supernatant with 0.9 volumes of 90% isopropanol.

- The tube is inverted 3–4 times to mix the components. Do not vortex or centrifuge.
- The mixture is incubated at −80°C for 30–40 min (or -20°C for 3–4 hr. or overnight).
- The mixture is centrifuged for 15 min, 11,000 rpm at room temperature.
- The isopropanol is discarded and the pellet obtained is transferred into 1.5 ml micro centrifuge tube.
- The pellet is washed with 1 ml of 70% ethanol followed by washing with 1 ml absolute ethanol until the clean pallet is obtained.
- The pellet is suspended in 50 µl of sterile double distilled water.
- The pellet is immediately used for RT-PCR or stored in -20°C until use.
- Reverse Transcription Polymerase Chain Reaction [69]:

5.2.1.3 Synthesis of cDNA

The extracted RNA is used to run RT-PCR. Reverse Transcription process is carried out in two steps to synthesis cDNA as follows;

Step 1: (1X)

Experimental RNA = 5 µl

Reverse primer = 1 µl

Double distilled water = 2.5 µl

Total Volume = 8.5 µl

The reaction is incubated at 80°C for 12 min then immediately transferred to ice for 5 min.

Step 2: (1X)

AMV-RT = 1 µl

dNTPs = 2 µl

RNAse Inhibitor = 0.5 µl

$MgCL_2$ = 4 µl

RT buffer = 4 µl

Total volume = 11.5 µl

The reaction is incubated at 55°C for 30 min. After 30 min, the process is stopped when it reaches to 10°C. The cDNA obtained is stored in −80°C freezer until use (or it can be used immediately).

5.2.1.4 PCR protocol

The final volume of PCR should be 25 µl which consists of 12.5 µl of PCR master mix, 5 µl of cDNA, 5.5 µl of sterile double distilled water, 1 µl of forward primer and 1 µl of reverse primer.

The conditions for PCR amplification (35 cycles) are as follows:

a. Denaturation:

94°C for 10 min

94°C for 30 seconds

60°C for 1 min

b. Annealing at 60°C for 10 seconds.

c. Extension at 72°C for 10 seconds and then 5 min.

The list of specific primers used is given in Table 7.

Table 7. List of specific primer for citrus viroids [69].

Viroid	Type	Sequence	Target (Product size)
CEVd	RT-Reverse	*5′ -CCGGGGATCCCTGAAGGACTT-3'*	371 bp
	PCR-Forward	*5′ -GGAAACCTGGAGGAAGTCGAG-3'*	
CVd-I	RT-Reverse	*5′ -TCGACGACGACCAGTCAGCT-3'*	233 bp
	PCR-Forward	*5′ -TCCCCTTCACCCGAGCGCTGC-3'*	
CVd-I-LSS	RT-Reverse	*5′ -ACGACCGCTCAGTCTCCTCT-3'*	247 bp
	PCR-Forward	*5′ -CTGTAACCGGACCGGTCTCCTTC-3'*	
CVd-II	RT-Reverse	*5′ -CCGGGGCTCCTTTCTCAGGTAAGT-3'*	302 bp
	PCR-Forward	*5′ -GGCAACTCTTCTCAGAATCCAGC-3'*	
CVd-III	RT-Reverse	*5'-TCACCAACTTAGCTGCCTTCGTC-3'*	271 bp
	PCR-Forward	*5′ -CTCCGCTAGTCGGAAAGACTCCGC-3'*	
CVd-IV	RT-Reverse	*5′ -TCTATCTCAGGTCGCGAAGGAAGAAGC-3'*	209 bp
	PCR-Forward	*5′ -TCTGGGGAATTTCTCTGCGGGACC-3'*	
CVd-VI	RT-Reverse	*5′ -GTCCGCTCGACTAGCGGCAGAGAGC-3'*	166 bp
	PCR-Forward	*5′ -CGTCGACGAAGGCATGTGAGCTT-3'*	

5.2.1.5 Agarose gel electrophoresis

The amplified RT-PCR product is separated using 2% agarose gel as follows [85];

- 2% agarose gel is prepared with 1x TBE buffer
- Samples are loaded in the gel and electricity is provided at 100 volts for 50 min.
- The gel is stained with Ethidium bromide for 10 min and washed with distilled water for 5 min.
- The gel is visualized under Trans UV and captured with Gel Doc XR system.

5.2.1.6 PCR product purification

- Positive PCR products with expected size are purified using MinElute® Gel Extraction Kit according to the standard protocol provided with Kit.
- The expected size of band is excised from the agarose gel with a sterile, sharp scalpel.

- The gel slice is put in a sterile 1.5 ml micro-centrifuge tube and weighed.
- QC buffer, provided with the kit, is added @ 3:1 volume of gel.
- The gel slice is incubated at 50°C for 10 min until the gel slice has completely dissolved.
- The mixture is vortexed every 2–3 min to facilitate the dissolution of gel slices.
- Then, 1 gel volume of isopropanol is added and mixed by inverting with pipette.
- The MinElute spin column is placed into 2 ml collection tube.
- The sample is transferred into the MinElute column and centrifuged at 13000 rpm for 1 min.
- The flow-through is discarded and put back the column into the same collection tube. 750 µl of Buffer PE is added to MinElute column and let it stand for 1–2 min.
- Centrifuged at 13000 rpm for 1 min and the flow-through is discarded.
- The process is repeated to remove Buffer PE completely.
- The ethanol residual left at the bottom of the column is discarded and MinElute column is placed into a sterile 1.5 ml micro-centrifuge tube.
- 30 µl of EB Buffer is added to the center of MinElute membrane to elute DNA. The mixture is let to stand for 1 min, and then is centrifuged at 13000 rpm for 1 min.
- MinElute column is discarded and the tube is stored in - 20°C.

5.2.1.7 Molecular Cloning (TOPO TA cloning kit, Invitrogen)

Positive PCR samples will be cloned using the TOPO TA cloning kit according to the standard protocol provided along with the Kit as follows;

Ligation

4 µl purified PCR products.

1 µl vector (pCR2.1-TOPO).

1 µl Salt solution.

Incubate in PCR machine/ heat block at 25°C for 30 min.

Add 2 µl of ligation mixture into competent cell *E. coli* - do not pipette up and down, just thaw a bit/ swirl.

Put 30 min in ice.

Transformation

Put 30 sec in 42°C water bath (heat shock).

Put in ice for 5 min (immediately after heat shock).

Add 250 µl SOC medium to mixture-seal competent cell tube with parafilm.

Put at 200 rpm in 37°C incubator shaker for 1 h 30 min.

Warm the petri dish in incubator for 20–30 min.

Spread 40 µl X-gal on petri dish (LBA media).

After spread the X-gal, put the petri dish in incubator for 20–30 min.

Finally, spread the sample mix on petri dish and incubate overnight at 37°C.

Note: Strictly follow the incubation time and temperature in the protocol during cloning.

5.2.1.8 Two Dimensional poly acrylamide gel electrophoresis (2D PAGE)

2D PAGE is carried out to for the detection and to check the circularity of Viroid RNA. Following is the recipe and protocol for PAGE.

Gel Ingredients

Acrylamide (A)

Bisacrylamide (B)

40% AB in 50 ml distilled water @ 19:1 ratio

Non Denaturing Gel:

Ingredients	8% GEL	5% GEL
40% AB	6 ml	6.25 ml
10X TBE	3 ml	5 ml
dH_2O	20.25 ml	37.7 8 ml
10% APS	750 µl	937.5 µl
TEMED	40 µl	43.75 µl
Total Volume	30 ml	50 ml

Mix the gel with magnetic bar.

Wash the glass with KOH and dH_2O.

KOH washing buffer includes 10 g KOH + 10 ml dH_2O + 90 ml and 99% ETOH.

Rinse the glass with dH_2O and let it dry.

Prepare the gel and cast into electrophoresis set.

Let the gel to polymerase for 30 min.

Pre-run empty gel for 20 min at 10 mA.

Pre-run sample for 10 min at 10 mA and then run sample for 1 hr. plus at 20 mA.

5% Non-denaturing Gel

40% AB: 8.7 ml

Urea: 25.2 g

10XTBE: 5.25 ml

10% APS: 750 µl

TEMED: 45 µl

dH2O: 17.25 ml

Total volume: 52.5 ml

Fixer 1:	
ETOH (10%, V/V)	10 ml
Acetic Acid (5%, V/V)	5 ml
dH_2O	85 ml
Fixer 2:	
ETOH (10%, V/V)	10 ml
Acetic Acid (5%, V/V)	0.5 ml
dH_2O	89.5 ml
Silver Solution:	
Silver Nitrate	0.3 g
dH_2O	150 ml
Developer Solution:	
3 mM $NaBH_4$	0.023 g
Formaldehyde	0.75 ml
0.375 M NaOH	3 g
dH_2O	200 ml

Silver Staining:

Fix the gel in fixer 1 for 10 min at room temperature in shaker.

Fix the gel in fixer 2 for 10 min at room temperature in shaker.

Dip the gel in silver stain solution for 1 hr. under dark condition.

Rinse with dH_2O twice with the interval of 1 min.

Add developer solution in the end.

Stop developing by adding 5% acetic acid.

5.3 Citrus tristeza virus (Ctv): a case study

5.3.1 Introduction

CTV belongs to the genus *Closterovirus* of the family *Clostervoiridae*. Virus particle is a monopartite, positive sense, comprising of ssRNA genome of approximately 20Kb in size. It is the largest known form of a plant virus and its genome is encapsulated in a flexuous rod 2000 nm long particles composed of coat protein subunits of 25KDA [86, 87, 88, 89]. ssRNA genome comprised of 19,296 nucleotides that encode for 12 open reading frames [90]. CTV probably originated in Asia and has been spread to all citrus growing areas by infected plant material movement and now is widely distributed to all major citrus growing areas as summarized in Table 8. Over the two decades *i.e.* 1950–1930, millions of citrus trees were destroyed due to CTV infection and citrus orchards were almost wiped out in Brazil, Spain, and Argentina. This virus was the killer of three million citrus trees grafted on sour orange rootstock alone in south California [91, 92, 93, 94]. The tristeza disease was first reported in Florida in 1959 and by 1980s became the serious threat to citrus industry [95]. By 1991, an estimation of total world loss of 100 million trees was recorded due to CTV in Argentina, Brazil, Spain, California, Venezuela and other areas [96, 97]. Several strains of CTV have been identified primarily on the basis of their biological reaction in several citrus species and indicator plant. The major groups of strains are mild that cause barely detectable clearing of leaf veins in Mexican lime; decline-inducing strains cause death of trees when propagated on sour orange rootstock. Stem pitting strains cause mild to severe pitting of stems and branches of grapefruit and orange resulting in low yield [95, 98]. Almost all the citrus varieties and hybrids have been infected with CTV [91]. Symptom expression of CTV in citrus hosts is highly variable and depends upon host species (rootstock and scion combination), virulence of CTV isolates and soil or environmental conditions. Characteristics symptoms of CTV are vein clearing, decline, stem pitting, seedling yellows, stunting and leaf corking on different citrus hosts like sweet orange, grapefruit, grafted on sour range root stock. Severity of infection and symptoms expression on cultivars vary from mild to severe isolates [99, 100, 101]. CTV is transmitted in nature by different species of aphids in a semi-persistent manner and through grafting [102, 103]. The most efficient vector involved in semi-persistent manner is *T. citricida* Kirkaldy (brown or black citrus aphid) when compared with other aphids.

Table 8. Geographical Distribution of Citrus tristeza closterovirus.

Regions	Countries	Status
EPPO	Israel, Spain, Turkey	Present
	France	Found but not established
	Algeria, Cyprus, Egypt, Italy, Morocco, Tunisia	Scattered infection
Asia	Brunei, China, Georgia	Present
	India	Widespread
	Indonesia. Iran, Japan	Present
	Jordan	Unconfirmed
	Korea, Malaysia, Nepal	Present
	Pakistan	Present (Scarce Information available)
	Philippines, KSA, Sri Lanka, Taiwan, Thailand, Viet Nam, Yemen	Present
Africa	Cameroon, Chad, Ethopia, Gabon, Ghana, Kenya, Mauritious, Mozambique, Nigeria, South Africa, Tanzania, Zaire, Zambia, Zimbabwe	Present
	Libya	Unconfirmed
North America	Bermuda, Mexico, USA	Present
Central America and Caribbean	Antigua, Barbuda, Bahamas, Belize, Costa Rica, El Salvador, Guatemala, Honduras, Jamaica, Netherlands Antilles, Nicaragua, Puerto Rico, St. Lucia, Trinidad, Tobago	Present
	Dominica	Unconfirmed

Regions	Countries	Status
South America	Argentina, Bolivia, Brazil	Present (wide spread)
	Chile	Found, not established
	Colombia, Ecuador, Guyana, Peru, Paraguay, Suriname, Uruguay	Present
Oceania	American Samoa, Australia, Fiji, New Zealand	Present

Source: Anonymous, 2004.

5.3.2 Indexing

Serological and biological indexing: Indexing includes biological, serological and molecular methods, which are the common procedures according to their reliability, sensitivity and duration to detect the CTV. During a survey in Spain, 22 CTV isolates were collected on the basis of geographical information, source tree and symptomology and then were characterized by biological indexing. Diversified symptoms were produced on 9 indicator species. Mexican lime was found to be a good indicator host [104].

In Morocco, 14 diverse isolates were selected from samples during survey and then characterized on the basis of reaction pattern. Among these 14 isolates, four were severe and two were mild isolates. Isolates were also indexed against a series of monoclonal antibodies [105]. DAS-ELISA was used to detect the CTV from the samples collected during a survey in Western and Midwestern development regions of Nepal [106]. One hundred and eighty-eight samples were analyzed through biological indexing and DAS-ELISA to detect tristeza, psorosis and similar diseases like-symptoms including viroids in orange varieties in all the regions and the cachexia was detected as the most important and widespread disease [107]. Biological indexing is still considered as an important tool using for the characterization of CTV isolates. Different strains were identified through symptoms expression on differential hosts, including Mexican lime and sweet orange. Moreover, they observed visual symptoms of different strains on Mexican lime and sweet orange through biological indexing followed by ELISA [108]. Detection of CTV in Spain was compared by indexing using monoclonal and polyclonal antibodies [109].

Molecular indexing: Different nucleic acid based indexing methods have been developed for the quick detection of CTV. The adaptability of these methods depends upon the reliability, time duration and sensitivity. Alteration in protein patterns in rootstock bark from CTV infected tree were analyzed through PAGE [110]. There was a clear modification in protein pattern but not in CTV free trees. Similarly, Northern blot technique was used to compare dsDNAs extracted from CTV infected and CTV free plants. Two out of the three CTV isolates were detected by this method [111]. CTV was also detected in the three aphid species through RT-PCR. IC-RT-PCR was used to amplify the coat protein gene [112]. Sensitivity of cDNA probe was slightly better than hybridization with ^{32}P-labeled probe. Similarly, hybridization with tissue print with DIG-probe could differentiate CTV isolates grown under green house or field conditions [113]. In Taiwan, RT-PCR was found to be a rapid and sensitive assay than other serological methods but one step RT-PCR, which is the combination of reverse transcriptase and polymerase chain reaction in one tube. It is more sensitive and detects the CTV when virus concentration is very low. Comparison between ELISA and RT-PCR revealed that

ELISA was better than RT-PCR at detecting mild CTV strains as the virus was detected in all parishes, while RT-PCR detected CTV in only 8 parishes. It would appear that the primers used for RT-PCR are more specific for severe CTV isolates [114]. Some modifications were introduced in PCR-ELISA to increase its sensitivity and reduced the costs of detection. PCR-ELISA is the immune-detection of PCR products and effective for detection and differentiation of plant viral nucleic acids. PCR-ELISA being a laborious and expensive method was modified and simplified by using asymmetric PCR. It made PCR-ELISA more sensitive than TaqMAN™, a fluorescence-based detection method.

Three microscopy procedures for detecting CTV were compared which provided additional alternatives for very rapid CTV indexing, including the use of EM, SSEM and light microscopy. In light microscopy, inclusions were found in young phloem tissues of all CTV-infected hosts examined. Similarly, in SSEM virus particles were found on grids prepared with antiserum and extracts from infected tissue. CTV particles could be detected in pooled samples representing one in 100. Similarly, virus particle fragments were observed infrequently in samples representing one infected plant in 1,000 samples [32].

6. Conclusion

Citrus is an important fruit crop of the world and has a great potential for local consumption, export purposes and industrial uses. Unfortunately, citrus orchards are facing the problem of low productivity due to citrus decline. This is mainly attributed, among other factors to the prevalence of graft-transmissible virus and virus-like diseases, unhygienic nursery operations and poor orchard management. However, most of the problems arise from nurseries. It is the time that the nurseries should operate on highly technical and scientific lines and should work on providing disease-free and certified plants to the citrus growers. To establish the disease free nurseries, indexing of virus and virus-like diseases are the major area that needs to be focused. Implication of traditional and modern high-throughput biological, serological and molecular indexing techniques, such as ELISA, RT-PCR, PAGE, should be put in practice for the detection and indexing of virus and virus-like diseases of citrus plants. Moreover, citrus nurseries should be registered and indiscriminate multiplication and sale of uncertified citrus plants should be prohibited.

References

1. Davies FS, Albrigo LG. History, distribution and uses of citrus fruit. Citrus, CAB international. Printed in Great Britain by Redwood books. Trowbridges, Wiltshine, 1994;1–254
2. Hooker JD. 1872. Flora of British India. Reeve and co. London
3. Liu P. World markets for organic citrus and citrus juices. Food and Agriculture Organization of the United Nations, 2003; 6p
4. Khan MM, Y. Iftikhar Y, Abbas M, Naqvi SA, Iqbal, MT, Jaskani MJ, Khan IA. Disease free citrus nursery production system. In Proc. "Workshop on production of disease free citrus nursery plants". Institute of Horticultural Sciences, University of Agriculture, Faisalabad, September 12–13, 2007; 29–35

5. Iftikhar Y. Serological, Biochemical and Molecular characterization of Citrus tristeza virus. Thesis submitted in department of Plant Pathology, University of Agriculture, Faisalabad, 2009

6. Bove JM. 1995. Virus and virus-like diseases of citrus in the near east region. F.A.O. Rome. pp: 239–266

7. Khan IA. 1992. Virus and virus like diseases of citrus. In: Khan, I. A (Ed.), Proceedings of 1st International Seminar on Citriculture in Pakistan. University of Agriculture, Faisalabad, Dec 2–5, pp 343–352

8. Roistacher CN (Ed.). 1991. Graft transmissible diseases of citrus. F.A.O, Rome, 1991: 266

9. Iftikhar Y. Some Biological and physical properties of yellow vein clearing virus of lemon. Thesis submitted in department of Plant Pathology, University of Agriculture, Faisalabad, 2004

10. Mughal SM. Symptomatology, detection, distribution and management of virus and virus-like diseases of citrus in Pakistan. In: Proc. Int. Symp. Citriculture, U.A.F. Pakistan. 2004;106–113

11. Arif, M., Ahmad, A., Ibrahim, M. and Hassan, S. (2005). Occurrence and distribution of virus and virus-like diseases of citrus in north-west frontier province of Pakistan. Pakistan Journal of Botany, 37(2), 407–421

12. Catara, A., Azzaro, A., Mughal SM., and Khan DA. (1988). Virus, viroids and prokaryotic diseases of citrus in Pakistan. In: Proceedings 6th Int. Citrus Cong. Mar.6-11. pp. 957-962

13. Iftikhar Y, Mughal SM, Khan MM, Khan MA, Nawaz MA, Hussain Z. Symptomatic expression of tristeza-infected citrus plants in Pakistan. Archives of Phytopathology and Plant Protection. 2013; 46: 98–104

14. Kluth, S., Kruess, A. and Tscharntke, T. (2002). Insects as vectors of plant pathogens: mutualistic and antagonistic interactions. Oecologia, 133(2), 193–199

15. Agrios, G.N. (2009). Transmission of plant diseases by insects. Available: http://entomology.ifas.ufl.edu/capinera/eny5236/pest1/content/03/3_plant_diseases.pdf

16. Wu, F., Qureshi, J. A., Huang, J., Fox, E. G. P., Deng, X., Wan, F., Liang, G. and Cen, Y. (2018). Host plant-mediated interactions between 'Candidatus Liberibacter asiaticus' and its vector Diaphorina citri Kuwayama (Hemiptera: Liviidae). Journal of Economic Entomology, 111(5), 2038–2045

17. Roistacher, C.N. (2004). Diagnosis and management of virus and virus like diseases of citrus. In: Diseases of Fruits and Vegetables (Volume I), Springer, Dordrecht. pp. 109–189

18. Mitchel, P.L. (2004). Heteroptera as vectors of plant pathogens. Neotropical Entomology, 33, 519–545

19. Raccah, B. and Fereres, A. (2009). Plant Virus Transmission by Insects; John Wiley and Sons, Ltd.: Chichester, UK

20. Heck, M. (2018). Insect transmission of plant pathogens: A systems biology perspective. Msystems, 3(2). e00168–17
21. Bar-Joseph, M. and Nitzan, Y. (1991). The spread and distribution of citrus tristeza virus isolates in sour orange seedlings. In: Proceedings of the 11th Conference International Organization of Citrus Virologists (IOCV), Riverside, CA. pp. 162–165
22. Catara, A., Azzaro, A., Davino, M., and Polizzi, G. (1993). Yellow vein clearing of lemon in Pakistan. In: Proceedings of the 12th Conference International Organization of Citrus Virologists (IOCV), Riverside, CA. pp. 364–367
23. Önelge, N. (2002). First report of yellow vein clearing of lemons in Turkey. Journal of Turkish Phytopathology, 32, 53–55
24. Alshami, A., Ahlawat, Y.S. and Pant, R.P. (2003). A hitherto unreported yellow vein clearing disease of citrus in India and its viral etiology. Indian Phytopathology, 56(4), 422–427
25. Zhou, Y., Chen, H., Cao, M., Wang, X., Jin, X., Liu, K.H. and Zhou, C.Y. (2017). Occurrence, distribution, and molecular characterization of Citrus yellow vein clearing virus in China. Plant Disease, 101(1), 137–143
26. Hashmian, S.B. and Aghajanzadeh, S. (2017). Occurrence of citrus yellow vein clearing virus in citrus species in Iran. Journal of Plant Pathology, 99(1), 290–297
27. Önelge, N., Satar, S., Elibuyuk, O., Bozan, O. and Kamberoolu, M. (2011). Transmission studies on citrus yellow vein clearing virus. In: International Organization of Citrus Virologists Conference Proceedings (1957–2010) (Vol. 18, No. 18)
28. Noordam, D. Identification of Plant viruses. Methods and Experiments. Centre of Agril. Publishing and documentation, Wageningen, 1973; 207p
29. Ashfaq M, Mughal SM, Iftikhar Y, Khan MA, Khan NA. Study on host range, serology and inclusion bodies of yellow vein clearing virus (YVCV) of lemon. Pakistan Journal of Phytopathology. 2004;16: 1–4
30. Iftikhar Y, Mughal SM, Ashfaq M, khan MA, Haq IU. Some biological and physical properties of yellow vein clearing virus of lemon. Pakistan Journal of Phytopathology. 2004; 16: 5–8
31. Ahlawat YS. Diagnosis of Plant viruses and Allied Pathogens. Studium Press (India) Pvt. Ltd. 2010; 224p
32. Garnsey SM, Christie RG, Derrick KS, Bar-Joseph M. Detection of Citrus tristeza virus II. Light and Electron microscopy of inclusions and viral particles. In: Calavan, EC, Garnsey, SM and Timmer, LW (Eds.), Proceeding of 8th International organization of citrus virologists, conference, California, Riverside, 1979, U.S.A. pp.9–16
33. Garnsey SM., and Cambra M. (1991). Enzyme-linked immunosorbent assay (ELISA) for citrus pathogens. In: Roistacher CN ed. Graft transmissible diseases of citrusHandbook for detection and diagnosis. Rome, Italy, FAO. Pp. 193-216
34. Semancik, JS., Duran-Vila, N. (1991). The Grouping of Citrus Viroids: Additional Physical and Biological Determinants and Relationships with Diseases of Citrus. UC Riverside International Organization of Citrus Virologists Conference Proceedings, (1957-2010), 11(11) ISSN 2313-5123. https://escholarship.org/uc/item/64r9q712

35. Iftikhar Y, Khan MA, Rashid A, Mughal SM, Iqbal Z, Batool A, Abbas M, Khan MM, Muhammad S, Jaskani MJ. Occurrence and distribution of Citrus tristeza Closterovirus in the Punjab and NWFP, Pakistan. Pakistan Journal of Botany. 2009; 41: 373–380

36. Wylie, SJ, Wilson, CR, Jones, RAC, Jones, MGK. A polymerase chain reaction assay for Cucumber mosaic virus in lupin seeds. Australian Journal of Agricultural Research, 1993; 44: 41–51

37. Olmos, A., Bertolini, E, Gil, M, Cambra, M. Real-time assay for quantitative detection of non-persistently transmitted Plum pox virus RNA targets in single aphids. Journal of Virological Methods, 2005; 128: 151–155

38. Bertolini, E, Olmos, A, Martínez, MC, Gorris, MT, Cambra, M. Single-step multiplex RT-PCR for simultaneous and colourimetric detection of six RNA viruses in olive trees. Journal of Virological Methods, 2001; 96:33–41

39. Olmos, A, Bertolini, E, Cambra, M. Isothermal amplification coupled with rapid flowthrough hybridisation for sensitive diagnosis of Plum pox virus. Journal of Virological Methods, 2007; 139:111–115

40. Bertolini, E, Olmos, A, López, MM, Cambra, M. multiplex nested reverse transcriptionpolymerase chain reaction in a single closed tube for sensitive and simultaneous detection of four RNA viruses and Pseudomonas savastanoi pv. savastanoi in olive trees. Phytopathology, 2003; 93:286–292

41. Walsh, K, North, J, Barker, I, Boonham, N. Detection of different strains of Potato virus Y and their mixed infections using competitive fluorescent RT–PCR. Journal of Virological Methods, 2001; 91: 167–173

42. López, MM, Bertolini, E, Marco-Noales, E, Llop, P, Cambra, M. Update on molecular tools for detection of plant pathogenic bacteria and viruses. In Molecular diagnostics: current technology and applications, J.R. Rao, C.C. Fleming, and J.E. Moore, eds. Horizon Bioscience, Wymondham, UK, 2006: 1–46

43. Rocha-Pena, MA, Lee, RF, Lastra, R, Niblett, CL, Ochoa-Corona, FM, Garnsey, SM, Yokomi, RK. Citrus tristeza virus and its aphid vector Toxoptera citricida: threats to citrus production in the carribean and central and North America. Plant Disease, 1995; 79(5): 437–445

44. Ahlawat, YS. Viruses, greening bacterium and viroids associated with citrus (Citrus species) decline in India. Indian Journal of Agricultural Science, 1997; 67:51–57

45. Bar-Joseph M, Marcus R, Lee RF. The continous challenges of citrus tristeza virus control. Annual Review of Phytopathology, 1989;27: 291–316

46. Biswas, KK. Molecular diagnosis of Citrus tristeza virus in mandarin (Citrus reticulata) orchards of Darjeeling hills of West Bengal. Indian Journal of Virology, 2008; 19: 26–31

47. López, MM, Llop, P, Olmos, A, Marco-Noales, E, Cambra, M, Bertolini, E. Are molecular tools solving the challenges posed by detection of plant pathogenic bacteria and viruses?. Current issues in molecular biology, 2009; 11(1): 13

48. Lukhtanov, EA, Lokhov, SG, Gorn, VV, Podyminogin, MA, Mahoney, W. Novel DNA probes with low background and high hybridization-triggered fluorescence. Nucleic Acids Research, 2007; 35: 5 e30. doi:10.1093/nar/gkl1136

49. Gasparic, MB, Cankar, K, Zel, J, and Gruden, K. Comparison of different real-time PCR chemistries and their suitability for detection and quantification of genetically modified organisms. BMC Biotechnology, 2008; doi:10.1186/1472-6750-8-26

50. Korimbocus, J, Coates, D, Barker, I, and Boonham, N. Improved detection of Sugarcane yellow leaf virus using a real-time fluorescent (TaqMan) RT-PCR assay. Journal of Virological Methods, 2002; 103: 109–120

51. Beuret, C. Simultaneous detection of enteric viruses by multiplex real-time RT-PCR. Journal of Virological Methods, 2004; 115: 1–8

52. Munford, RA, Skelton, A, Metcalfe, E, Walsh, K, and Boonham, N. The reliable detection of Barley yellow and mild mosaic viruses using realtime PCR (TaqMan). Journal of Virological Methods, 2004; 117: 153–159

53. Varga, A, and James, D. Use of reverse transcription loop-mediated isothermal amplification for the detection of Plum pox virus. Journal of Virological Methods, 2006; 138: 184–190

54. Agindotan, BO, Shiel, PJ, and Berger, PH. Simultaneou detection of potato virus, PLRV, PVA, PVX and PVY from dormant potato tubers by TaqMan(®) real-time RT-PCR. Journal of Virological Methods, 2007; 142: 1–9

55. Kogovsek, P, Gow, ., Pompe-Novak, M, Gruden, K, Foster, GD, Boonham, N, and Ravnikar, M. Single-step RT real-time PCR for sensitive detection and discrimination of Potato virus Y isolates. Journal of Virological Methods, 2008; 149: 1–11

56. Agrios, G.N. (2005). Plant Pathology 5th Edition. Department of Plant Pathology, University of Florida.

57. Hadidi, A., Flores, R., Randles, J.W., and Semancik, J.S. (2003). Viroids. Science Publisher, Inc.

58. Bostan H, Nie X, Singh RP (2004). An RT-PCR primer pair for the detection of Pospiviroid and its application in surveying ornamental plants for viroids. J. Virol. Methods, 116: 189–193

59. Bar-Joseph, M. (1996) A contribution to the natural history of viroids.Proc. 13th IOCV Conf. (Riverside, CA, USA), pp. 226–229

60. Pagliano, G., Umaña, R., Pritsch, C., Rivas, F. and Duran-Vila, N. (2013). Occurrence, prevalence and distribution of citrus viroids in Uruguay. Journal of Plant Pathology. pp. 631-635

61. Bani Hashemian, S.M., Taheri, H., Duran-Vila, N. and Serra, P. (2010). First Report of Citrus viroid V in Moro Blood Sweet Orange in Iran. Plant Disease: An International Journal of Applied Plant Pathology. pp. 129-129

62. Semancik, J.S., Morris, T.J., Weathers, L.G., Rordorf, G.F., and Kearns, D.R. (1975). Physical properties of a minimal infectious RNA (viroid) associated with the exocortis disease. Virology 63: 160-167

63. Duran-Vila, N., Pina, J.A., Ballester, J.F., Juarez, J., Roistacher, C.N., Rivera-Bustamante, R., and Semancik, J.S. (1988). The Citrus Exocortis Disease: A complex of viroid-RNAs. Tenth IOCV.

64. Ashulin, L., Lachman, O., Hadas, R., and Bar-Joseph, M. (1991). Nucleotide sequence of a new viroid species, Citrus bent leaf viroid (CBLVd) isolated from grapefruit in Israel. Nucleic Acids Res

65. Cao, M. J., Atta, S., Liu, Y. Q., Wang, X. F., Zhou, C. Y., Mustafa, A. and Iftikhar, Y. (2009). First report of Citrus bent leaf viroid and Citrus dwarfing viroid from citrus in Punjab, Pakistan. Plant Disease 2009 Vol. 93 No. 8 pp. 840

66. Rakowski, A.G., Szychowski, J.A., Avena, Z.S. and Semancik, J.S. (1994). Nucleotide sequence and structural features of the group III citrus viroids. J Gen Virol 75:3581-3584

67. Putcha, H., Ramm, K., Luckinger, R., Haddas, R., Bar-Joseph, M. and Sanger, H.L. (1991). Primary and secondary structure of citrus viroid IV, a new chimeric viroid present in dwarfed grapefruit in Israel. Nucleic Acids Res 19:6640

68. Serra, P., Eiras, M., Bani-Hashemian, S. M., Murcia, N., Kitajima, E. W., Daròs, J. A., Flores, R., and Duran-Vila, N. (2008). Citrus viroid V: Occurrence, host range, diagnosis, and identification of new variants. Phytopathology 98:1199-1204

69. Ito, T., Ieki, H. and Ozaki, K. (2002). Simultaneous detection of six citrus viroids and Apple stem grooving virus from citrus plants by multiplex reverse transcription polymerase chain reaction. Journal of Virological Methods 106, 235–239

70. Srivastava, S., & Prasad, V. (2020). Viroids: small entities with a mean punch. In Applied Plant Virology (pp. 209–226). Academic Press

71. Garnsey, S.M, Zies, D.L., Irey, M., Sieburth, J.S., Semancik, J.S., Levy, L. and Hilf, M.E. (2002). Practical field detection of citrus viroids in Florida by RT-PCR. Fifteenth IOCV Conference.

72. Eiras, M., Silva, S.R., Stuchi, E.S., Carvalho, S.A., and Garcez, R.M. (2013). Identification and characterization of viroids in 'Naveline ISA 315' sweet orange. Tropical Plant Pathology, 38: 058–062

73. Abubaker, M. Y. A., & Elhassan, S. M. (2010). Survey and molecular detection of two citrus viroids affecting commercial citrus orchards in the Northern part of Sudan. Agric. Biol. JN Am, 1(5), 930–937

74. Pagliano, G., Peyrou, M., Campo, R.D., Orlando, L., Gravina, A., Wettstein, R., and Francis, M. (2000). Detection and characterization of citrus viroids in Uruguay. Fourteenth IOCV Conference.

75. Nakahara, K., Hataya, T., Uyeda, I., and Ieki, H. (1998). An Improved Procedure for Extracting Nucleic Acids from Citrus Tissues for Diagnosis of Citrus Viroids. Ann. Phytopathol. Soc. Jpn. 64: 532–538

76. Murcia, N., Serra, P., Olmos, A., and Duran-Vila, N. (2009). A novel hybridization approach for detection of citrus viroids. Molecular and Cellular Probes 23

77. Schumacher, J., Randles, J.W., and Riesner, D. (1983). A two dimensional electrophoretic technique for detection of circular viroids and virusoids. Anal. Biochem. 135:288

78. Nakaune, R. and Nakano, M. (2006). Efficient methods for sample processing and cDNA synthesis by RT-PCR for the detection of grapevine viruses and viroids. Journal of Virological Methods 134: 244–249

79. Jiang, D., Hou, W., Kang, N., Qin, L., Wu, Z., Li, S., and Xie, L. (2012). Rapid detection and identification of viroids in the genus Coleviroid using a universal probe. Journal of Virological Methods.

80. Kunta, M., Gracxa, J.V.D., and Skaria, M. (2007). Molecular Detection and Prevalence of Citrus Viroids in Texas. Hortscience 42(3):600–604

81. Papayiannis, L.C. (2013). Diagnostic real-time RT-PCR for the simultaneous detection of Citrus exocortis viroid and Hop stunt viroid. Virological Methods 196: 93–99

82. Targon, M.L.P.N., Carvalho, S.A.D., Stuchi, E.S., Souza, J.M., Muller, G.D. and Machado, K.M.B.M.A. (2005). Hybridization techniques for indexing of citrus viroids in Sao Paulo State, Brazil. LARANJA, Cordeiropolis: Volume 26, p25–38

83. Narayanasamy, P. (2010). Microbial Plant Pathogens-Detection and Disease Diagnosis: Viral and Viroid Pathogens.Vol 3. Springer Science & Business Media.

84. Ito, T., Ieki, H. and Ozaki, K. (2000). A population of variants of a viroid closely related to citrus viroid-I in citrus plants. Archives of Virology 145: 2105–2114

85. Bernard, L., and Duran-Vila, N. (2006). A novel RT-PCR approach for detection and characterization of citrus viroids. Mollecular and Cellular Probes 20: 105–113

86. Mehta P, Brlansky RH, Gowda S. Reverse Transcription Polymerase Chain Reaction Detection of Citrus Tristiza Virus in Aphids. Plant Disease, 1997; 81:1066–1069

87. Mathews DM, Riley KR, Dodds JA. Comparison of detection methods for citrus tristeza virus in field trees during months of non-optimal titer. Plant Disease, 1997; 81: 525–529

88. Niblett CL, Genc H, Cevik B, Halbert S, Brown I, Nolasco G, Bonaçalza B, Manjunath KL, Febres VJ, Pappu HR, Lee RF. Progress on strain differentiation of citrus tristeza virus and its application to the epidemiology of citrus tristeza disease. Virus Research, 2000; 71:97–106

89. Suastika G, Natsuaki T, Terui H, Kano T, Leki H, Okuda S. Nucleotide sequence of citrus tristeza virus seedlig yellows isolates. Journal of General Plant Pathology, 2001; 67: 73–77

90. Bar-Joseph M, Lee RF. 1989. Citrus tristeza virus. AAB Descriptions of Plant Viruses no. 353. AAB, Wellesbourne (GB). http://www.ncbi.nlm.nih.gov/ICTVdb/ICTVdB/

91. Anonymous. 2004. Citrus Tristeza Closterovirus. Data sheets on Quarantine pests. EPPO A2 list, No.93. www.eppo.org/QUARANTINE/virus/Citrus_tristeza/

92. Bar-Joseph M, Marcus R, Lee RF. The continous challenges of citrus tristeza virus control. Annual Review of Phytopathology, 1989;27: 291–316

93. Kallsen C. 2002. Controlling citrus tristeza virus in the San Joaquin valley of California. Citrus subtropical Horticulture/Pistachios. http://cekern.ucdavis.edu/Custom_Program143/Controlling_Citrus_Tristeza_Virus_in_the_SJ_Valley_of_California.htm

94. Mooney P, Harty A. 1992. Citrus tristeza virus. The Orchardist. http://www.hortnet.co.nz/publications/science/kk0992.htm

95. Futch SH, Brlansky RH. 2005. Field diagnosis of citrus tristeza virus. HS996, one of a series of the Horticultural services department, Florida cooperative extension service. IFAS, University of Florida, U.S.A. http://edis.ifas.ufl.edu

96. Bar-Joseph M, Roistacher CN, Garnsey SM, Gumpf DJ. A review of tristeza, an ongoing threat to citriculture. In: Proceeding of International Society of Citriculture, Tokyo, Japan, 1981; 419–423

97. Mooney P, Dawson T, Harty A. 1994. Citrus tristeza virus preimmunization strategies. The Orchadist. http://www.hortnet.co.nz/publications/science/kk0894.htm

98. Chung KR, Brlansky RH. Citrus diseases exotic to Florida: Citrus Tristeza Virus-Stem Pitting (CTV-SP). Fact Sheet. pp 227. Plant Pathology Department, Florida Cooperative Extension Service, Institute of Food and Agricultural Sciences, University of Florida, 2006, Florida. http://edis.ifas.ufl.edu

99. Brlansky RH, Damsteegt VD, Howd DS, Roy A. Molecular analysis of citrus tristeza virus subisolates separated by aphid transmission. Plant Disease, 2003;87:397–401

100. EPPO Bulletin. Protocol for the diagnosis of quarantine organism, citrus tristeza closterovirus. 2004; 34: 239–246

101. Lbida B, Bennani A, Serrhini MN, Zemzami M. Biological, serological and molecular characterization of three isolates of citrus tristeza closterovirus introduced into Morocco. OEPP/EPPO Bulletin 2005; 35:511–517

102. Brown LG, Denmark HA, Yakomi RK. Citrus Tristeza Virus and its vectors. In; Florida. Plant Pathology circular no.311. Florida Department of Agriculture and Consumer Service. Division of Plant Industry. 1988

103. Komazaki S. Biology and virus transmission of citrus aphid. Akitsu Branch, Fruit and Tree Research Station, Ministry of Agriculture, Forestry and Fisheries, Hiroshima, Japan, 1993. http://www.fftc.agnet.org/library/tb/136/

104. Ballester OJF, Pina JA, Carbonell EA, Moreno P, Hermoso de Mandoza A, Cambra M, Navarro L. Biological diversity of citrus tristeza virus (CTV) isolates in Spain. Plant Pathology, 1993;42:219–229

105. Zemzami M, Garnsey SM, Nadori EB, Hill JH. Biological and serological characterization of citrus tristeza virus (CTV) isolates from Morocco. Phytopathologia-Mediterranea, 1999; 38: 95–100

106. Malla S, Sah DN. Severity and prevalence of citrus tristeza virus (CTV) in the western and Mid-Western regions of Nepal. Working paper lumle Agricultral Research Centre, 2001; 14

107. Besoain XA, Valenzuela M, Castro M, Ballester Olmos JF. Current status of some virus and virus like diseases of citrus in Chile. Fitopatologia. 2000;35: 98–104

108. Polek M, Gumpf DJ, C. M. Wallen CM, Riley KM. Biological Characterization of Naturally Occurring Citrus tristeza virus Strains in California Citrus. In: In: M.E. Hilf, N. Duran-Vila and M.A. Rocha-Pena (Eds.), Proceeding of the 16th International Organization of Citrus Virologist Conference, California, Riverside, 2005, USA:68–74

109. Cambra M, Serra J, Bonet JC, Moreno P. Present status of the citrus tristeza virus in the valencian community. In: L. W. Timmer, S. M. Garnsey and L. Navarro. (Eds), Proceeding of 10th International Organization of Citrus Virologists, California Riverside, 1998; 1–7

110. Moreno, P., J. Guerri and J. Ortiz. Alteration of bark proteins associated with citrus tristeza virus (CTV) infection on susceptible citrus species and scion-rootstock combinations. Phytopathology, 1989; 125:55–66

111. Acikgoz S. Detection of citrus tristeza virus (CTV) isolates with northern blot hybridization. Doga Turk Tarim-ve-Ormancilik Dergisi, 1991;15:836–840

112. Anfoka GH, Abhary MK, Fattash I, Nakhla MK. Occurrence and distribution of citrus tristeza virus in the Jordan valley. Phytopathologia Mediterranea, 2007;44:17–23

113. Narvaez G, Skander BS, Ayllon MA, Rubio L, Guerri J, Moreno P. A new procedure to differentiate citrus tristeza virus isolates by hybridization with digoxigenin-labelled cDNA probes. Journal of Virological Methods, 2000; 85: 83–92

114. Fisher L, Tennant P, Molaughlin W. 2005. Detection and differentiation of Citrus Tristiza Virus (CTV) in Jamaica using ELISA, RT-PCR, DNA hybridization and RFLP. Ministry of Agriculture's citrus replanting project-research services. Conference, The Environment: Biodiversity, O-23, Faculty of Pure and Applied Sciences, University of West Indies. www.mona.uwi.edu/fpas/conference/fpas7

Chapter-7

Xanthomonas *Citri* ssp. Citri Pathogenicity, A Review

1. Introduction

The surface of the plants is one of the most hostile environments, prevailing factors at the phyllosphere such as: the low availability of nutrients, the high incidence of UV rays, the fluctuating periods of temperature and humidity, mechanical disruption by winds, antibacterial compounds produced by the host plant or by microorganisms member of leaf microbiome, among others, make the bacterial persistence and survival itself a pathogenicity strategy. Due the symptoms development ceases when one pathway involved in the bacterial epiphytic survival is seriously threatened [1]. In phytopathogenic bacteria whose infection route is the phyllosphere, it is important to understand how phenotypic traits upset to ensure survival and surface fitness, and how these traits interact with the phyllosphere microbiome in order to secure the onset of infection (Figure 1). Besides, over the time, on a large-scale, plant leaves will age and fall, thus, the phyllosphere bacteria must have to anticipate living outside of the leaf, for example in the air, soil or reach to young leaves [2].

Bacterium *Xanthomonas citri* ssp. *citri* (*Xcc*) is the etiological agent of bacterial citrus canker. This bacterium is equipped with a huge arsenal of cellular structures that allow its survival in the phyllosphere before it reaches the target mesophyll tissue. *Xcc* secretes toxins that directly affect the survival of its competitors. Once in the mesophilic tissue *Xcc* produces effectors that are responsible by the appearance of spongy and corky pathognomonic lesion of citrus canker. In this chapter we will review the both bacterial life style outside and inside of the host.

2. *Xanthomonas citri* ssp. *citri* taxonomy

The bacterium *Xantomonas citri* ssp. *citri* is a gram negative rod shape bacteria with a single polar flagellum. *Xcc* belongs to the *Xanthomonas* genus from the gamma proteobacteria group. This genus is constituted by 28 species and more than 150 pathovars [3]. In the early 1900s, due to pathogenicity experiments, the bacterium was classified as *Pseudomonas citri* [4]. Subsequently, the bacterium was classified into different genus such as: *phytomonas bacterium* and finally at late 1930 classified as *Xanthomonas citri* [5]. The bacterium continued in *X. citri* until 1978, when it was classified in *X. campestris* pv. *citri* in order to reserve citri at the specific level [6]. In 1989 Gabriel suggest the replaced of bacterium as *X. citri* [7]. Using DNA–DNA hybridization approach and based on renaturation rates, the bacterium was classified as *X. axonopodis* pv. *citri* by Vauterin [8]. Lately, It was suggested major changes to Xanthomond taxonomy, it which were based on multilocus sequence analysis (MLSA) and digital DNA–DNA hybridization of whole genome nucleotide, the author has been recommend the names *Xanthomonas citri ssp citri* for the etiological agent of citrus cancer type A [9].

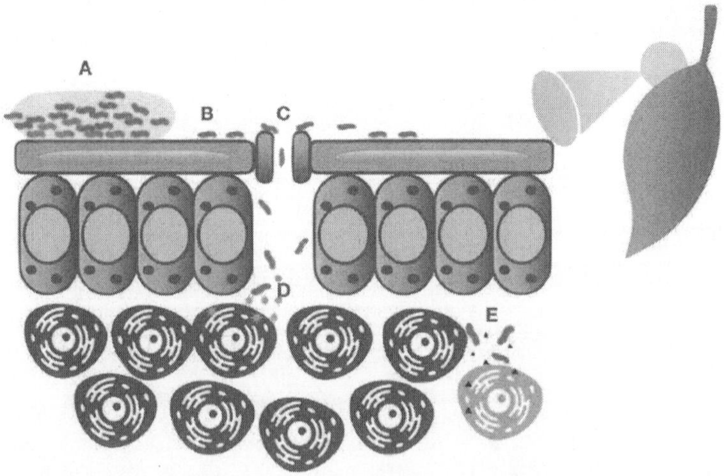

Figure 1. Phytopathogenic bacteria plant infection. A. Surface leaf survivor and biofilm formation. B. Bacterial movement to natural opening on leaf. C. Phytotoxins secretion to modulate stomatal closure. D. Effector secretion that affect the cell host behavior. E. Degrading cell wall plant protein secretion.

3. Microbe- host interaction

3.1 Microbe -host interaction outside the susceptible host "epiphytic life style"

Bacterial citrus canker disease cycle begins with the deposit of inoculum of *XCC* at the leaf surface by rain splash. Subsequently, the bacteria move toward the natural opening of leaves, the stomata, then, the bacteria reach the apoplastic space and start the infection process inside the host "endophytic lifestyle". In this section we are going to focus on the structures, toxins, molecules and extracellular substances that favor and promote the epiphytic interaction between XCC and susceptible citrus host.

3.1.1 Type IV pili

Several bacterial genera are endowed with filamentous appendages called pili. These filamentous organelles include the chaperon- Usher pili, type IV pili (T4P) and gram-positive pili. All types of pili are homopolymers ensembled of thousands of units of pilin protein. The outstanding function of pili is the attachment to surfaces, besides, in *Xcc* pili type IV is also responsible for the twitching motility and biofilm formation [10]. Type 4 pili is unique in its dynamism, since, it polymerizes and depolymerizes in very fast cycles, which leads to instantaneous extension and retraction cycles producing considerable mechanical force [11], as a consequence, this organelle could attract several substrates like DNA or bacteriophages in order to internalize to periplasmic space, as well as to secrete protein across the membrane [12]. Twitching motility is a bacterial displacement that able to cell to move over humid on organic and inorganic surfaces on a fashion independent of flagella [13]. In the process of biofilm development, the T4P contributes in the initial steps exactly in the reversible attachment phase and subsequently, in the formation of mushroom microcolonies. Contribution of T4P in the pathogenicity in XCC is not completely demonstrated, however, the mutation of *pilM* gene responsible to encode a membrane protein that participate in the T4P pili ensemble reduce drastically the bacterial virulence [10].

3.1.2 Type V secretion system *(non fimbrial adhesins)*

Xanthomonads encode type V secretion system (T5SS), it which has a function as non fimbrial adhesins [14]. Compared with the other bacterial secretion systems, the secretion system 5 is one of the simplest complexities from the structural point of view; it is smaller and has only presence at the outer membrane of gram negative bacteria [15]. This T5SS do not have a direct energy source, there is no ATP accessible in the periplasm space neither proton gradient. Consequently, the name of autotransporter has been coined for the this T5SS [16]. The T5SS is comprised of two domains: the β barrel that is located at the out membrane and a secreted passenger. There are five subtypes of T5SS from Va-Ve and recently a new subtype the Vf has been discovered [17]. The bacterium *Xcc* is endowed with three subclasses of T5SS: Va, Vb and Vc. (Figure 2). Va is a classical auto-transporter, it which transport proteases, lipases and adhesins. The type Vb is a secretion system knowing as Two-Partner Secretion System (TPS), which is composed by a translocator protein and a cognate passenger protein. Translocation from the cytoplasm to the periplasm space occurs by Sec translocase pathway once the perception of amino terminal from signaling peptide is done. The passenger protein has effector function and is termed TpsA. It is transported by TpsB, which forms a pore in the outer membrane in order to enable the TpsA translocation. TpsB also comprise two periplasmic domains. TpsB typically contains a 16-stranded beta-barrel domain that forms the outer membrane pore and two periplasmic POTRA (Polypeptide transport associated). Its function is the recognition of the cognate partner via binding to a TPS domain in TpsA.

The T5SS subclass Vc have a trimeric transporter adhesin conformation, this surface exposed adhesin assembles as homotrimeric structure at the outer membrane [18]. Proteomic and functional studies involving T5SS have revealed roles in pathogenicity to host primarily implicated in the adhesion, especially in the initial steps of pathogenicity process [19, 20].

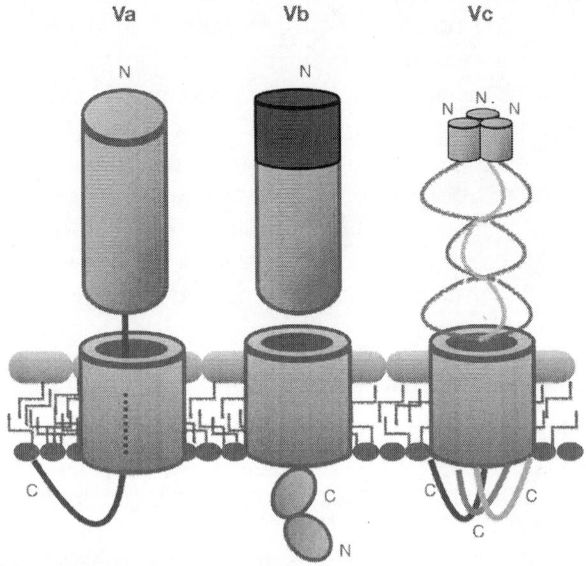

Figure 2. Schematic representation of T5SS present in Xcc. β barrel domain and POTRA are characterized with blue, linker, passengers transported are represented in green and two partner secretion system domain are characterized with red.

3.1.3 Xanthomonadin pigment

Xanthomonads bacteria produce a yellow pigment membrane bound known as Xanthomonadins. Several studies have shown that Xanthomonadin has a pivotal role in a epiphytic survival and in plant-pathogen interaction [21, 22]. In the early years this yellow pigment was associated with the carotenoids. However, it was only until its full characterization was achieved that this pigment represents a unique group of aryl-polyene, water insoluble new type of pigment [23]. Genomic analysis shows that a region near to 25.4 kb contains seven transcriptional units (*pigA, pigB, pigC, pigD, pigE, pigf* and *pigG*). This gene cluster encodes necessary elements for Xanthomonadin biosynthesis [24]. Biological roles of xanthomonadin in a pathogenicity context are: (i). *favor the bacterial epiphytic survival,* since, Xanthomonadin avoid the photodamage produced by UV light irradiation that results in ROS production. Similar as structural related carotenoids, Xanthomonadin absorbs wavelengths between UV-C to red light. This pigment gives the bacteria additional advantages against the other phyllosphere colonizer bacteria as it is to deal with stress related factors such as UV irradiation and consequently the photo oxidative damage. Xanthomonadin also offers protection against visible light in the presence of exogenous photosensitizers. Cellular location of Xanthomadin (outer membrane) strongly suggests that this pigment stabilizes cell membrane in the epiphytic phase of this phytopathogenic bacterium. Previous studies in which *Xcc* deletion mutants of the *pig* genes were used and which were inoculated using the needleless syringe pressure technique did not show a significant reduction in virulence compared to the wild type phenotype inoculated using the same technique. Instead, when using the spray infection method, that resembles the natural infection method, it which involve the epiphytic fitness stage, the *Xcc* pig mutant strains display great reduction in the virulence compare with the wild type phenotypes [25]. (ii) *Antioxidant activity*, the oxidative stressors as ROS and H_20_2 injury the membranes, DNA and proteins, the carotenoids pigments could efficiently quench the ROS.

3.1.4 EPS xanthan and LPS

The EPS in Xanthomonas is named as xanthan, this polysaccharide surrounds the outer membrane through non-covalent ligations [26]. Pathogenicity roles in Xanthomonas genus differ greatly depending on specie, e.g. in *Xanthomonas campestris*, xanthan suppresses induced innate immunity by calcium chelation [27]. In addition xanthan increases the plant susceptibility to *X. campestris* due to avoiding the callose deposition [28]. In Xac there is controversy regarding the direct participation of xanthan in the pathogenicity process, while some authors find just a discrete participation in the epiphytic survival [29], another study shows that xanthan deletion mutants reduce the surface leave colonization ability and consequently the severity of citrus canker disease was deeply reduced [30]. Xanthan is a key component in the biofilm formation. The gene cluster *gum* is responsible for the xanthan production and exportation. This gene cluster comprise 12 successive genes with one operon-like identical direction of transcription i.e. *gumB* to *gumM*. The first two genes of cluster *gumB* and *gumC* encode components of channel than spans the outmembrane and the periplasmic space and enable the xanthan secretion [31].

The LPS is the major component of the outer leaflet of the outer membrane. The LPS in *Xcc* have a classic conformation being a tripartite glycoconjugate forming by: lipid A that carries a core oligosaccharide and polysaccharide the O- antigen. LPS that lack the O-antigen are named as lipooligosaccharide (LOS) or rough-type LPS. LPS has an essential role in bacterial growth acting as a barrier for antibacterial compounds and delivering protection against stress as well contributing to the structural proprieties of outer membrane. Lipid A is fairly conserved in most gram-negative bacteria, however, in Xanthomonas genus there is variation in the core oligosaccharide and O antigen structures, there may even be variation between the different species of Xanthomonas [32]. Nowadays is has been established that LPS has a double role in plant-microbial interaction; (i) elicitor of immunity plant response and (ii) It has a role in the promotion of virulence, because it acts as a barrier against antimicrobial activity compounds produced by root hair. *Xcc* is able to overwhelming plant defense responses induced by LPS.

3.1.5 Quorum sensing and biofilm formation

One discovery in microbiology that completely changed the conception of microbial ecology in the last two decades was the establishment of cooperative behavior in bacterial populations. This social behavior allows members of the bacterial community to adapt to new ecological niches, colonize new habitats, gain a competitive advantage against potential competitors and resist or avoid the host defense [33]. This cooperative behavior is based on a cell to cell communication system known as Quorum Sensing. Quorum sensing (QS) is a system of bacterial cell–cell communication that enables the microorganism to sense a minimum number of cells (quorum) in order to respond to external stimuli in a concerted fashion [34]. The process of QS relies upon the production, release and detection of small signaling molecules called auto-inducers. Each bacterial cell produces a basal quantity of auto-inducers, which are exported to the extracellular environment and reflect bacterial population density. At high cell densities, the auto-inducers reach a critical concentration, at which point they are recognized by their cognate receptor, triggering a cascade of biological functions [35].

The autoinducer in *Xcc* is a short chain fatty acid molecule known as DSF (Diffusible Signal Factor). Once this DSF accumulates at the extracellular space up to a critical level, it is sensed by its cognate receptor and triggers a cascade of biological function via the internal second messenger cyclic di-GMP, which is involved in virulence, resistance and biofilm formation. The encoding genes for quorum sensing components in *Xcc* form a cluster termed as *rpf* (Regulation of Pathogenicity Factors). For detailed revision of DSF quorum sensing circuit in *Xcc* [36].

Once *Xcc* reaches a leaf surface, it begins the initial adhesion process that was mention above. This attachment is followed by the formation of biofilm-like structures. Biofilm classical definition is an aggregated composed by several bacterial communities, which are embedded in a self-produced matrix of EPS, these bacterial cells are attached to each other or/and to a surface [37]. Biofilm is composed by polysaccharides, nucleic acids (eDNA), proteins, and have a pivotal role in attachment and protection against biotic and abiotic factors. In *Xcc* the biofilm formation in leaf and fruit surfaces is a main virulence factor in the early stage of development of citrus canker disease. In *Xcc* biofilm formation and dispersion is modulated by the quorum sensing autoinducer molecule DSF. How it was mention before DSF autoinducer promotes the biofilm formation because it stimulates the EPS production and pilus ensemble. On the other hand, DSF negatively regulates the biofilm formation because; it upregulates β 4–1 mannanase, ManA, leading to EPS dispersion and disassembly of biofilm [38]. Our previous study shown that quorum sensing signaling plays an essential role in the epiphytic stage survival, which is crucial at the early phase of pathogenicity development. Since, quorum quenchers bacteria belonging to genus Pseudomonas and Bacillus, it which were isolated from leaves of susceptible citrus host, which displayed the ability to disrupt the DSF pathway in *Xcc* and reduce citrus canker severity in a high susceptible citrus host [34].

3.1.6 T4SS and T6SS potentiates the *Xcc* antagonism with bacteria inhabiting the phylloplane and the soil amoeba

Nutrient limitation in the phyllosphere additional to environmental changes conditions, make the surface of the leaves one of the most hostile, restrictive and competitive habitats [38]. The type IV protein secretion system is used by bacteria to inject proteins and/or DNA into the prokaryotes and eukaryotes targets. Xanthomonas are endowed with genes that encode components of T4SS, the encoding genes VirB7, VirB8 and VirB9 responsible for the outer membrane pore formation. Genes that encode for VirB3, VirB4, VirB6, VirB8, Vir11, VirD4 and VirB10, responsible for the pore formation at the inner membrane. Finally, the gene that encodes for the subunits VirB2 and VirB5 that form the extracellular pilus structure. Besides, the encoding gene for VirB1 subunit predicted as a periplasmic lytic transglycosylase that plays a role in peptidoglycan alteration throughout T4SS biogenesis [39].

A recent study shows that in *Xcc* there are near to 12 proteins that interact with inner membrane associated ATPase VirD4, that is responsible for the recognition of substrates to be secreted [40]. These proteins share a C terminal domain termed XVIPCDs (Xanthomonas VirD4-interacting proteins conserved domains). These proteins are translocated into the target bacteria cell resulting in the dead of the receptor cells [41]. This bactericidal T4SS is knowing as X-T4SS and the effectors secreted by this nanomachine are termed X-Tfes (Xanthomonadales likeceeae t4SS effectors). Finally, a recent study reported that T6SS protect *Xcc* against the predatory amoeba Dyctiostelium [42].

3.2 Microbe -host interaction inside the susceptible host

Once the bacterium *Xcc* reaches the mesophilic tissue, after of epiphytic fitness and survival events mention before, must have to face the host defense response and parallel to express the pathogenicity factors;

3.2.1 T3SS the main pathogenicity determinant

The type 3 Secretion systems T3SS is the main protein secretion system widely studied in relationship to the pathogenicity. This secretion system is shared with several pathogenic bacteria ranging from animal to plants. This system is known as the "needle" and it works by delivering effector proteins directly to the target cells and modifying their behavior. Effectors from *Xcc* strains determine the host range. i.e. avirulence factors limit the specificity at the pathogen race/cultivar level by triggering immunity reactions in hosts with a related specific resistance gene. [43]. The effector delivered by the T3SS in *Xcc* belongs to the AvrBs3/PthA family. *Xcc* contains four PthA genes that encode transcription activator-like effector (TALE); of these four genes, pthA4 is responsible for the formation of citrus canker lesions. In citrus host the gene known as CsLOB is targeted by the TALE encoded by the *Xcc* gene pthA4; this gene was assessed in two susceptible host to *Xcc* infection, i.e., grape fruit and sweet orange [44]. CsLOB1-specific function still remains unclear; some previous studies suggest that CsLOB1 is involved in the regulation of development of lateral organ and metabolism of nitrogen and anthocyanin. Some plant hormones such as auxin, gibberellin, and cytokines also have proven to exert an effect on CsLOB1 gene [45]. Therefore, TALE have been shown to promote host cell transcriptional reprogramming as a virulence strategy [46].

4. Conclusions

The bacterium *Xcc* uses various adaptation and colonization strategies, it which are mainly aimed at guaranteeing its epiphytic survival, either by overcoming stress factors of biotic origin (predators, competitors, nutrient limitation) and abiotic origin (UV radiation, humidity and temperature variability). Because, this phase of epiphytic adaptation is crucial for the subsequent development of citric cancer symptoms in the susceptible host. Despite, these mechanism not having a direct effect on the health of the host, they become virulence factors, since its abolition avoid the subsequent development of the characteristic symptoms of citric cancer. Already inside the host, the bacterium uses as the main direct pathogenicity factor, the inoculation of effector proteins TALE, this effector is responsible for inducing cell hyperplasia, leading to rupture of the leaf epidermis and resulting in raised corky and spongy lesions surrounded by a water-soaked margin, the pathognomonic lesson of bacterial citrus canker.

References

1. Pfeilmeier S, Caly DL, Malone JG. Bacterial pathogenesis of plants: future challenges from a microbial perspective: Challenges in Bacterial Molecular Plant Pathology. Mol Plant Pathol. 2016 Oct;17(8):1298-1313. doi: 10.1111/mpp.12427
2. Vorholt JA. Microbial life in the phyllosphere. Nat Rev Microbiol. 2012 Dec;10(12):828-840. doi: 10.1038/nrmicro2910. PMID: 23154261
3. Bull CT, De Boer SH, Denny TP et al (2012) List of new names of plant pathogenic bacteria (2008 2010). J Plant Pathol 94(1):21-27
4. Hasse, C.H. (1915) Pseudomonas citri, the cause of citrus canker – a preliminary report. J. Agric. Res. 4, 97-100
5. Dowson, W.J. (1939) On the systematic position and generic names of the gram negative bacterial plant pathogens. Zentr. Bakteriol. Parasitenk. Abt. II. 100, 177-193
6. Young, J.M., Dye, D.W., Bradbury, J.F., Panagopoulos, C.G. and Robbs, C.F. (1978) Proposed nomenclature and classification for plant pathogenic bacteria. N. Z. J. Agric. Res. 21, 153-177
7. Gabriel, D.W., Kingsley, M.T., Hunter, J.E. and Gottwald, T. (1989) Reinstate- ment of Xanthomonas citri (ex Hasse) and Xanthomonas phaseoli (ex Smith) to species and reclassification of all Xanthomonas campestris pv citri strains. Int. J. Syst. Bacteriol. 39, 14-22
8. Vauterin, L., Hoste, B., Kersters, K. and Swings, J. (1995) Reclassification of Xanthomonas. Int. J. Syst. Bacteriol. 45, 472-489
9. Constantin, E.C., Cleenwerck, I., Maes, M., Baeyen, S., Van Malderghem, C., De Vos, P. and Cottyn, B. (2016) Genetic characterization of strains named as Xanthomonas axonopodis pv. dieffenbachiae leads to a taxonomic revision of the X. axonopodis species complex. Plant Pathol. 65, 792-806
10. German Dunger, Cristiane R. Guzzo, Maxuel O. Andrade, Jeffrey B. Jones, and Chuck S. Farah (2014) Xanthomonas citri subsp. citri Type IV Pilus Is Required for Twitching Motility, Biofilm Development, and Adherence Molecular Plant-Microbe Interactions 27:10, 1132-1147
11. Ribbe, J., Baker, A. E., Euler, S., O'Toole, G. A. & Maier, (2017) B. Role of cyclic Di-GMP and exopolysaccharide in type IV pilus dynamics. J. Bacteriol. 199, e00859–e00816
12. Ellison, C. K. et al (2018). Retraction of DNA-bound type IV competence pili initiates DNA uptake during natural Sheetz, M. pili enables transformation in Vibrio cholerae. Nat. Microbiol. 3, 773-780
13. Henrichsen, J. 1983. Twitching motility. Annu. Rev. Microbiol. 37:81-93
14. Moreira LM, de Souza RF, Almeida Jr NF, Setubal JC, Oliveira JC, Furlan LR, Ferro JA, da Silva AC (2004) Comparative genomics analyses of citrus-associated bacteria. Annu Rev Phytopathol 42:163-184
15. Leo, J. C., Grin, I., and Linke, D. (2012). Type V secretion: mechanism(S) of autotransport through the bacterial outer membrane. Philos. Trans. R. Soc. B Biol. Sci. 367, 1088-1101. doi: 10.1098/rstb.2011.0208

16. Drobnak, I., Braselmann, E., Chaney, J. L., Leyton, D. L., Bernstein, H. D., Lithgow, T., et al. (2015). Of linkers and autochaperones: an unambiguous nomenclature to identify common and uncommon themes for autotransporter secretion. Mol. Microbiol. 95, 1-16. doi: 10.1111/mmi.12838

17. Grijpstra, J., Arenas, J., Rutten, L., and Tommassen, J. (2013). Autotransporter secretion: varying on a theme. Res. Microbiol. 164, 562-582. doi: 10.1016/j. resmic.2013.03.010

18. Fan E, Chauhan N, Udatha DBRKG, Leo JC, Linke D. Type V Secretion Systems in Bacteria. Microbiol Spectr 2016;4. https://doi.org/10.1128/microbiolspec. VMBF-0009-2015

19. Bogdanove AJ, Koebnik R, Lu H, Furutani A, Angiuoli SV, Patil PB, et al. Two new complete genome sequences offer insight into host and tissue specificity of plant pathogenic Xanthomonas spp. J Bacteriol 2011;193:5450-5464

20. Mhedbi-Hajri N, Darrasse A, Pigné S, Durand K, Fouteau S, Barbe V, et al. Sensing and adhesion are adaptive functions in the plant pathogenic xanthomonads. BMC Evol Biol 2011;11:67

21. Park, Y.J., Song, E.S., Noh, T.H., Kim, H., Yang, K.S., Hahn, J.H. et al. (2009) Virulence analysis and gene expression profiling of the pigment-deficient mutant of Xanthomonas oryzae pathovar oryzae. FEMS Microbiol Lett 301: 149-155

22. Rajagopal, L., Sundari, C.S., Balasubramanian, D., and Sonti, R.V. (1997) The bacterial pigment xanthomonadin offers protection against photodamage. FEBS Lett 415: 125-128

23. Andrewes, A.G., Jenkins, C.L., Starr, M.P., Shepherd, J., and Hope, H. (1976) Structure of xanthomonadin I, a novel di- brominated aryl-polyene pigment produced by the bacterium Xanthomonas juglandis. Tetrahedron Lett 17: 4023-4024

24. Poplawsky, A.R., and Chun, W. (1997) pigB determines a diffusible factor needed for extracellular polysaccharide slime and xanthomonadin production in Xanthomonas campestris pv. campestris. J Bacteriol 179: 439-444

25. Poplawsky, A. R., Urban, S. C., & Chun, W. (2000). Biological role of xanthomonadin pigments in Xanthomonas campestris pv. campestris. Applied and environmental microbiology, 66(12), 5123-5127.doi.org/10.1128/aem.66.12.5123-5127.2000

26. BeckerA, Vorholter F-J. Xanthan biosynthesis by Xanthomonas bacteria: an overview of the current biochemical and genomic data. In: Bernd H. A. Rehm (ed). Microbial Production of Biopolymers and polymer Precursors: Applications and Perspectives. UK: Caister Academic Press, 2009, 1-12

27. Aslam SN, Newman MA, Erbs G et al. Bacterial polysaccharides suppress induced innate immunity by calcium chelation. Curr Biol 2008;18:1078-1083

28. Yun MH. Xanthan induces plant susceptibility by suppressing callose deposition. PLANT Physiol 2006;141:178-187

29. Dunger G, Relling VM, Tondo ML et al. 2007. Xanthan is not essential for pathogenicity in citrus canker but contributes to Xanthomonas epiphytic survival. Arch Microbiol;188:127-135

30. Rigano LA, Siciliano F, Enrique R et al. 2007. Biofilm Formation, Epi- phytic Fitness, and Canker Development in Xanthomonas axonopodis pv. citri. Mol Plant-Microbe Interact;20:1222-30

31. Bianco MI, Jacobs M, Salinas SR et al. 2014. Biophysical characterizaion of the outer membrane polysaccharide export protein and the polysaccharide co-polymerase protein from Xanthomonas campestris. Protein Expr Purif;101:42-53

32. Molinaro A, Silipo A, Lanzetta R et al. 2003. Structural elucidation of the O-chain of the lipopolysaccharide from Xanthomonas campestris strain 8004. Carbohydr Res;338:277-281

33. Ng WL, Bassler BL, 2009. Bacterial quorum-sensing network architectures. Annual Review of Genetics 43, 197-222

34. Caicedo JC, Villamizar S, Ferro MIT, Kupper KC, Ferro JA. (2016). Bacteria from the citrus phylloplane can disrupt cell–cell signalling in Xanthomonas citri and reduce citrus canker disease severity. Plant Pathology.;65:782-791

35. Federle MJ, Bassler BL, 2003. Interspecies communication in bacteria. Journal of Clinical Investigations 112, 1291-1299

36. Juan Carlos Caicedo, Sonia Villamizar and Jesus Aparecido Ferro (2017). Quorum Sensing, Its Role in Virulence and Symptomatology in Bacterial Citrus Canker, Citrus Pathology, Harsimran Gill and Harsh Garg, IntechOpen, DOI: 10.5772/66721

37. Vert, M. et al. Terminology for biorelated polymers and applications (IUPAC Recommendations 2012). Pure Appl. Chem. 84, 377-410 (2012)

38. Lindow SE, Brandl MT, 2003. Microbiology of the phyllosphere. Applied and Environmental Microbiology 69, 1875-1883

39. Ilangovan A, Connery S, Waksman G. (2015). Structural biology of the Gram-negative bacterial conjugation systems. Trends Microbiol;23:301-310

40. Sgro GG, Oka GU, Souza DP, Cenens W, Bayer-Santos E, Matsuyama BY, et al. 2019. Bacteria-killing type IV secretion systems. Front Microbiol;10:1078

41. Alegria MC, Souza DP, Andrade MO, Docena C, Khater L, Ramos CHI, et al. 2005. Identification of new protein-protein interactions involving the products of the chromosome- and plasmid-encoded type IV secretion loci of the phytopathogen Xanthomonas axonopodis pv. citri. J Bacteriol;187:2315-25

42. Bayer-Santos E, Lima L dos P, Ceseti L de M et al. Xanthomonas citri T6SS mediates resistance to Dictyostelium predation and is regulated by an ECF σ factor and cognate Ser/Thr kinase. Environ Microbiol 2018;20:1562-1575

43. He YQ, Zhang L, Jiang BL et al. 2007. Comparative and functional genomics reveals genetic diversity and determinants of host specificity among reference strains and a large collection of Chinese isolates of the phytopathogen Xanthomonas campestris pv. campestris. Genome Biol;8:R218

44. Yang H, Junli Z, Hongge J, Davide S, Ting L, Wolf BF, et al (2014). Lateral organ boundaries 1 is a disease susceptibility gene for citrus bacterial canker disease. Proceedings of the National Academy of Sciences of the United States of America.;111:E521-E529

45. Majer C, Hochholdinger F. (2011). Defining the boundaries: Structure and function of LOB domain proteins. Trends in Plant Science.;16(1):47-52. DOI: 10.1016/j.tplants.2010.09.009

46. Peng Z, Hu Y, Zhang J, Huguet-Tapia JC, Block AK, Park S, et al. (2019). Xanthomonas translucens commandeers the host rate-limiting step in ABA biosynthesis for disease susceptibility. Proc Natl Acad Sci;116:20938-46. https://doi. org/10.1073/pnas.1911660116

Chapter-8

Climate Change and Citrus

1. Introduction

Citrus and its related genera i.e., Poncirus, Eremocitrus, Fortunella, and Microcitrus belong to the family Rutaceae [1, 2]. Citrus is a prominent fruit tree of tropical and sub-tropical regions that require a suitable climate for quality production. Citrus fruit quality and quantity are inclined by multiple factors including climatic conditions [3]. Change in optimum climate elements like low temperature/freezing, heat stress/heatwaves, CO_2 assimilation, drought/water scarcity, intensive rainfall, and relative humidity, may affect directly and indirectly citrus production [4].

Citrus tree (rootstock and scion) growth, development, fruit production, and fruit quality is reduced under the biotic and abiotic stresses [5]. Citrus with tolerant rootstocks against biotic and abiotic factors improve the growth and productivity of the trees [6]. The potential citrus yield is 18–20 tones ha^{-1}, which goes up to 25 tones ha^{-1} in the developed world; however, the citrus average yield in Pakistan is 12–10 tones ha^{-1} and is affected by abiotic and biotic stresses [7]. The yield gap is due to biotic, abiotic, and general factors, like agronomic practices in countries of climate risk [8].

The productivity and growth of plants are affected by climate change especially drought and high temperatures collectively [9]. Reactive oxygen species accumulate superoxide and hydrogen peroxide [10] is due to water stress and high-temperature stress which reduce the biochemical, physiological, and molecular regulation. Reduction in carbohydrate accumulation affects the flowering, fruit set, and fruit yield. However, to reduce the negative plant physiological stresses, there should be good management practices in citrus orchards. Choice of better scion enhances citrus trees to produce higher yield with good fruit quality [11].

Citrus has a phenological life cycle of the whole year, starting from February to next year January. Flowering starts during February–March in subtropical regions and is generally considered a critical period for citrus production. An increase in temperature and water stress after pollination inhibits ovule fertilization [12], which in return reduces tree fruit set, increases June fruit drop, and reduces tree yield [13, 14, 15]. Fruit growth phases from button size to mature fruit are more sensitive to heat stress and deficit irrigation. Citrus under water deficit conditions faces reduced fruit growth and ripening, which is associated with a decrease in fruit size, an increase in fruit acidity [16], and low tree yield

[13]. Water stress at the pre-harvest stage in oranges develops fruit peel wrinkles [17]. An increase in optimum temperature at fruit ripening causes pre-harvest fruit drop and reduced yield (Figure 1).

Figure 1. Key phenological stages and management activities [18].

To deal with heat and water deficit stress, there is a need to improve agronomic management practices and adopt breeding and molecular approaches. Agronomic management practices encompass factors like irrigation, nutrition, pruning, pests, diseases, and other injuries which have a key role in citrus fruits quantity and quality [19]. Breeding approaches need to search out/develop rootstocks that are tolerant/resistant against abiotic and biotic stresses. Based on breeding techniques, better rootstocks can be developed that can mitigate climate risk and other major biotic factors [20]. Molecular approaches are very helpful to deal with heat and water deficit conditions. Modulation in genes and gene expression related to stress help plants to cope and mitigate stress adversity. Henceforth, somatic hybridization, mutation, somaclonal variation, and genetic transformation techniques help to improve trees for thermotolerance [21].

Thus, in this chapter, we present an overview of climate change i.e., heat and drought stress impact on citrus and its management through agronomic, breeding, and molecular approaches.

2. Climate change

The main reason behind climate change is greenhouse gasses; especially carbon dioxide accumulation in the environment. Fossil fuel burning is the primary source of greenhouse gasses emission. The use of pesticides in agriculture and cutting of forests are also contributing to the proliferation of such gasses that cause climate change. An optimum amount of these gasses is necessary for controlling the earth's temperature, but now their concentration is increasing dramatically. From the expected beginning of human civilization about thousand years ago to 1900, the carbon dioxide concentration in the environment was 0.03%, but now due to climate change, it has been reached to 0.04%, the highest in history [22].

2.1 What is the effect of climate change?

The earth's mean temperature has risen for the past hundred years [23]. The increase in temperature of the earth due to climate change can affect the environment adversely. Today, the average temperature is 4°F more in comparison to the last Ice Age [24]. Global warming is causing the melting of polar caps and warming the ocean's water, which is leading to greater storms and frequent floods along with heavy winds and rains. A heat rise is also enhancing the incidences of wildfires, which damage natural habitat and creatures [25]. Climate change threatens the world's population. The world is severely facing the issue of climate change, especially the third world countries. American and European countries are prepared well against climate change [26]; however, the countries of the Middle East, Asia, and Africa are more exposed to environmental changes due to less preparedness and technology to tackle these issues [27]. Norway would be the country likely to survive climate change due to its low vulnerability against climate change. The neighboring countries: Finland (third), Sweden (fourth), Denmark (sixth), and Iceland (eighth) are well prepared. The countries least likely to survive global warming change include the Central African Republic, South Africa, Eritrea, Chad, Somalia, and the Democratic Republic of the Congo. These countries have poor infrastructure, unstable governance, poor health, and food and water scarcity (Figure 2).

Figure 2. The map shows the countries most at risk and least at risk against climate change [28].

2.2 Rise in temperature

NASA center graph associated with climate (Figure 3) indicates the average global surface temperature during the era of 1880–2018. After 1940, an abrupt increase in temperature was noted for a duration of two years and then continuous high temperature was witnessed after 1980–2016 [28]. The researchers believe that global temperature will rise continuously over the next few decades, mainly due to humans generated greenhouse gases. The IPCC (The Intergovernmental Panel on Climate Change) predicts a rise of 2.5–10°F over the next century [29].

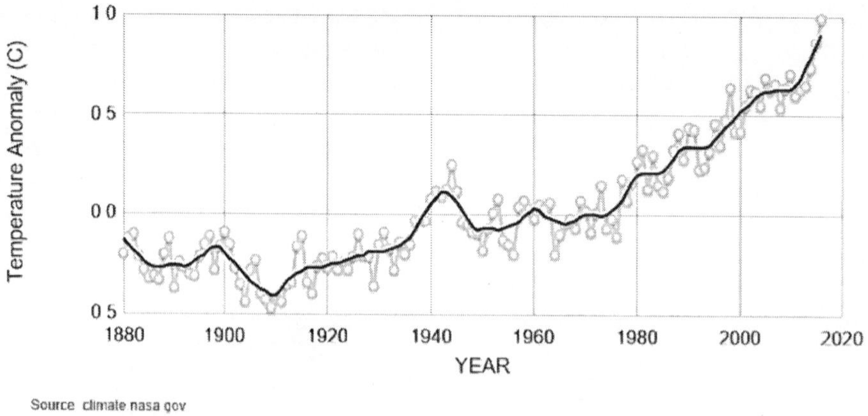

Figure 3. Global surface temperature anatomy 1880–2018 [28].

2.3 Water scarcity

Water is highly important for plants and its global importance is not difficult to understand. There is a frequent rise in water scarcity due to changes in climatic patterns. It is expected that the world will face a decrease of 66% in water availability up till 2050. The water cycle is adversely affected by climate change. Due to the changing climate, several areas are getting dry. There are 17 nations under the extremely high risk of water scarcity; out of which 12 are in North Africa and Middle East [29]. India and Pakistan, two Asian countries, fell in the list of 17 countries having a risk of water scarcity (Figure 4).

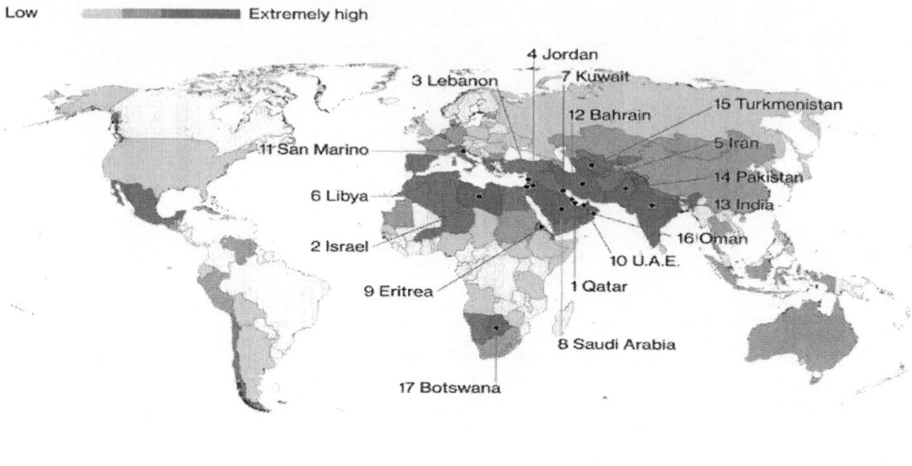

Figure 4. Top 17 countries facing the risk of extremely high water stress [29].

3. Effect of heat and water deficit on tree health

The yield of any crop begins to decrease when the temperature exceeds the ideal temperature range and the water level falls below the ideal water demand of the crop. Temperature and precipitation variables of climate are described as diagrammatic sketch alternatively for intensity and duration of drought, which show a small portion of the climate space presently exceeding tree mortality threshold (Figure 5). It is predicted that there will be high temperature and drought due to extreme climate change, which can cause severe damage to agriculture and could become a risk for tree populations [31].

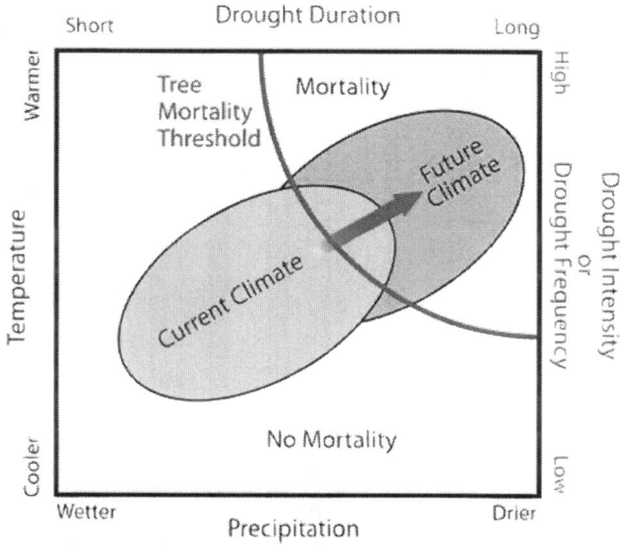

Figure 5. Climatic variables, temperature, and precipitation, with a range of variability [30].

3.1 Tree physiology

Plant physiology includes all the dynamic processes of growth, metabolism, reproduction, defense, and communication responsible for plant survival [32, 33]. Heatwaves affect the plant's physiological processes and responses, their ability to tolerate heat, as well as the effectiveness of strategies used for thermotolerance improvement [34]. In citrus fruits, the temperature above than optimum causes a big difference in the leaf to air vapor ratio as well as high leaf temperature, but the shade conditions can relieve the water pressure and lower the temperature of the leaves [35]. Stomatal conductance and carbon dioxide uptake are reduced by superoptimal leaf temperature and water tension [36].

The gas exchange activity is reduced badly in citrus trees under deficient irrigation [37, 38]. CO_2 assimilation, conductivity, and transpiration rate decrease under water stress, so these gas exchange parameters are a water stress indicator [39, 40, 41]. Citrus trees under water deficit conditions reduce the conductivity of stomata and increase photorespiration [42], which reduces the yield, size, and quality of fruits [43]. Under groundwater deficiency, citrus trees lead to stomata closure associated with high ABA content and result in a sudden decrease in photosynthesis [16, 44, 45] and production losses [14, 35]. Chlorophyll *a* is easily damaged compared to chlorophyll *b* due to lack of water. Genotypes/species that maintain stomata conduction under dry conditions also maintain chlorophyll fluorescence and high growth levels [46].

3.2 Tree morphology

Altered temperatures and water deficit conditions affect citrus leaf-to-air vapor pressure during the day in the early morning and midday [8], indicating a vapor pressure of 4.3 kPa at 37–40°C and 6.2 kPa at ≥40°C, respectively. Citrus exposed to the temperature above than optimum (37°C) and vapor pressure deficit (3.6 kPa) at 330 μmols^{-1} CO_2 concentration during midday, shows depression in carbon dioxide exchange rate [47]. Swingle citrumelo, a citrus species increased its total biomass when kept under slightly high temperature [8]. Drought stress affects both plant vegetative and reproductive growth parameters [48, 49]. Citrus under water deficit lessens vegetative growth, fruit size and quality and orchards face a major economic loss [14]. Orange trees exposed to prolonged or excessive water deficit can lead to leaf drop, gradual drying of the tips of the branches, and a drastic decline in fruit production due to severe flower and fruit drop [50]. Young lemons exposed to water stress showed a decrease in daily stem diameter and water flow [51]. The growth of plant roots is dependent on the soil water availability. The roots of irrigated soils are well distributed and widespread compared to roots with less irrigation. Valencia orange roots on the Swingle citrumelo rootstock and a significant difference in root distribution between irrigated and non-irrigated trees were observed [52]. Water stress decreases the growth and metabolism of citrus fruits [43, 53] and increases the cost of extracting juice [21]. Dryness also reduces the thickness of fruit peel, making citrus fruits more sensitive to damage during handling and transportation [54]. Irrigation stoppage during initial and final growth phases of Lane Late orange (*Citrus sinensis* Osbeck) significantly reduced yield [55]. General studies of water stress in citrus show that the extent and duration of water stress at critical development stages are more vulnerable to the production of citrus. On the other hand, the cultivar and properties of orchards *i.e.*, soil, climate, and cultivation also play important role in success under deficit irrigation [56].

3.3 Tree water status

The transport of water is determined inside the plant by the availability of soil water and relative humidity. Plant physiological adjustments under changing environmental conditions maintain the turgor pressure of the plant cell. In perennial species, seasonal variations in environmental conditions can affect water relationship. In citrus fruits, the large crown and low hydraulic conductivity of the trunk and roots contribute to severe water scarcity [57]. Transient water deficit in citrus at midday [58] reduces photosynthetic rates [47]. Root hydraulic conductivity decreases under drought stress to prevent the plants from mortality. High temperatures can also increase the loss of root moisture to a harmful level [59]. In plants, heat stress appears as the supply of water is insufficient to meet the evaporation requirement. Heat stress is linked to drought as the plant and soil quickly lose water at high temperatures. It is known that heat stress and drought reduce nutrient uptake and photosynthetic efficiency in plants [60].

3.4 Tree biochemistry

Citrus leaf water potential and leaf abscisic acid (ABA) are the indicator of water stress. Citrus rootstock Rangpur lime grafted with scion Pera orange resulted in decreased leaf water potential and decreased leaf ABA concentration when subjected to water stress [61]. Citrus trees produce endogenous hormones and their regulation by promoting synthesis and accumulation under severe water stress [43]. Plant phytohormones are found in a minor quantity but drought stress accumulates jasmonic acid [62]. Drought synthesizes roots ABA and leaves by transpiration stream [43, 63]. The amount of sugar, like non-reducing sugar (sucrose) and reducing sugar (fructose and glucose), in contrast to the sorbitol content, decreases dramatically over the drought period. Water stress accumulates proline contents, an important osmoprotective agent, and its concentration increases with increasing water deficit conditions in citrus orchard [64]. Proline levels in leaf were recorded in Gada dahi citrus rootstock on day 24 of the water stress in comparison to tolerant rootstocks, which indicated that the accumulation of proline was greater in susceptible genotypes than in tolerant genotypes due to higher stress. Lower accumulation of proline was due to its protective function, removing radicals, maintaining the redox balance, and reducing cell damage [65]. The total phenol content also increases in plants under drought stress, compared to normal irrigated plants [66, 67]. Proteins are involved in several processes that change the plant metabolism under stress conditions and activate the plant defense signal [19, 68]. The protein content of drought-tolerant genotypes is generally higher than that of drought-sensitive genotypes. Carrizo citrange, a tolerant genotype, shows notable soluble proteins in leaves and roots [69, 70]. Higher MDA and H_2O_2 contents observed in plants under water stress indicate greater oxidative damage, which determines the severity of the plant and indicates low efficacy of antioxidant machinery of Carrizo citrange drought-tolerant rootstock [43]. Plants produce several antioxidant enzymes, such as CAT, SOD, and POD to treat the cell damage caused by stress at the oxidative level. SOD is the main enzyme that is expressed under stress, especially under water stress conditions. Carrizo citrange has shown an excellent defense mechanism under water stress with high activity of CAT, SOD, and POD in roots and leaves [19, 68].

3.5 Tree anatomy

Alteration in anatomy by applying heat stress is established. Stress treatment at 40–45°C was given to similar size plants and anatomical changes (size of the epidermis, size of pith, cortex, leaf thickness, epidermal cells, parenchyma tissue) in root, stem, and rhizomes were studied. The thickness of mesophyll, epidermis, and cortex was increased in stressed plants [71]. Some common anatomical changes include increased densities of stomata and trachomatous, cell size reduction, stomata enclosure, and higher xylem vessels in roots and shoots [72]. It has been demonstrated that grafted plant size is reduced on dwarfing rootstock, and such plants are unable to maintain drought or water-deficient conditions [73]. The researchers explain that the vessel density of root and stem are decreased with tree height [74]. The rootstock growth ability is dramatically affected by the number of xylem traits, xylem phloem ratio, vessel size, and vessel density [73, 75]. Maintaining hydraulic conductance of stem, root [74, 76], vessel size and number is the basic factor in hydraulic conductance maintenance [77]. Fewer small vessels may decrease hydraulic conductance, as a result, growth decrease in fruit trees [78]. Water stressed leaf spongy cells have a dense arrangement and reduce the conductivity of leaf diffusion. These results give an idea to understand the direct relationship between mesophilic conductivity and the porosity of the soils [79].

3.6 Tree genetics

Stress-related genes are activated through high temperature and drought, [80], and sugars, different functional proteins, amino acids, and amines are synthesized through these genes [81]. HSPs are the heat shock proteins consisting of a group of genes relevant to heat stress in plants and animals [82, 83]. Heat shock proteins play an important role in maintaining/removing ROS, cell membrane integrity, producing antioxidants, and osmolytes [84, 85]. Heat shock proteins protect plant cells/tissues from drought and heat stress [84]. Citrus HSP70 expression has been examined against water scarcity and high-temperature stress in the *Poncirus trifoliata* rootstock. In *P. trifoliata* HSP70 and HSP90 genes against abiotic stress are upregulated [86]. HSP90s play a vital role in signal transduction, cell cycle regulation, protein breakdown, genomic mutation, and protein trade [81, 87]. Aquaporins are transmembrane channel proteins found in tonoplasts, plasma membranes, and other intracellular membranes and are abundantly expressed in plant roots [88, 89]. Major intrinsic proteins (MIPs) are a superfamily of aquaporins that regulate intracellular water passage [90]. The plasma membrane proteins (PIPs) are the most important group of natural proteins that respond to water transport. Overexpression of PIP under abiotic stress conditions confirms the importance of PIP for heat and water stress tolerance [91] as the combination of heat stress and the scarcity of groundwater generally limits the physiology, growth, and productivity of plants [92].

3.7 Tree productivity

The citrus phenological cycle starts from February to next year January in subtropical regions. Flowering starts during February–March and is generally considered a critical period for fruit production. An increase in temperature and water stress after pollination inhibits ovule fertilization [12] which reduces tree fruit set, increases fruit drop, and reduces tree yield [13]. Phases of fruit growth from button size to mature fruit are more sensitive to deficit irrigation and heat stress. Hence, reduced fruit growth, and delayed ripening occur which are associated with a decrease in fruit size, increase in fruit acidity, and low tree yield [13]. Drought at a pre-harvest stage in oranges develops wrinkle on fruit peel [17]. An increase in optimum temperature at fruit pre-harvest causes fruits drop and reduced yield. Citrus under different phenological stages respond to deficit irrigation or water stress and contribute negatively to yield/production and fruit quality. In an experiment, eleven-year-old sweet orange scion grafted on Carrizo citrange were evaluated against water stress and revealed 10–12% relative yield decline. Gonzalez [37] compared Clementina (*Citrus clementina*) tree performance under 25–50% deficit irrigation during initial fruit enlargement and pre-maturation phases and recorded a significant negative effect on fruit yield [13]. Navelina sweet orange (*Citrus sinensis* Osbeck) yield reduced significantly when irrigation was reduced at %55 with respect to crop water requirement during flowering and fruit set.

4. Management of citrus under climate change

4.1 Agronomic management

Implementation of proper orchard management practices decreases the adverse effects of heat and drought stress. The management includes trees requirement based nutrition and irrigation techniques, organic and synthetic mulches, as well as selecting the most suitable cultivars/rootstocks that are resistant to various stresses.

The selection and development of new rootstocks tolerant to biotic and abiotic stress is inevitable for the stable production of citrus under the scenario of climate change. New and known diseases and environmental conditions also help to force developing new citrus rootstocks according to the demand [93]. Citrus rootstocks like Volkamer lemon (*C. volkameriana*), Rangpur lime (*C. limonia*), and Rough lemon (*C. jambhiri* Lush.) resist water stress and increase the production of cultivars grafted on these.

Fertilizer application can also be helpful to manage plants against abiotic stresses [94]. Application of Ca and K macronutrients and B and Mn micronutrients modify the function of stomata under heat/high-temperature stress [95]. K, Ca, B and Mn activates physiological and metabolic processes that help maintain a high water potential in tissues, which increases tolerance to heat stress [96]. The use of N, K, Ca and Mg also reduces the toxicity of ROS, thereby increase the levels of antioxidant enzymes in plant cells [96].

Plant growth regulators (PGRs) in managing water and heat stress also play an important role. PGRs like cytokinins, abscisic acid (ABA), and salicylic acid play role in resistance to heat and drought. The application of PGRs increases the water potential and chlorophyll content in citrus trees [97]. The exogenous use of ABA increases productivity in the absence of water [98]. ABA formulations are available with commercial manufacturers to improve the drought tolerance of trees [99, 100].

Mulching underneath the trees is often used as a technique for water conservation [101, 102]. Mulches are used to maintain moisture levels high in the soil, control soil temperature, and evaporation [93, 103]; thereby reduce the need for irrigation during growing seasons [104]. The need for water in the soil is decreased and the ability to withstand drought and heat is increased by using mulches [105, 106]. Plastic films are more effective than organic compost for groundwater protection [107].

4.2 Breeding strategies

Citrus rootstock breeding programs are aimed to combine biotic and abiotic tolerance/resistance in new rootstocks. However, conventional plant breeding (Figure 6) in mitigating the abiotic stresses has limited success against plant productivity [108]. Similarly, developing better rootstocks through breeding by the conventional method is a long-term approach due to many difficulties, particularly the complexity of citrus biology (high heterozygosity, long juvenility, polyembryony) [109, 110]. Typically, from a breeding program, it takes at least 15 years for a new standard variety to emerge in the citrus industry. Moreover, a sexual hybrid is difficult to identify at an early stage. In this case, trifoliate leaf (a morphological marker) is used as a male parent and unifoliate as a female for identification of sexual hybrid at the seedling stage [111]. The trifoliate trait is dominant, and the seedlings showing the parental trifoliate pollen phenotype are considered hybrids. In absence of a trifoliate pollen parent, the hybrids can be identified by using SSR and RAPD molecular markers [112].

Figure 6. Traditional cross-hybridization in citrus. (A) a large unopened bud, (B) emasculation, (C) pollination of the emasculated flower, (D) bagging of the pollinated flower, (E) bagged twig, (F) general view of the seed parent after crossing.

Besides citrus biological constraints, valuable traits like resistance to low/high temperature, root rot, viruses, nematodes, salinity, and drought are important to incorporate in rootstocks. A list of some important biotic and abiotic traits of citrus rootstocks are presented in Table 1, which can help in the development of new tolerant/ resistant rootstocks.

Table 1. Some biotic and abiotic traits of selective citrus rootstocks [113].

Rootstocks	Alkalinity	Chloride	Drought	Tristeza	Nematodes	Phytophthora	Fruit quality
Cleopatra mandarin	T	TT	pT	T	S	pT	G
Chios mandarin	T	S	pT	T	S	SS	—
Sweet orange	T	pT	T	T	S	SS	G
Macrophylla	TT	TT	TT	S	S	T	L
Pomeroy Poncirus	SS	SS	S	R	T	T	H
C-35 citrange	S	S	T	R	T	T	H
4475 citrumelo	S	pT	pT	R	T	T	H
Citrandarin	pT	pT	—	R	pT	T	G

Abbreviations in the table: T; tolerant, TT; very tolerant, pT; poorly tolerant, S; susceptible, SS; very susceptible, G; good, L; low, H; high.

Several rootstock hybrids have been released to the citrus industry worldwide. Swingle citrumelo rootstock is a hybrid of Duncan grapefruit and Trifoliate orange, crossed by Swingle in 1907 and released in 1974. Since then, it has been used successfully as the standard rootstock for better traits *i.e.*, moderate drought tolerance and high fruit quality. Citranges are hybrids of Washington Navel orange and *Poncirus trifoliata*, out of which Carrizo and Troyer are the two main citranges. These can tolerate water shortages and produce excellent quality fruits. Benton citrange is a cross of Ruby blood orange and trifoliate orange developed in late 1940 and is more tolerant to heat and water scarcity [114]. Brazilian sour orange also shows tolerance against heat, drought, and their combined stress [85].

4.3 Biotechnological interventions

Hybridization by somatic approaches is a protoplast fusion process that has become an important tool for plant production, combine (partially or totally) desired cultivars somatic cells, species, or genera, resulting in the development of new genetic combination. In addition to intergenerational mixtures in somatic hybridization, more emphasis has been placed on interspecific mixtures between *C. reticulata* and *C. maxima* [109, 110] to meet the specific needs of the citrus industry. Poncirus is drought-prone, while Citrange C-35 is more drought-tolerant. Among these rootstocks, 4475 citrumelo have the best ability to adapt to the environment. Cleopatra mandarin + *Poncirus trifoliata* and Cleopatra mandarin + C35- Citrange somatic hybrids have resistance to CTV, tolerance to nematodes, and phytophthora. The Sweet orange + *Poncirus* and the Sweet orange + C35-, as well as the Sweet orange + Citrumelo 4475, can adapt to low moisture soils and tolerate biotic stresses. Macrophylla is a productive rootstock and adapts well to saltwater, limestone, and water stress [115].

In vitro mutagenesis and somaclonal variation are important tissue culture techniques being used in citrus improvement. Somaclonal variation, genetic and phenotypic variation between plants, can be used to improve citrus cultivars under conditions of water and heat stresses. Genetic improvement by *in vitro* selection of Satsuma mandarins (*Citrus unshiu* Marc.) has been made successfully; however, the frequency of somaclonal variation by factors, including genotypes, explant culture length, sources, and environmental composition [116, 117]. Cell lines success stories of some salt-tolerant cultivars are *C. sinensis* cv. "Shamouti" [118], *C. limonium* [119] and "Troyer citrange" [120].

Genetic transformation is an alternate technique for citrus genetic improvement. PEG-mediated genetic transformation of citrus fruits is a direct DNA transfer method [120] that seeks to express an aminoglycoside phosphotransferase II gene in isolated protoplasts from sweet orange (*Citrus sinensis* Osbeck) culture for suspension. The genetic transformation of citrus fruits has mainly been carried out from young materials such as embryogenic cells from the epicotyl segment of *in vitro* germinated seedlings. Excess protein for late embryogenesis (OHL), heat shock proteins, and certain transcription factors that affect the expression of various stress-related target genes have also been used to improve drought tolerance in transgenic plants. Drought-induced genes with different functions have been identified through molecular and genomic analyses in a variety of plant species such as the C/CBF family (Shinozaki) [121]. By regulating stress gene expression and signal transformation, plants indirectly become more stress-resistant [97, 122]. The TDF genes have been identified as drought-induced and the proteins encoded include fructose aldols bisphosphate, a cold-like protein found in WCOR413. The PIP2 protein, an aquaporin specializing in a water channel to transport water across the plasma membrane, and the tonoplast have been observed in sweet orange. TDF21, TDF38 and TDF80 are involved in the regulation of signal transduction and expression of genes. These are sensitive to stress and also regulate the expression of stress-induced genes, possibly induced by drought.

5. Future research strategies

Soil microbiome research offers the opportunity to improve abiotic stress in plants. The mechanisms by which plants recover from drought and/or heat stress can be mediated by microbes surrounding the plant, particularly the roots, and these are involved in various stages of plant growth. Advances in the application of new molecular and genomic tools and technologies have paved the way for the study of plant microbiota, and these promising advances enable the study of the biological functions of various microorganisms both inside and outside the host tissue.

Significant advances in the characterization of the plant genome and the optimization of techniques for manipulating the plant genome have contributed and will further improve our knowledge and ability to develop stress-resistant plants. Ultimately, genetic engineering or transgenic methods must be combined with conventional breeding activities and supported by markers in order to obtain the desired improved varieties.

Plants have developed complex adaptive mechanisms to withstand diverse and complex abiotic stresses. With the advent of new technologies such as genomics and genetic transformation, significant advances have been made in understanding these complex traits in higher plants. However, the commercial application of successful research results requires additional validation of the products or prototypes in the field.

These efforts will lead to tangible practical outcomes that may help mitigate the effects of climate change, especially concerning drought and heat stresses, and will contribute to improved crop productivity and food security.

6. Conclusion

Plants adaptation is considered a striking strategy to manage the impacts of climate change. In climate change, the most important factors are fluctuating patterns of temperature and drought which have an adverse effect on plants physiology, morphology, water status, biochemistry, anatomy, genetics, and productivity. Hence the emphasis should be on the development of production systems for improved water-use efficiency and to adapt to the hot and dry conditions through agronomic practices. Development of climate-resilient citrus rootstocks and scion through genomics and biotechnology are essentially required.

References

1. Ghada B, Amel O, Aymen M, Aymen A, Amel SH. Phylogenetic patterns and molecular evolution among 'True citrus fruit trees' group (Rutaceae family and Aurantioideae subfamily). Sci Hortic (Amsterdam). 2019;253:87-98. https://doi.org/10.1016/j.scienta.2019.04.011

2. Ristow M. Unraveling the truth about antioxidants: mitohormesis explains ROS-induced health benefits. Nat Med. 2014;20:709-711. https://doi.org/10.1038/nm.3624
3. Nawaz R, Abbasi NA, Hafiz IA, Khalid A, Ahmad T, Aftab M. Impact of climate change on kinnow fruit industry of Pakistan. Agrotechnology. 2019;8:2. https://doi.org/10.35248/2168-9881.19.8.186
4. Abobatta WF. Drought adaptive mechanisms of plants–a review. Adv Agr Env Sci. 2019;2:42-45
5. Liu G-H, Yang J-Y. Content-based image retrieval using color difference histogram. Pattern Recognit. 2013;46:188-198. https://doi.org/10.1016/j.patcog.2012.06.001
6. Rouphael NG, Stephens DS. Neisseria meningitidis: biology, microbiology, and epidemiology. In: Neisseria meningitidis. Springer; 2012. p. 1-20. https://doi.org/10.1007/978-1-61779-346-2_1
7. Naeem A, Ghafoor A, Farooq M. Suppression of cadmium concentration in wheat grains by silicon is related to its application rate and cadmium accumulating abilities of cultivars. J Sci Food Agric. 2015;95:2467-2472. https://doi.org/10.1002/jsfa.6976
8. Rosenzweig C, Jones JW, Hatfield JL, Ruane AC, Boote KJ, Thorburn P, et al. The agricultural model intercomparison and improvement project (AgMIP): protocols and pilot studies. Agric For Meteorol. 2013;170:166-182. https://doi.org/10.1016/j.agrformet.2012.09.011
9. Mittler R, Finka A, Goloubinoff P. How do plants feel the heat? Trends Biochem Sci. 2012;37:118-125. https://doi.org/10.1016/j.tibs.2011.11.007
10. Jaspers P, Kangasjärvi J. Reactive oxygen species in abiotic stress signaling. Physiol Plant. 2010;138:405-413. https://doi.org/10.1016/j.tibs.2011.11.007
11. Zekri M. Factors affecting citrus production and quality. Trees. 2011
12. Hsiao TC. Effects of drought and elevated CO 2 on plant water use efficiency and productivity. In: Interacting stresses on plants in a changing climate. Springer; 1993. p. 435-465. https://doi.org/10.1007/978-3-642-78533-7_28
13. García-Tejero I, Romero-Vicente R, Jiménez-Bocanegra JA, Martínez-García G, Durán-Zuazo VH, Muriel-Fernández JL. Response of citrus trees to deficit irrigation during different phenological periods in relation to yield, fruit quality, and water productivity. Agric Water Manag. 2010;97:689-699. https://doi.org/10.1016/j.agwat.2009.12.012
14. Guerra P, Romero PR. Pistol grip style elongated dissecting and dividing instrument. 2006
15. García O, Johnson SI, Seltzer K. The translanguaging classroom: Leveraging student bilingualism for learning. Caslon Philadelphia, PA; 2017. https://doi.org/10.21283/2376905x.9.165
16. Pérez-Pérez JG, Robles JM, Botía P. Influence of deficit irrigation in phase III of fruit growth on fruit quality in 'lane late'sweet orange. Agric Water Manag. 2009;96:969-974. https://doi.org/10.1016/j.agwat.2009.01.008

17. Goldhamer DA, Salinas M. Evaluation of regulated deficit irrigation on mature orange trees grown under high evaporative demand. In: Proceedings of internat soc citriculture IX congress. 2000. p. 227-231

18. Khurshid T. Project, the enhancement of citrus value chains production in Pakistan and Australia through improved orchard management practices. 2015. https://research.aciar.gov.au/aik-saath/sites/_co-lab.aciar.gov.au.pdf

19. Zandalinas SI, Balfagón D, Arbona V, Gómez-Cadenas A. Modulation of antioxidant defense system is associated with combined drought and heat stress tolerance in citrus. Front Plant Sci. 2017;8:953. https://doi.org/10.3389/fpls.2017.00953

20. Cimen B, Yesiloglu T. Rootstock breeding for abiotic stress tolerance in citrus. In: Abiotic and Biotic Stress in Plants-Recent Advances and Future Perspectives. IntechOpen; 2016. https://doi.org/10.5772/62047

21. Hasanuzzaman M, Nahar K, Fujita M. Extreme temperature responses, oxidative stress and antioxidant defense in plants. Abiotic Stress responses Appl Agric. 2013;13:169-205. https://doi.org/10.5772/54833

22. Buis A. A Degree of Concern: Why Global Temperatures Matter. NASA's Glob Clim Chang Website, June. 2019;19

23. Steffen W, Leinfelder R, Zalasiewicz J, Waters CN, Williams M, Summerhayes C, et al. Stratigraphic and Earth System approaches to defining the Anthropocene. Earth's Futur. 2016;4:324-345. https://doi.org/10.1002/2016EF000379

24. Wang H, Ni J, Prentice IC. Sensitivity of potential natural vegetation in China to projected changes in temperature, precipitation and atmospheric CO 2. Reg Environ Chang. 2011;11:715-727. https://doi.org/10.1007/s10113-011-0204-2

25. Pausas JG, Marañón T, Marañón T, Marañón T, Caldeira MC, Pons J. Natural regeneration. 2009

26. Gössling S, Scott D, Hall CM, Ceron J-P, Dubois G. Consumer behaviour and demand response of tourists to climate change. Ann Tour Res. 2012;39:36-58. https://doi.org/10.1016/j.annals.2011.11.002

27. Bone RC, Balk RA, Cerra FB, Dellinger RP, Fein AM, Knaus WA, et al. Definitions for sepsis and organ failure and guidelines for the use of innovative therapies in sepsis. Chest. 1992;101:1644-1655. https://doi.org/10.1378/chest.101.6.1644

28. Tomsits E, Pataki M, Tölgyesi A, Fekete G, Rischak K, Szollár L. Safety and efficacy of a lipid emulsion containing a mixture of soybean oil, medium-chain triglycerides, olive oil, and fish oil: a randomised, double-blind clinical trial in premature infants requiring parenteral nutrition. J Pediatr Gastroenterol Nutr. 2010;51:514-521. https://doi.org/10.1378/chest.101.6.1644

29. Ford JD, Cameron L, Rubis J, Maillet M, Nakashima D, Willox AC, et al. Including indigenous knowledge and experience in IPCC assessment reports. Nat Clim Chang. 2016;6:349-353

30. Garcia ES, Swann ALS, Villegas JC, Breshears DD, Law DJ, Saleska SR, et al. Synergistic ecoclimate teleconnections from forest loss in different regions structure global

ecological responses. PLoS One. 2016;11:e0165042. https://doi.org/10.1371/journal.pone.0165042

31. Allen GL. Principles and practices for communicating route knowledge. Appl Cogn Psychol Off J Soc Appl Res Mem Cogn. 2000;14:333-359. https://doi.org/10.1002/1099-0720(200007/08)14:4<333::aid-acp655>3.0.co;2-c
32. Brenner ED, Stahlberg R, Mancuso S, Vivanco J, Baluška F, Van Volkenburgh E. Plant neurobiology: an integrated view of plant signaling. Trends Plant Sci. 2006;11:413-419. https://doi.org/10.1016/j.tplants.2006.06.009
33. Scott WR. Approaching adulthood: the maturing of institutional theory. Theory Soc. 2008;37:427. https://doi.org/10.1007/s11186-008-9067-z
34. Wahid A. Physiological implications of metabolite biosynthesis for net assimilation and heat-stress tolerance of sugarcane (Saccharum officinarum) sprouts. J Plant Res. 2007;120:219-228. https://doi.org/10.1007/s10265-006-0040-5
35. Syvertsen JP, Lloyd JJ. Citrus. Handb Environ Physiol fruit Crop. 1994;2:65-99. https://doi.org/10.17660/ActaHortic.2012.928.44
36. Kiefer AK, Sanchez DT, Kalinka CJ, Ybarra O. How women's nonconscious association of sex with submission relates to their subjective sexual arousability and ability to reach orgasm. Sex Roles. 2006;55:83-94. https://doi.org/10.1007/s11199-006-9060-9
37. Ginestar C, Castel JR. Responses of young clementine citrus trees to water stress during different phenological periods. J Hortic Sci. 1996;71:551-559. https://doi.org/10.1080/14620316.1996.11515435
38. González-Altozano P, Castel JR. Regulated deficit irrigation in'Clementina de Nules' citrus trees. I. Yield and fruit quality effects. 1999. https://doi.org/10.17660/actahortic.2000.537.89
39. Sinclair TR, Allen Jr LH. Carbon dioxide and water vapour exchange of leaves on field-grown citrus trees. J Exp Bot. 1982;33:1166-1175. https://doi.org/10.1093/jxb/33.6.1166
40. 40.Shalhevet J, Levy Y. Citrus trees. Agron. 1990. https://doi.org/10.21273/HORTTECH.15.1.0095
41. Chattha WS, Atif RM, Iqbal M, Shafqat W, Farooq MA, Shakeel A. Genome-wide identification and evolution of Dof transcription factor family in cultivated and ancestral cotton species. Genomics. 2020;112:4155-4170. https://doi.org/10.1016/j.ygeno.2020.07.006
42. Flexas J, Medrano H. Drought-inhibition of photosynthesis in C3 plants: stomatal and non-stomatal limitations revisited. Ann Bot. 2002;89:183-189. https://doi.org/10.1093/aob/mcf027
43. Arbona V, Iglesias DJ, Jacas J, Primo-Millo E, Talon M, Gómez-Cadenas A. Hydrogel substrate amendment alleviates drought effects on young citrus plants. Plant Soil. 2005;270:73-82. https://doi.org/10.1007/s11104-004-1160-0
44. Lopez-Medina J, Moore JN. Chilling enhances cane elongation and flowering in primocane-fruiting blackberries. HortScience. 1999;34:638-640. https://doi.org/10.21273/hortsci.34.4.638

45. Adamis PDB, Gomes DS, Pinto MLCC, Panek AD, Eleutherio ECA. The role of glutathione transferases in cadmium stress. Toxicol Lett. 2004;154:81-88. https://doi.org/10.1016/j.toxlet.2004.07.003

46. Blum A. Drought resistance–is it really a complex trait? Funct Plant Biol. 2011;38:753-757. https://doi.org/10.1071/fp11101

47. Brakke M, Allen LH. Gas exchange of Citrus seedlings at different temperatures, vapor-pressure deficits, and soil water contents. J Am Soc Hortic Sci. 1995;120:497-504. https://doi.org/10.21273/jashs.120.3.497

48. Xie Z, Miller GM. Trace amine-associated receptor 1 is a modulator of the dopamine transporter. J Pharmacol Exp Ther. 2007;321:128-136. https://doi.org/10.1124/jpet.106.117382

49. Wu K, Horejsi T, Byrum J, Bringe N, Yang J, Pei D, et al. Agronomically elite soybeans with high beta-conglycinin content. US Patent Application 11/517186. Date issued: 11 September. Agron Elit soybeans with high beta-conglycinin content US Pat Appl 11/517186 Date issued 11 Sept. 2007

50. Syvertsen T. The many uses of the "public service" concept. Nord Rev. 1999;20:5-12

51. Fauchere J-L, Ortuno J-C, Levens N, Chamorro S, Boutin J. Novel aminotriazolone compounds, method for preparing same and pharmaceutical compositions containing same. 2004

52. Carvalho-Junior O, Birolo AB, Macedo-Soares LCP de. Ecological aspects of neotropical otter (Lontra longicaudis) in Peri lagoon, south Brazil. IUCN Otter Spec Gr Bull. 2010;27:105-115. http://dx.doi.org/10.1590/S0001-37652013005000014

53. Perez-Campo R, Lopez-Torres M, Cadenas S, Rojas C, Barja G. The rate of free radical production as a determinant of the rate of aging: evidence from the comparative approach. J Comp Physiol B. 1998;168:149-158. https://doi.org/10.1007/s003600050131

54. Takeyama K, Dabbagh K, Lee H-M, Agustí C, Lausier JA, Ueki IF, et al. Epidermal growth factor system regulates mucin production in airways. Proc Natl Acad Sci. 1999;96:3081-3086. https://doi.org/10.1073/pnas.96.6.3081

55. Adie BAT, Pérez-Pérez J, Pérez-Pérez MM, Godoy M, Sánchez-Serrano J-J, Schmelz EA, et al. ABA is an essential signal for plant resistance to pathogens affecting JA biosynthesis and the activation of defenses in Arabidopsis. Plant Cell. 2007;19:1665-1681. https://doi.org/10.1105/tpc.106.048041

56. Panigrahi P, Sharma RK, Hasan M, Parihar SS. Deficit irrigation scheduling and yield prediction of 'Kinnow'mandarin (Citrus reticulate Blanco) in a semiarid region. Agric Water Manag. 2014;140:48-60. https://doi.org/10.1016/j.agwat.2014.03.018

57. Moreshet S, Huck MG, Hesketh JD, Peters DB. Relationships between sap flow and hydraulic conductivity in soybean plants. 1990. https://doi.org/10.1051/agro:19900504

58. Shapiro L, Cohen S. Novel alginate sponges for cell culture and transplantation. Biomaterials. 1997;18:583-590. https://doi.org/10.1016/s0142-9612(96)00181-0

59. Berger B, Parent B, Tester M. High-throughput shoot imaging to study drought responses. J Exp Bot. 2010;61:3519-3528. https://doi.org/10.1093/jxb/erq201

60. Lamaoui M, Jemo M, Datla R, Bekkaoui F. Heat and drought stresses in crops and approaches for their mitigation. Front Chem. 2018;6:26. https://doi.org/10.3389/fchem.2018.00026

61. Kavar T, Maras M, Kidrič M, Šuštar-Vozlič J, Meglič V. Identification of genes involved in the response of leaves of Phaseolus vulgaris to drought stress. Mol Breed. 2008;21:159-172. https://doi.org/10.1007/s11032-007-9116-8

62. Fonseca S, Chini A, Hamberg M, Adie B, Porzel A, Kramell R, et al. (+)-7-iso-Jasmonoyl-L-isoleucine is the endogenous bioactive jasmonate. Nat Chem Biol. 2009;5:344-350. https://doi.org/10.1038/nchembio.161

63. ZHANG J, Davies WJ. Does ABA in the xylem control the rate of leaf growth in soil-dried maize and sunflower plants? J Exp Bot. 1990;41:1125-1132. https://doi.org/10.1093/jxb/41.9.1125

64. Hayat S, Alyemeni MN, Hasan SA. Foliar spray of brassinosteroid enhances yield and quality of Solanum lycopersicum under cadmium stress. Saudi J Biol Sci. 2012;19:325-335. https://doi.org/10.1016/j.sjbs.2012.03.005

65. Naliwajski MR, Skłodowska M. Proline and its metabolism enzymes in cucumber cell cultures during acclimation to salinity. Protoplasma. 2014;251:201-209. https://doi.org/10.1007/s00709-013-0538-3

66. Olmstead MA, Lang NS, Ewers FW, Owens SA. Xylem vessel anatomy of sweet cherries grafted onto dwarfing and nondwarfing rootstocks. J Am Soc Hortic Sci. 2006;131:577-585. https://doi.org/10.21273/jashs.131.5.577

67. Djoukeng JD, Arbona V, Argamasilla R, Gomez-Cadenas A. Flavonoid profiling in leaves of citrus genotypes under different environmental situations. J Agric Food Chem. 2008;56:11087-11097. https://doi.org/10.1021/jf802382y

68. Sarkar C, Guenther AB, Park J-H, Seco R, Alves E, Batalha S, et al. PTR-TOF-MS eddy covariance measurements of isoprene and monoterpene fluxes from an eastern Amazonian rainforest. Atmos Chem Phys. 2020;20:7179-7191. https://doi.org/10.5194/acp-20-7179-2020

69. Mathews H, Litz RE, Wilde HD, Merkle SA, Wetzstein HY. Stable integration and expression of β-glucuronidase and NPT II genes in mango somatic embryos. Vitr. 1992;28:172-178. https://doi.org/10.1007/bf02823312

70. Chakraborty U, Pradhan B. Oxidative stress in five wheat varieties (Triticum aestivum L.) exposed to water stress and study of their antioxidant enzyme defense system, water stress responsive metabolites and H2O2 accumulation. Brazilian J Plant Physiol. 2012;24:117-130. https://doi.org/10.1590/s1677-04202012000200005

71. Chaidir L, Annisa J, Dian S, Parwati I, Alisjahbana A, Purnama F, et al. Microbiological diagnosis of adult tuberculous meningitis in a ten-year cohort in Indonesia. Diagn Microbiol Infect Dis. 2018;91:42-46. https://doi.org/10.1016/j.diagmicrobio.2018.01.004

72. Ribotta PD, Perez GT, Leon AE, Anon MC. Effect of emulsifier and guar gum on micro structural, rheological and baking performance of frozen bread dough. Food Hydrocoll. 2004;18:305-313. https://doi.org/10.1016/s0268-005x(03)00086-9

73. Trifilò P, Gullo MA Lo, Nardini A, Pernice F, Salleo S. Rootstock effects on xylem conduit dimensions and vulnerability to cavitation of Olea europaea L. Trees. 2007;21:549-556. https://doi.org/10.1007/s00468-007-0148-9

74. Zach C, Klopschitz M, Pollefeys M. Disambiguating visual relations using loop constraints. In: 2010 IEEE Computer Society Conference on Computer Vision and Pattern Recognition. IEEE; 2010. p. 1426-1433. https://doi.org/10.1109/cvpr.2010.5539801

75. Meland M. Efficacy of chemical bloom thinning agents to European plums. Acta Agric Scand Sect B-Soil Plant Sci. 2007;57:235-242. https://doi.org/10.1080/09064710600914236

76. Tombesi F, Cappi M, Reeves JN, Palumbo GGC, Yaqoob T, Braito V, et al. Evidence for ultra-fast outflows in radio-quiet AGNs-I. Detection and statistical incidence of Fe K-shell absorption lines. Astron Astrophys. 2010;521:A57. https://doi.org/10.1051/0004-6361/200913440

77. Tyree MT, Ewers FW. The hydraulic architecture of trees and other woody plants. New Phytol. 1991;119:345-360. https://doi.org/10.1111/j.1469-8137.1991.tb00035.x

78. Olmstead RG, Bohs L. A summary of molecular systematic research in Solanaceae: 1982-2006. In: VI International Solanaceae Conference: Genomics Meets Biodiversity 745. 2006. p. 255-268. https://doi.org/10.17660/actahortic.2007.745.11

79. Loreto F, Harley PC, Di Marco G, Sharkey TD. Estimation of mesophyll conductance to CO2 flux by three different methods. Plant Physiol. 1992;98:1437-1443. https://doi.org/10.1104/pp.98.4.1437

80. Shinozaki K, Yamaguchi-Shinozaki K. Gene networks involved in drought stress response and tolerance. J Exp Bot. 2007;58:221-227. https://doi.org/10.1093/jxb/erl164

81. Wang W, Vinocur B, Altman A. Plant responses to drought, salinity and extreme temperatures: towards genetic engineering for stress tolerance. Planta. 2003;218:1-14. https://doi.org/10.1007/s00425-003-1105-5

82. Kotak S, Vierling E, Bäumlein H, von Koskull-Döring P. A novel transcriptional cascade regulating expression of heat stress proteins during seed development of Arabidopsis. Plant Cell. 2007;19:182-195. https://doi.org/10.1105/tpc.106.048165

83. Shalchi A, Döring H. Velocity correlation functions of charged test particles. J Phys G Nucl Part Phys. 2007;34:859. https://doi.org/10.1088/0954-3899/34/5/007

84. Khan Z, Shahwar D. Role of Heat Shock Proteins (HSPs) and Heat Stress Tolerance in Crop Plants. In: Sustainable Agriculture in the Era of Climate Change. Springer; 2020. p. 211-234. https://doi.org/10.1007/978-3-030-45669-6_9

85. Shafqat S, Kishwer S, Rasool RU, Qadir J, Amjad T, Ahmad HF. Big data analytics enhanced healthcare systems: a review. J Supercomput. 2020;76:1754-1799. https://doi.org/10.1007/s11227-017-2222-4

86. Alves MS, Dadalto SP, Gonçalves AB, De Souza GB, Barros VA, Fietto LG. Plant bZIP transcription factors responsive to pathogens: a review. Int J Mol Sci. 2013;14:7815-7828. https://doi.org/10.3390/ijms14047815

87. Kang B, Rancour DM, Bednarek SY. The dynamin-like protein ADL1C is essential for plasma membrane maintenance during pollen maturation. Plant J. 2003;35:1-15. https://doi.org/10.1046/j.1365-313x.2003.01775.x
88. Javot H, Maurel C. The role of aquaporins in root water uptake. Ann Bot. 2002;90:301-313. https://doi.org/10.1093/aob/mcf199
89. North RJ, Jung Y-J. Immunity to tuberculosis. Annu Rev Immunol. 2004;22:599-623. https://doi.org/10.1146/annurev.immunol.22.012703.104635
90. Banerjee A, Samanta S, Singh A, Roychoudhury A. Deciphering the molecular mechanism behind stimulated co-uptake of arsenic and fluoride from soil, associated toxicity, defence and glyoxalase machineries in arsenic-tolerant rice. J Hazard Mater. 2020;390:121978. https://doi.org/10.1016/j.jhazmat.2019.121978
91. Kumar V, Kumar M, Sharma S, Varma A, Bhalla-Sarin N. Procedural Insights on In Vitro Propagation of Litchi chinensis (Sonn.). In: Lychee Disease Management. Springer; 2017. p. 217-35. https://doi.org/10.1007/978-981-10-4247-8_13
92. Barnabás B, Jäger K, Fehér A. The effect of drought and heat stress on reproductive processes in cereals. Plant Cell Environ. 2008;31:11-38. https://doi.org/10.1111/j.1365-3040.2007.01727.x
93. Yang N, Sun Z-X, Feng L-S, Zheng M-Z, Chi D-C, Meng W-Z, et al. Plastic film mulching for water-efficient agricultural applications and degradable films materials development research. Mater Manuf Process. 2015;30:143-154. https://doi.org/10.1080/10426914.2014.930958
94. Prasad RN. Varietal Evaluation of Pomegranate under Arid Conditions. Ann Arid Zone. 2000;39:427-430. http://krishi.icar.gov.in/jspui/handle/123456789/2370. Accessed 4 Feb 2020. http://doi.org/10.123456789/2370
95. Samoliński B, Raciborski F, Lipiec A, Tomaszewska A, Krzych-Fałta E, Samel-Kowalik P, et al. Epidemiologia chorób alergicznych w Polsce (ECAP). Alergol Pol J Allergol. 2014;1:10-18. https://doi.org/10.1016/j.alergo.2014.03.008
96. Waraich EA, Ahmad R, Halim A, Aziz T. Alleviation of temperature stress by nutrient management in crop plants: a review. J soil Sci plant Nutr. 2012;12:221-244. https://doi.org/10.4067/s0718-95162012000200003
97. Zhang B, Schmoyer D, Kirov S, Snoddy J. GOTree Machine (GOTM): a web-based platform for interpreting sets of interesting genes using Gene Ontology hierarchies. BMC Bioinformatics. 2004;5:1-8. https://doi.org/10.4067/10.1186/1471-2105-5-16
98. Huang D. Dietary antioxidants and health promotion. 2018. https://dx.doi.org/10.3390%2Fantiox7010009
99. Hu ZQ, Huang XM, Chen HB, Wang HC. Antioxidant capacity and phenolic compounds in litchi (Litchi chinensis Sonn.) pericarp. In: III International Symposium on Longan, Lychee, and other Fruit Trees in Sapindaceae Family 863. 2008. p. 567-574. https://doi.org/10.17660/actahortic.2010.863.79
100. Waterland NL, Finer JJ, Jones ML. Benzyladenine and gibberellic acid application prevents abscisic acid-induced leaf chlorosis in pansy and viola. HortScience. 2010;45:925-933. https://doi.org/10.21273/hortsci.45.6.925

101.Chakraborty D, Nagarajan S, Aggarwal P, Gupta VK, Tomar RK, Garg RN, et al. Effect of mulching on soil and plant water status, and the growth and yield of wheat (Triticum aestivum L.) in a semi-arid environment. Agric water Manag. 2008;95:1323-1334. https://doi.org/10.1016/j.agwat.2008.06.001

102.Zribi W, Aragüés R, Medina E, Faci JM. Efficiency of inorganic and organic mulching materials for soil evaporation control. Soil Tillage Res. 2015;148:40-45. https://doi.org/10.1016/j.still.2014.12.003

103.Kader MA, Senge M, Mojid MA, Ito K. Recent advances in mulching materials and methods for modifying soil environment. Soil Tillage Res. 2017;168:155-166. https://doi.org/10.1016/j.still.2017.01.001

104..Zhang H, Huang Z, Xie B, Chen Q, Tian X, Zhang X, et al. The ethylene-, jasmonate-, abscisic acid-and NaCl-responsive tomato transcription factor JERF1 modulates expression of GCC box-containing genes and salt tolerance in tobacco. Planta. 2004;220:262-270. https://doi.org/10.1007/s00425-004-1347-x

105.Li Q, Li H, Zhang L, Zhang S, Chen Y. Mulching improves yield and water-use efficiency of potato cropping in China: A meta-analysis. F Crop Res. 2018;221:50-60. https://doi.org/10.1016/j.fcr.2018.02.017

106.Kader MA, Singha A, Begum MA, Jewel A, Khan FH, Khan NI. Mulching as water-saving technique in dryland agriculture. Bull Natl Res Cent. 2019;43:1-6. https://doi.org/10.1186/s42269-019-0186-7

107.Li SX, Wang ZH, Li SQ, Gao YJ, Tian XH. Effect of plastic sheet mulch, wheat straw mulch, and maize growth on water loss by evaporation in dryland areas of China. Agric water Manag. 2013;116:39-49. https://doi.org/10.1016/j.agwat.2012.10.004

108.Parmar N, Singh KH, Sharma D, Singh L, Kumar P, Nanjundan J, et al. Genetic engineering strategies for biotic and abiotic stress tolerance and quality enhancement in horticultural crops: a comprehensive review. 3 Biotech. 2017;7:239. https://doi.org/10.1007/s13205-017-0870-y

109.Shen X, Gmitter FG, Grosser JW. Immature embryo rescue and culture. In: Plant Embryo Culture. Springer; 2011. p. 75-92. https://doi.org/10.1007/978-1-61737-988-8_7

110.Ananthakrishnan G, Ćalović M, Serrano P, Grosser JW. Production of additional allotetraploid somatic hybrids combining mandarings and sweet orange with pre-selected pummelos as potential candidates to replace sour orange rootstock. Vitr Cell Dev Biol. 2006;42:367-371. https://doi.org/10.1079/ivp2006784

111.De Oliveira AC, Garcia AN, Cristofani M, Machado MA. Identification of citrus hybrids through the combination of leaf apex morphology and SSR markers. Euphytica. 2002;128:397-403. https://doi.org/10.1007/s10722-014-0188-0

112.Kepiro JL, Roose ML. AFLP markers closely linked to a major gene essential for nucellar embryony (apomixis) in Citrus maxima× Poncirus trifoliata. Tree Genet genomes. 2010;6:1-11. https://doi.org/10.1007/s11295-009-0223-z

113.Dambier D, Benyahia H, Pensabene-Bellavia G, Kaçar YA, Froelicher Y, Belfalah Z, et al. Somatic hybridization for citrus rootstock breeding: an effective tool to solve some important issues of the Mediterranean citrus industry. Plant Cell Rep. 2011;30:883-900. https://doi.org/10.1007/s00299-010-1000-z

114. Uzun A, Yesiloglu T, Aka-Kacar Y, Tuzcu O, Gulsen O. Genetic diversity and relationships within Citrus and related genera based on sequence related amplified polymorphism markers (SRAPs). Sci Hortic (Amsterdam). 2009;121:306-312. https://doi.org/10.1016/j.scienta.2009.02.018

115. Grosser JW, Gmitter FG. Protoplast fusion for production of tetraploids and triploids: applications for scion and rootstock breeding in citrus. Plant Cell, Tissue Organ Cult. 2011;104:343-357. https://doi.org/10.1007/s11240-010-9823-4

116. Skirvin RM, McPheeters KD, Norton M. Sources and frequency of somaclonal variation. HortScience. 1994;29:1232-1237. https://doi.org/10.21273/hortsci.29.11.1232

117. Duncan RR, Waskom RM, Nabors MW. In vitro screening and field evaluation of tissue-culture-regenerated sorghum (Sorghum bicolor (L.) Moench) for soil stress tolerance. In: The Methodology of Plant Genetic Manipulation: Criteria for Decision Making. Springer; 1995. p. 373-380. https://doi.org/10.1007/978-94-011-0357-2_46

118. Kochba J, Ben-Hayyim G, Spiegel-Roy P, Saad S, Neumann H. Selection of stable salt-tolerant callus cell lines and embryos in Citrus sinensis and C. aurantium. Zeitschrift für Pflanzenphysiologie. 1982;106:111-118. https://doi.org/10.1016/s0044-328x(82)80073-1

119. Piqueras Castillo A. Selección mediante cultivo" in vitro" y caracterización de una linea de callo embriogenico de (Citrus Limonum R.). 1992. http://hdl.handle.net/10201/31365

120. El Yacoubi H, Rochdi A, Ayolie K, Rachidai A. Sélection et évaluation de lignées de cals stables et tolérantes vis-à-vis du stress salin chez le citrange 'Troyer' [Citrus sinensis (L.)× Poncirus trifoliata (L.) Raf.]. Fruits. 2004;59:325-337. https://doi.org/10.1051/fruits:2004031

121. Matsukura S, Mizoi J, Yoshida T, Todaka D, Ito Y, Maruyama K, et al. Comprehensive analysis of rice DREB2-type genes that encode transcription factors involved in the expression of abiotic stress-responsive genes. Mol Genet Genomics. 2010;283:185-196. https://doi.org/10.1007/s00438-009-0506-y

122. Forner-Giner MA, Llosá MJ, Carrasco JL, Perez-Amador MA, Navarro L, Ancillo G. Differential gene expression analysis provides new insights into the molecular basis of iron deficiency stress response in the citrus rootstock Poncirus trifoliata (L.) Raf. J Exp Bot. 2010;61:483-490. https://doi.org/10.1093/jxb/erp328

SECTION-3
CITRUS NUTRITIONAL AND NUTRACEUTICAL IMPORTANCE

Chapter-9

Citrus Fruits: Nutritive Value and Value-Added Products

1. Introduction

Citrus is an evergreen shrub that belongs to the Rutaceae family from South Asia, China, India and the Malay Archipelago, which is native to the subtropical and tropical regions of Asian regions [1]. The genus of citrus includes sweet orange (*C. sinensis*: 61.1 % of world citrus production), tangerine (*C. reticulata*: 19.9 %), limon and lime (*C. limon* and *C. aurantifolia*: 12.1 %) and grapefruit (*C. paradisi*: 5%). Minor types of citrus, which constitute much of the remaining 2.0%, include sour orange (*C. quarantium*), shaddocks (*C. grandis*), citrus (*C. medica*), which seem to be promising sources for many beneficial human nuts [2]. Citrus fruit is divided into two sections like peel and flesh (Figure 1). Peel is made from epicarp or flavedo (colored peripheral surface) and mesocarp or albedo (white soft middle layer). The peel (60–65%), internal tissues (30–35%), and seeds (0–10%) comprise citrus fruits [3]. Citrus fruits provide carbohydrates, such as sucrose, glucose, and fructose mostly. Fresh citrus fruits are also an immeasurable source of dietary fiber associated with gastrointestinal disease prevention and lowered circulating cholesterol. Citrus fruits also have a distinct aroma and delicious taste along with low protein and fat content.

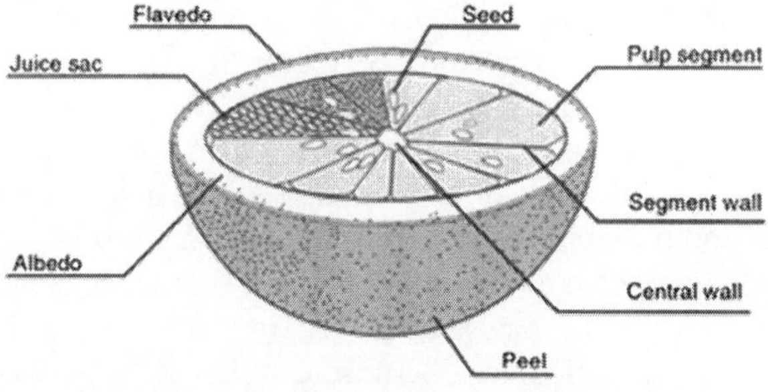

Figure 1. Structure of the citrus fruit [2].

Citrus fruits also provide the most potent source of vitamins C and B (thiamines, pyridoxines, niacins, riboflavin, pantothenic acids, and folate). The fruit also leads to the use of phytochemicals, such as carotenoids, flavonoids, and limonoids [1]. Citrus phytochemicals contain antibacterial, antiviral, antifungal, anti-carcinogenic, anti-thrombotic, or anti-inflammatory agents [4]. Several studies have proposed citrus fruit evaluation as a healthy and delicious diet [5]. Prior research suggested that citrus and citrus products are rich sources of vitamins, minerals, and dietary fibers [6]. However, the bioactive and non-nutrient compounds in citrus are appreciated to reduce the risk of various chronic diseases [7].

Citrus fruits are eaten as fresh goods and juice throughout the world. Peel is discarded as waste containing many secondary components with significant antioxidant activity related to other fruit portions [8]. In recent years, flavonoids such as polymethoxy flavones (PMFs), which are present in citrus fruits, have been attracted growing attention by their antioxidants [9] and anti-cancer properties [10]. Various bioactive compounds in citrus peel extract and powder may reduce overall cholesterol, triglycerides, LDL, and glucose levels [11]. Citrus by-products produce a range of value-added products, including essential oils, pectin, enzymes, single-cell collagen, natural antioxidants, ethanol, organic acids, and prebiotics. Orange, lemon, mandarin, and grapefruit contained essential oils show antifungal activity upon the fungi *A. niger*, *A. flavus, P. chrysogenum,* and *P. verrucosum.* The essential oil may be regarded as acceptable for the food industry as alternatives to chemicals [12]. Pectin extracted from *Citrus* peel is used in various industrial food processes as gelling agents, including jam, jellies, and as thickener, texturizer, emulsifier, and stabilizer in dairy products. Pectin is also used to jellify properties in the pharmaceutical, dental, and cosmetic industries [13]. Therefore, this chapter highlighted the nutritional values of major essential nutrients such as Vitamin C, carotenoids and vitamin A, Folate, Dietary fiber, flavonoids, and limonoids, as well as value-added products such as food ingredients, pectin, essential oil, enzymes, a natural antioxidant, and packaging film retrieved from citrus especially lemon and pomelo fruit.

2. Characteristics of citrus fruits (lemon and pomelo) and their chemical compositions

2.1 Lemon

Lemon (*Citrus limon L.* from Rutaceae) is one of the most common globally and ranks third among the Citrus species globally by 4,200,000 metric tons after orange and mandarin [14]. Lemon fruits typically consist of three parts: pulp, skins (albedo and flavedo), and seeds. It offers an extensive supply of natural compound products such as citric acid, ascorbic acid, minerals, flavonoids and essential oils [15]. Lemon bioactive compounds like flavonoids, vitamins, minerals, dietary fiber (Table 1), and essential oils are used in the food, cosmetic, and pharmaceutical industries. Most by-products of the lemon juice industry can provide functional foods with nutritional substances such as non-digestible carbohydrates, dietary fiber and bioactive (flavonoids and ascorbic acid). Lemon fruits can function against photo-oxidamage because carotenoids exist. Lemon fruit, rich in flavonoids, has a significant role in the healthy diet, particularly in preventing diseases such as obesity, diabetes, lowering blood lipids, cardiovascular disease, and some forms of cancer [18]. The citrus fruits used for direct consumption or converted into juices, jam, jelly, molasses, lemoncello beverage and more in addition to the lemon skin are added value products such as pectins, essential oil and functional ingredients [12, 18].

Table 1. Chemical composition of Citrus fruits as [15, 16, 17].

Component	Lemon (*Citrus limon*)	Pomelo (*Citrus maxima*)
Moisture (g/100g)	84.2	87.0
Fiber (g/100g)	1.6	1.60
Carbohydrate (g/100g)	10.8	11.5
Protein (g/100g)	0.9	0.6
Fat (g/100g)	0.8	0.20
Vitamin (mg/100g)		
β-carotene	50.0	120
Thiamine	0.02	0.03
Riboflavin	0.01	0.03
Vitamin C	37	20.0
Mineral (mg/100g)		
Ca	70.0	10.0
Mg	12.0	21.6
Na	1.50	2.70
P	10.0	20
K	148	106
Fe	0.23	0.40
Zinc	0.12	0.15
Cu	0.20	0.19
Total phenol (mg GAE/g)	204.40	70.56
Total flavonoids (mg QUE/g)	27.50	13.06
Carotenoid (mg/100 mL)	0.31-0.35	0.72-0.73

2.2 Pomelo

Pomelo is one of the most commonly grown and eaten citrus fruits and orange, mandarin, lemon, and grapefruit [19]. Pomelo (Table 1) is a promising source of carbohydrates, proteins, fiber, vitamins and minerals originating in warm tropical climates in south-eastern Asia [20]. The presence of bioactive (carotenoids, lycopene, polyphenols, flavonoids, limonoids, fiber and vitamin C) contributes to their protection against oxidative stress, hyperglycemia, and high blood pressure. Due to its essential health promotion properties, pomelo segments in food products are growing in importance in producing functional foods [21]. Pomelo is eaten fresh or made into juice [19], or pomelo fortified noodles help the diabetic population [21]. On the other hand, researchers have investigated alternative ways of restoring pomelo peels to the advantage of value-added products such as pectin, essential oils, polysaccharides, phytochemicals [19]. Production of juice and consumption of fresh fruit create large quantities of agricultural waste. The main components of wet Pomelo Peel waste, like other citrus fruits, include water, cellulose and hemicellulose, soluble sugars, lipids (mainly D-limonene), and bioactive compounds (i.e., polyphenols, mostly flavonoids).

3. Nutritional values of citrus fruits

Citrus has many natural plant compounds such as vitamin C, carotenoids (some can convert to vitamin A), folic acid, flavonoids, and fiber. Table 2 shows the amount of vitamin and mineral consumption in lemon and pomelo fruits.

Table 2. The number of nutrients and the percent of the recommended daily allowance or adequate intake met from the consumption of 100 g of selected citrus fruit [22].

	Vitamin C	Vitamin A	Folate	Fiber
Oranges	**53-88 mg**	**17 μg**	**30 μg**	**2.4 g**
Children under 9 y (%)	213-589	3-6	15-20	10-13
Persons 9+ y	59-195	2-4	8-10	6-11
Pregnant/Lactating women	44-110	2-3	5-6	8-9
Grapes	**31-61 mg**	**58 μg**	**13 μg**	**1.6 g**
Children under 9 y	125-244	12-15	7-9	6-8
Persons 9+	35-135	6-10	3-4	4-8
Pregnant/lactating women	26-76	4-8	2-3	6
Tangerines	**27-72 mg**	**46-144 μg**	**16 μg**	**1.8 g**
Children under 9	107-480	9-36	8-11	7-9
Persons 9+	30-160	5-24	4-5	5-9
Pregnant/lactating women	21-90	4-19	3-4	6
Lemons/limes	**29-61 mg**	**2-22 μg**	**11-16 μg**	**1.8-2.8 g**
Children under 9	116-407	0.4-6	4-7	11-15
Persons 9+	32-135	0.2-4	2-4	9-13
Pregnant/lactating women	24-76	0.2-3	1-2	10

3.1 Vitamin C (ascorbic acid)

Citrus is a valuable source of vitamin C. By consuming a moderate amount of citrus fruits each day, an individual can achieve 100 percent Vitamin C level. Vitamin C is an essential water-soluble vitamin essential for the body's defense [22]. It is transmitted through muscle fibers, carnitine biosynthesis, neurotransmitters, collagen, and bones because these particles connect the fibers. The immune system can be effectively stimulated by consuming vitamin C, which boosts white blood cells [23]. When Vitamin C is taken for pregnancy, it can decrease pre-eclampsia risk [24]. Some studies indicate that vitamin C supplementation can reduce the severity of colds symptoms or duration [23]. Anti-oxidants such as Vitamin C could reduce the risk of artery stiffening and cardiovascular diseases [25]. Above 200 mg of vitamin C daily is a healthy intake, and citrus fruits are a huge source of this vitamin. Lemon provides 37 mg of ascorbic acid per 100 g of fruit [16]. Pomelos have 52.3 mg of ascorbic acid in 100 g of the flesh [26].

3.2 Carotenoids and Vitamin A

There are many types of carotenoids, including terpenes responsible for pigments commonly found in plants, and there are about 600 carotenoids in foods and 50 in human bodies [27, 28]. The highest carotenoid levels, such as lutein, zeaxanthin, lycopene, and vitamin A, are found in fruits and vegetables, including orange and carotene. Benefits of carotenoids in foods include improving immune function, promoting bone formation, promoting eye health, and maintaining visual quality [22]. There is a large amount of data supporting that carotenoids reduce the risk of cancer, macular degeneration, cataracts, skin damage to the sun, and cardiovascular diseases [29]. Higher consumption of β-carotene is linked to a lower breast cancer risk [30]. Beta carotene, lycopene, or lutein may decrease the rate of UV-induced lipid peroxidation in human skin fibroblast cells [30]. Lutein is inversely related to colorectal cancer in both men and women [31]. The levels of lutein, zeaxanthin, β-cryptoxanthin, and β-carotene in the lemon and pomelo, were around 2.95, 0.81, 0.81 and 10.3 (μg/g, db), respectively [32]. The content of carotenoids in pummelos' peel was 0.012-0.015 mg/gdb [33].

3.3 Folate (folic acid)

Folic acid, which is a water-soluble vitamin, and its derivatives are collectively called folate or folacin. The most notable folate compounds in Citrus are the reduced 5-methyl tetrahydrofolate (monoglutamate) and polyglutamate compounds [34]. Folate plays a vital role in DNA, which is involved in homocysteine regulation and protein production primarily through the methylation transfer reactions [22]. Because there is a high DNA production during pregnancy, a folate deficiency is significantly linked to birth defects such as neural tube defects [35]. Lack of folic acid caused higher levels of homocysteine, raising heart disease and atherosclerosis [22]. Previous studies show that citrus fruits' daily consumption can help improve folate levels, which will subsequently decrease blood homocysteine (tHcy), thus reducing cardiovascular disorder and neural tube defects [36]. Citrus is a parallel source of dietary folate that can help to cover up to 10% to 20% of the recommended daily allowance of adults, children, and infants with a consumption of 100 g of citrus fruits. The consumption of citrus fruit is an easy way to obtain vitamin C and dietary folate, which is vital for absorption in the body. Lemon, a citrus fruits representative, has eleven to sixteen micrograms of Folate in 100 grams [22]. According to El-Otmani and Ait-Oubahou [37] Citrus limon contained 11mg of folic acid per 100 g of citrus.

3.4 Dietary fibre

The fiber is found in vegetables and fruits cannot be digested and absorb in the small intestine. There are two kinds of dietary fiber; soluble and insoluble fiber. Insoluble fibers are highly fermentable and connected with carbohydrate and lipid metabolism, while soluble fibers contribute to fecal bulk and reduce transit time [34]. Although pectin, cellulose, and hemicellulose comprise the most abundant dietary fiber on the plants, they also contain only trace amounts of lignin. Pectin is citrus' primary fiber, which occurs primarily in citrus peels and rinds. Consumption of citrus fruit can contribute significant quantities of pectin in a diet. Dietary incorporation of pectin appears to affect several metabolic and digestive processes; principal interest affects glucose absorption and cholesterol level [38, 39]. There is a significant benefit in consuming citrus fruit because of its pectin content. Dietary incorporation of pectin appears to have many implications for metabolic, digestive, and health affairs. One way fiber can reduce colon cancer is by diluting and trapping the harmful chemicals in the colon from bile-absorption and bile-excretion [34]. Scientific studies have proven that fiber can help promote laxation and satiety, the uptake and reabsorption of glucose, fat, cholesterol, and bile acids, thereby lessening heart disease risk and possibly enhancing healthy intestinal microbial fermentation [40, 41]. Citrus fruits significantly reduce cholesterol levels depending on the esterification degree of fiber consumption, viscosity, and molecular mass [22]. A fiber-rich diet has a low risk of deadly chronic diseases such as diabetes, heart disease, weight, and cancer and lowers cholesterol levels and blood sugar [42]. Several epidemiological studies reported that citrus peel support reducing plasma liver cholesterol, total serum cholesterol, serum triglyceride levels, and total liver lipids [43].

3.5 Flavonoids and limonoids

Citrus pulp, peel are rich sources of flavonoids. Toh et al. [44] found that pomelo peeled (356.95 mg/QE) had higher total flavonoid content than pomelo pulp (13.06 mg/QE). Makni et al. [15] found the amount of quercetin in lemon flesh (56.16 mg Eq Quercetin/g dry weight) was higher than in peel (27.50 mg Eq Quercetin/g dry weight). Citrus fruits are also rich in flavonoids such as hesperidin, hesperetin, naringin, naringenin, diosmin, quercetin, rutin, nobiletin, tangeretin, and others [45]. Citrus flavonoids have both antioxidant and anti-inflammatory properties, and because of that, it can increase the antioxidant capacity and effect reducing cholesterol and triglycerides levels and provide more excellent bone health [22]. Preclinical studies and clinical trials demonstrated that flavonoids' effects in the forms of hesperidin and its aglycone hesperetin prevent various types of diseases, including neurological, psychiatric, and cardiovascular disorders [46]. Over the years, naringin and hesperidin are gaining attention for their great antioxidant capacity, contributing sweet flavor to foods and beverages [47]. Some known naturally occurring flavonoids have potency in defending against certain types of RNA and DNA viruses [48].

Limonoids are also known as flavonoids, which are compounds found in citrus fruits. In citrus fruits, there are two groups of limonoids: aglycones and their corresponding glucosides. Bitter taste in citrus results from limonoids present. Limoninin's most important constituents are glycosides called limonin and nomilin [22]. In animal and human cell lines, limonoids slow down the development of aggressive cancers like the pancreas, colon, stomach, and breasts. On the other hand, limonoids are also reported to reduce skin cancer in animal models. Limonoids are known for their medicinal or health beneficial effects like anti-cancer, anti-microbial and antimalarial activities [49]. Limonoids have antibacterial and antiviral effects. Some limonoids are known to stimulate the in vivo production of the detoxifying enzyme glutathione S-transferase in the liver and inhibit the formation of chemically induced tumor cells in the oral cavity, forestomach, small intestine, colon, lung, and skin of animals [50, 51]. Limonoids found in citrus fruits decreased the spread of cancer cells in animal studies [52]. Studies showed that lemon-lime oil in the form of obacunone appeared to prevent breast cancer by inhibiting aromatase enzyme and anti-inflammatory pathways [53] (Figure 2). Several studies have revealed that limonin and nomilin are found in fruits, pulp, and seeds. Pummelo juice contained 18 ppm of limonin and 29 ppm of total limonoid glucosides [54]. Wattanasiritham et al. [55] reported limonin content of 18 ppm in the juice of pomelo cultivars. Limonin levels in extracted juice from seven pummelo cultivars from Florida ranged from 10.07 to 29.62

ppm. As shown by analysis, mature lemon seeds contain 1300 µg/g of different limonoid glucosides on a fresh weight basis [56]. The mature seeds contain much higher amounts of glucosides than commercialized juice, therefore. Fong et al. [57] documented that the commercial lemon juice had only 82 µg/g of glucosides.

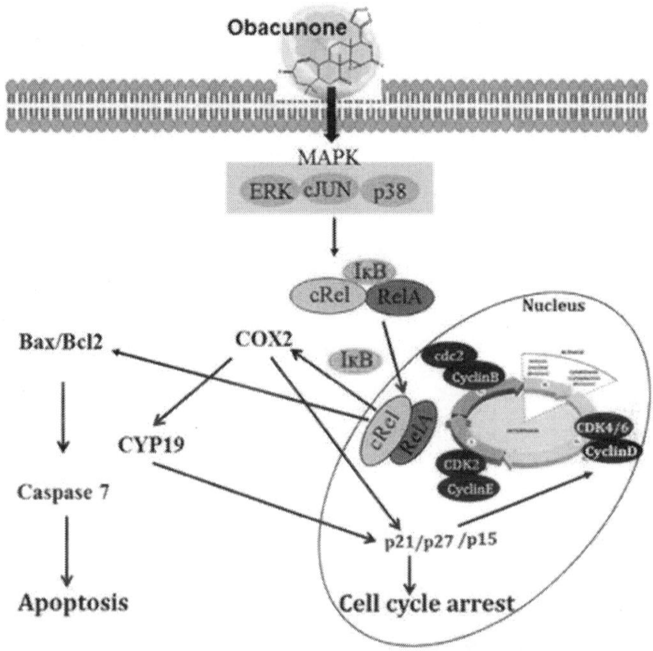

Figure 2. Proposed model for signaling pathways leading to growth inhibition by obacunone in estrogen-responsive breast cancer (MCF-7) cells [53].

4. Value-added products

Citrus fruits are known for being highly fragrant, with a tart taste and higher vitamin C content. The world has a wider variety of citrus fruit because of the continued type changes, such as sour oranges, oranges, pummelos, lemons, and limes, among others [22]. Nowadays, citrus pulp/pomace, seed, and peel are used for various commercially valuable products such as food ingredients, pectin, essential oils, enzyme production, a natural antioxidant, and packaging film formation.

4.1 Food ingredient or food products

The nutritional supplement of pomelo fruit segments has been added to products to be developed like noodles prepared with 30% new segments and 5% dry. These noodles can satisfy those with regular diabetes and the general public [21]. A high dietary fiber food was created by reducing the dietary fiber-rich pomelo peel to a powder that contained nearly 50% of dietary fiber. Lemon fruit is usually eaten fresh, but it is also processed to make juices, jams, jellies, molasses, candies and much more [18]. Lemon juice has been used as a coagulant during the manufacture of wara cheese [58]. Another innovation implemented in the beef burger is to use lemons for "enhancing the cooking properties of the burger" [59]. Lario et al. [60] had reported that the high-fiber lemon powder extract from lemon peel debris by-products is an ideal additive for food products (as meat, dairy, and bakery products).

4.2 Pectin

Pectin is an agent of gelling, emulsifying, stabilizing, texturizing, which appears as a white to light brown powder broadly accepted as a functional ingredient [61]. Fruit peels are a highly desirable pectin source because they cover up to 20% of the fruit's total Pectin [62]. Pomelo is a highly valued source of natural Pectin. About 20.75% pectin is derived from lemon peel for jams. In the study, the high extraction (36.71%) of Pectin from lemon peel has something to contribute to this industry [63]. Moneim et al. [64] recommended utilizing 20.75% of the lemon peels' total weight in making pumpkin jam. The researchers were added 16.740% of Pectin from pomelo peel to the pressed carrots before storage [65]. Studies by Methacanon et al. [66] have shown that pectin yield was 23.19% for pomelo peel. On the other hand, Roy et al. [65] were found that pomelo peels are a good source of Pectin, and then carrot jam made by extracting Pectin from pomelo peel.

4.3 Essential oil

Essential oils (Eos) are volatile, complex, natural mixture of aromatic oils obtained from plants [67]. Citrus essential oil is commonly known to produce a good fragrance and has been officially approved for healthy public consumption. All over the world, EOs are used in cosmetics, perfumery, toiletries, flavoring, beverage, pharmaceuticals and other personal hygiene products [68, 69, 70]. Lemon oil is often used on the skin because of its antimicrobial and antifungal properties. Pomelo peel (PP) has approximately 299 recognized volatile compounds. A significant number of these volatile compounds is considered as terpenoids (189 volatiles, 63.2%). Various kinds of chemicals released are monoterpenoids, monocyclic monoterpenoids, bicyclic monoterpenoids, diterpenoids, acyclic sesquiterpenoids, monocyclic sesquiterpenoids, bicyclic sesquiterpenoids, and tricyclic sesquiterpenoids. Another major volatile present in PP EO is nonterpenoid alcohols (4.7%), nonterpenoid aldehydes (6.0%), nonterpenoid hydrocarbons (5.7%), and esters (8.7%). The unknown volatiles covered 11.7% (35 volatiles) of total volatile compounds. The structures of widely known terpenoids in Polypropylene EO has shown in Figure 3a,b. The most critical monoterpenoids (1 to 16) and sesquiterpenoids (17 to 29) are present in PP EO [19]. According to studies conducted, lemon essential oils retain aroma in foods because of their natural preservative and flavoring properties [71]. The effect of lemon essential oils on the cheesemaking process dramatically reduces microorganisms' population, especially those of the Enterobacteriaceae family [58]. In the food and pharmaceutical industries, citrus EOs can be employed to inhibit mold and fungal growth. Lemon EO (Citrus limon) was used as a possible fungicide to manage the pathogenic fungi attacking grapevines, namely Eutypa sp., Botryospaeria dothidea, and Fomitiporia mediterranea. The antifungal activity was observed for EO against all the three fungi with the highest action against strain Eutypa sp. (82% inhibition) and the lowest tolerance (33.1%) towards F. Mediterranean [72]. These essential oils repress the growth of mold and yeast. The raspberries coated with alginate and lemon EO (0.2 percent) or orange EO (0.1 percent) halted bacteria, yeast, and mold and also reduced the quality deterioration right after harvest [73]. The lemon EOs mixed in the chitosan films can be used to control L. Multicellular pathogens in refrigerated foods Researchers Rahmawati et al. [74] observed that an edible coating applied with lemon essential oil only delayed aging of tofu and fresh strawberry.

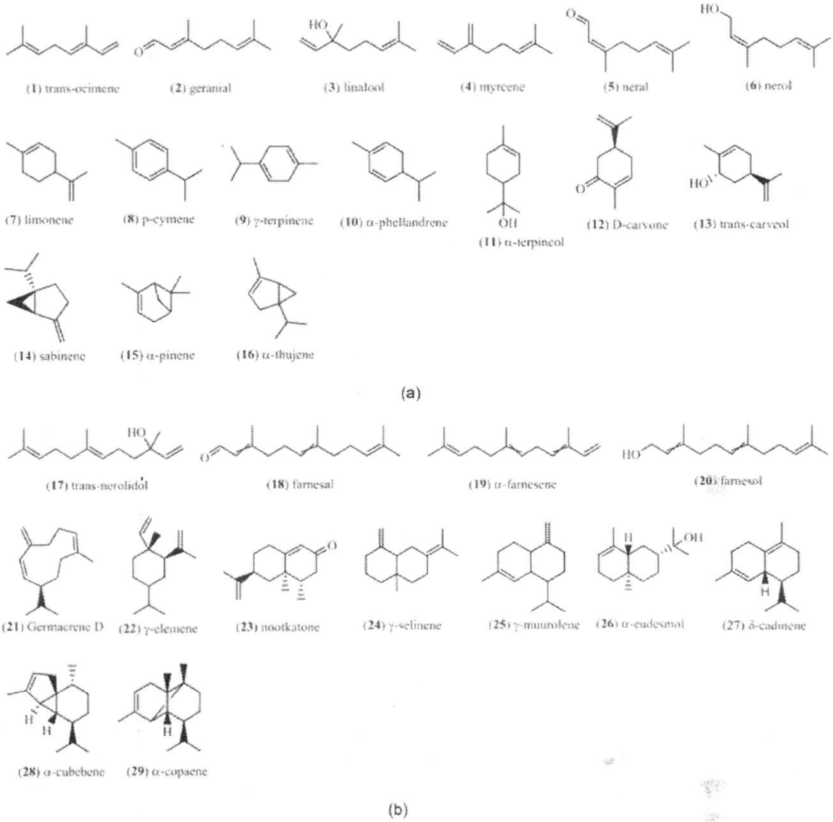

Figure 3. (a) Major monoterpenoids in pomelo peel essential oils (1-16); (b) Major sesquiterpenoids in pomelo peel essential oils (17-29).

4.4 Enzymes

The most basic usage of citrus peels is to produce the pectinolytic enzyme for beneficial purposes. Larios et al. [75] studied endo-polygalacturonase production by Aspergillus sp. CH-Y-1043 using untreated lemon peel and citrus pectin as carbon sources. Lemon peel being employed as a substrate in submerged cultures to obtain high pectinase titers by A. flavipes FP-500 and A. terreus FP-370 [76]. A. niger produces approximately 2,181.23 U/L pectinases from lemon, peel pomace in a solid-state reactor [77]. Seyis and Aksoz [78] have shown that lemon pomace and peel are suitable substrates for heterotrophic xylanases enzyme production using fungus Trichoderma harzianum. Aspergillus niger LFP-1 was studied in solid-state fermentation (SSF) using pomelo (Citrus grandis) peels as a substrate [79]. Maller, et al. [80] determined that lemon peels are extremely capable of triggering the production of Polygalacturonase in the aspergillus niveus. Pectin lyase yield increased through fungal strain Aspergillus oryzae process derived from lemons peel and used in solid-state fermentations [81]. Studies said that Polemo pericarp powder utilized as a substrate for Aspergillus oryzae JMU316 has Naringinase enzyme [82]. Pectinase enzyme produced from pomelo peels by By Aspergillus niger through Solid State Fermentation [83]. Lemon peels could be a good source of naringin, which could be used as a carbon source in submerged fermentation for naringinase production using Aspergillus niger [84]. Naringinase is essential for the production of sweetener precursors, preparation of prunin, aroma enhancement in winemaking, biotransformation of antibiotics, and rhamnose manufacturing [82].

4.5 Natural antioxidant

Antioxidants are chemical substances that can reduce or prevent the damage caused by free radicals in the body, thus reducing the risks of cardiovascular disease and cancer [85]. Results of studies showed that lemon peel contained almost 75.9% of antioxidant content. The unique ability of Paneer was derived from compounds found in the Peel of orange, lemon and pomegranate [85]. Peel taken from Tambun White pomelo type contains higher levels of antioxidants and is also a rich source of natural antioxidants [44]. Lemon peel and flesh had the highest antioxidant capacity, and they had a significant impact on the prevention of cardiovascular diseases and other diseases [86].

4.6 Packaging film formation

More sustainable, biodegradable plastic has gained popularity among environmental scientists. The researcher created plastic films from citrus peels. By applying the peels of citrus, biodegradable packaging material could be made [87]. Wu et al. [88] prepared fruit peel as the edible packaging film with high content biopolymer to form film for packaging. The film was designed to incorporate tea polyphenols, which causes interacting molecules to become more closely crowded. Soy protein with essential oil of lemon peel was used to create a degradable film and create cheese curdorants for preservation [89]. Dias et al. [90] reported that the use of citrus essential oil and its aroma significantly improved consumers' health and significantly increased the acceptance of biscuits' packaging. Das et al. [91] demonstrated that chicken feather keratin combined with pomelo peel pectin to form biodegradable composite film and wrapping of fried fish fillets resulted in less weight loss, hardness value, and reduction in the surface microbial count.

5. Conclusion

Citrus has positive effects on human health, and it could be an essential raw material to the biotechnological industry. Citrus is a mighty source of vitamins, minerals, and dietary fibers. Bioactive and non-nutrient compounds in citrus are valuable for controlling chronic diseases such as diabetes, cholesterol, obesity, cardiovascular disease, and some forms of cancer. Besides citrus, vitamin C also has other benefits, including fighting diseases such as cardiovascular disease, boosting white blood cells, immune function, and symptoms or duration of colds. Peel, flesh/pomace, and seed from the citrus fruit are employed in making different novel foods like noodles, extract pectin, enzyme extracts, and essential oil. Therefore this information might be necessary for the readers because it gives facts about the popular citrus fruit. Also, choosing the best citrus for an edible ingredient can be beneficial for citrus processors.

References

1. Liu Y, Heying E, Tanumihardjo SA. History, Global Distribution, and Nutritional Importance of Citrus Fruits. Comprehensive Reviews in Food Science and Food Safety. 2012; 11: 530-545. DOI: 10.1111/j.1541-4337.2012.00201.x
2. Mamma D, Christakopoulos P. Biotransformation of Citrus By-Products into Value Added Products. Waste and Biomass Valorization. 2014; 5: 529-549. DOI: 10.1007/s12649-013-9250-y
3. Alnaimy A, Gad AE, Mustafa MM, Atta MAA, Basuony HAM. Using of Citrus By-products in Farm Animals Feeding. Open Access Journal of Science. 2017; 1: 58-67. DOI: 10.15406/oajs.2017.01.00014
4. Abobatta FW. Nutritional Benefits of Citrus Fruits. American Journal of Biomedical Science & Research. 2019; 3: 303-306. DOI: 10.34297/ajbsr.2019.03.000681
5. Economos C, Clay WD. Nutritional and health benefits of citrus fruits. Food, Nutrition and Agriculture – FAO. 2005; 24: 11-18. http://www.fao.org/3/a-x2650t/X2650t03.pdf
6. Ghanem N, Mihoubi D, Kechaou N, Mihoubi NB. (2012). Microwave dehydration of three citrus peel cultivars: Effect on water and oil retention capacities, color, shrinkage and total phenols content. Industrial Crops and Products. 2012; 40: 167-177. DOI: 10.1016/j.indcrop.2012.03.009
7. Bamise CT, Oziegbe EO. Laboratory Analysis of pH and Neutralizable Acidity of Commercial Citrus Fruits in Nigeria Department of Child Dental Health, Faculty of Dentistry. 2013; 7: 72-76. DOI: 10.5829/idosi.abr.2013.7.4.73175
8. Manthey JA, Grohmann K. Phenols in Citrus Peel Byproducts. Concentrations of Hydroxycinnamates and Polymethoxylated Flavones in Citrus Peel Molasses. Journal of Agricultural and Food Chemistry. 2001; 49:3268-3273. DOI: 10.1021/jf010011r
9. Chen XM, Tait AR, Kitts DD. Flavonoid composition of orange peel and its association with antioxidant and anti-inflammatory activities. Food Chemistry. 2017; 218:15-21. DOI: 10.1016/j.foodchem.2016.09.016
10. Wang L, Wang J, Fang L, Zheng Z, Zhi D, Wang S, Li S, Ho CT, Zhao H. Anticancer activities of citrus peel polymethoxyflavones related to angiogenesis and others. BioMed Research International. 2014; 2014:1-10. DOI: 10.1155/2014/453972
11. Ashraf H, Butt MS, Iqbal MJ, Suleria HAR. Citrus peel extract and powder attenuate hypercholesterolemia and hyperglycemia using rodent experimental modeling. Asian Pacific Journal of Tropical Biomedicine. 2017; 7: 870-880. DOI: 10.1016/j.apjtb.2017.09.012
12. Viuda-Martos M, Ruiz-Navajas Y, Fernández-López J, Pérez-Álvarez J. Antifungal activity of lemon (Citrus lemon L.), mandarin (Citrus reticulata L.), grapefruit (Citrus paradisi L.) and orange (Citrus sinensis L.) essential oils. Food Control. 2008; 19: 1130-1138. DOI: 10.1016/j.foodcont.2007.12.003
13. Pagán J, Ibarz A, Llorca M, Pagán A, Barbosa-Cánovas GV. Extraction and characterization of pectin from stored peach pomace. Food Research International. 2001; 34: 605-612. DOI: 10.1016/S0963-9969(01)00078-3

14. Boluda-Aguilar M, López-Gómez A, Production of bioethanol by fermentation of lemon (Citrus limon L.) peel wastes pretreated with steam explosion. Industrial Crops and Products. 2013; 41: 188-197. DOI: 10.1016/j.indcrop.2012.04.031

15. Makni M, Jemai R, Kriaa W, Chtourou Y, Fetoui H. Citrus limon from Tunisia: Phytochemical and Physicochemical Properties and Biological Activities. BioMed Research International. 2018; 2018: 1-10. DOI: 10.1155/2018/6251546

16. Paul DK, Shaha RK. Nutrients, Vitamins and Minerals Content in Common Citrus Fruits in the Northern Region of Bangladesh. Pakistan Journal of Biological Sciences. 2004; 7: 238-242. DOI: 10.3923/pjbs.2004.238.242

17. Kumar D, Ladaniya MS, Gurjar M. Underutilized Citrus sp. Pomelo (Citrus grandis) and Kachai lemon (Citrus jambhiri) exhale in phytochemicals and antioxidant potential. Journal of Food Science and Technology. 2019; 56: 217-223. DOI: 10.1007/s13197-018-3477-3

18. González-Molina E, Domínguez-Perles R, Moreno DA, García-Viguera C. Natural bioactive compounds of Citrus limon for food and health. Journal of Pharmaceutical and Biomedical Analysis. 2010; 51: 327-345. DOI: 10.1016/j.jpba.2009.07.027

19. Tocmo R, Pena-Fronteras J, Calumba KF, Mendoza M, Johnson JJ. Valorization of pomelo (Citrus grandis Osbeck) peel: A review of current utilization, phytochemistry, bioactivities, and mechanisms of action. Comprehensive Reviews in Food Science and Food Safety. 2020; 19: 1969-2012. DOI: 10.1111/1541-4337.12561

20. Mäkynen K, Jitsaardkul S, Tachasamran P, Sakai N, Puranachoti S, Nirojsinlapachai N, Chattapat V, Caengprasath N, Ngamukote S, Adisakwattana S. Cultivar variations in antioxidant and antihyperlipidemic properties of pomelo pulp (Citrus grandis [L.] Osbeck) in Thailand. Food Chemistry. 2013; 139: 735-743. DOI: 10.1016/j.foodchem.2013.02.017

21. Reshmi SK, Sudha ML, Shashirekha MN. Noodles fortified with Citrus maxima (pomelo) fruit segments suiting the diabetic population. Bioactive Carbohydrates and Dietary Fibre. 2020; 22:100213. DOI: 10.1016/j.bcdf.2020.100213

22. Turner T, Burri B. Potential Nutritional Benefits of Current Citrus Consumption. Agriculture. 2013; 3: 170-187. DOI:10.3390/agriculture3010170

23. Wintergerst ES, Maggini S, Hornig DH. Immune-enhancing role of vitamin C and zinc and effect on clinical conditions. Annals of Nutrition and Metabolism. 2006, 50: 85-94. DOI: 10.1159/000090495

24. Chappell LC, Seed PT, Briley AL, Kelly FJ, Lee R, Hunt BJ, Parmar K, Bewley SJ, Shennan AH, Steer PJ, Poston L. Effect of antioxidants on the occurrence of pre-eclampsia in women at increased risk: A randomized trial. Lancet 1999, 354: 810-816. DOI: 10.1016/S0140-6736(99)80010-5

25. Wang YC, Chuang YC, Ku YH. Quantitation of bioactive compounds in citrus fruits cultivated in Taiwan. Food Chemistry. 2007; 102: 1163-1171. DOI: 10.1016/j.foodchem.2006.06.057

26. Haque MN, Saha BK, Karim MR, Bhuiyan MNH. Evaluation of Nutritional and Physico-Chemical Properties of Several Selected Fruits in Bangladesh. Bangladesh Journal of Scientific and Industrial Research. 2009; 44: 353-358. DOI: 10.3329/bjsir.v44i3.4410

27. Britton G. Structure and properties of carotenoids in relation to function. The FASEB Journal. 1995; 9: 1551-1558. DOI:10.1096/fasebj.9.15.8529834

28. Khachik F, Christopher JS; Smith JC, Louise MC, Steck A, Pfander H. Identification, quantification, and relative concentrations of carotenoids and their metabolites in human milk and serum. Analytical Chemistry. 1997; 69: 1873-1881. DOI:10.1021/ac961085i

29. Aust O, Sies H, Stahl W, Polidori MC. Analysis of lipophilic antioxidants in human serum and tissues: tocopherols and carotenoids. Journal of Chromatography A. 2001; 936: 83-93. DOI: 10.1016/s0021-9673(01)01269-9

30. Fraser PD, Bramley PM. The biosynthesis and nutritional uses of carotenoids. Progress in Lipid Research. 2004; 43: 228-265. DOI: 10.1016/j.plipres.2003.10.002

31. Slattery ML, Benson J, Curtin K, Ma KN, Schaeffer D, Potter JD. Carotenoids and colon cancer. The American Journal of Clinical Nutrition. 2000; 71: 575-582. DOI: 10.1093/ajcn/71.2.575

32. Wang YC, Chuang YC, Hsu HW. The flavonoid, carotenoid and pectin content in peels of citrus cultivated in Taiwan. Food Chemistry. 2008; 106: 277-284. DOI: 10.1016/j.foodchem.2007.05.086

33. Xu CJ, Fraser PD, Wang WJ, Bramley PM. Differences in the carotenoid content of ordinary citrus and lycopene-accumulating mutants. Journal of Agricultural and Food Chemistry. 2006; 54: 5474-5481. DOI: 10.1021/jf060702t

34. Silalahi J. Anticancer and health protective properties of citrus fruit components. Asia Pacific Journal of Clinical Nutrition. 2002; 11:79-84. DOI: 10.1046/j.1440-6047.2002.00271.x

35. Green NS. Folic acid supplementation and prevention of birth defects. The Journal of nutrition. 2002; 132: 2356S–2360S. DOI: 10.1093/jn/132.8.2356S

36. Brouwer IA, Van Dusseldorp M, West CE, Meyboom S, Thomas CMG, Duran M, Van Het Hof KH, Eskes TKAB, Hautvast JGAJ, Steegers-Theunissen RPM. Dietary folate from vegetables and citrus fruit decreases plasma homocysteine concentrations in humans in a dietary controlled trial. Journal of Nutrition. 1999; 129: 1135-1139. DOI: 10.1093/jn/129.6.1135

37. El-Otmani M, Ait-Oubahou A. Citrus spp.: orange, mandarin, tangerine, clementine, grapefruit, pomelo, lemon and lime. Elhadi MY, editor. In Postharvest Biology and Technology of Tropical and Subtropical Fruits. Woodhead Publishing Limited: 2011; p.437-514. DOI: 10.1533/9780857092762.437

38. Baker RA. Potential Dietary Benefits of Citrus Pectin and Fibre. Food Technology. 1994; 11: 133-139

39. Silalahi J. Hypocholesterolemic Factors in Foods. A Review. Indonesian Food and Nutrition Progress. 2000; 7: 26-35

40. Brown L, Rosner B, Willett WW, Sacks FM. Cholesterol-lowering effects of dietary fiber: A meta-analysis. The American Journal of Clinical Nutrition. 1999, 69: 30-42. DOI: 10.1093/ajcn/69.1.30

41. Dikeman CL, Fahey GC. Viscosity as related to dietary fiber: A review. Critical Reviews in Food Science and Nutrition. 2006; 46: 649-663. DOI: 10.1080/10408390500511862

42. Anderson JW, Baird P, Davis RH Jr, Ferreri S, Knudtson M, Koraym A, Waters V, Williams CL. Health benefits of dietary fiber. Nutrition Reviews. 2009; 67:188-205. DOI: 10.1111/j.1753-4887.2009.00189.x

43. Terpstra AHM, Lapré JÁ, de Vries HT, Beynen AC. The hypocholesterolemic effect of lemon peels, lemon pectin, and the waste stream material of lemon peels in hybrid F1B hamsters. European Journal of Nutrition. 2002; 41: 19-26. DOI:10.1007/s003940200002

44. Toh JJ, Khoo HE, Azrina A. Comparison of antioxidant properties of pomelo [Citrus Grandis (L) Osbeck] varieties. International Food Research Journal. 2013; 20: 1661-1668

45. Mahmoud AM, Hernández Bautista RJ, Sandhu MA, Hussein OE. Beneficial effects of citrus flavonoids on cardiovascular and metabolic health. Oxidative Medicine and Cellular Longevity. 2019; 2019: 1-19. DOI: 10.1155/2019/5484138

46. Li C, Schluesener H. Health-promoting effects of the citrus flavanone hesperidin. Critical Reviews in Food Science and Nutrition. 2015; 57: 613-631. DOI:10.1080/10408398.2014.906382

47. HungPV,NhiNHY,TingLY,PhiNTL.Chemical composition and biological activities of extracts from pomelo peel by-products under enzyme and ultrasound-assisted extractions.Journal of Chemistry.2020;2020:1-7. DOI: 10.1155/2020/1043251

48. Puri M, Banerjee UC. Production, purification, and characterization of the debittering enzyme naringinase. Biotechnology Advances. 2000; 18: 207-217. DOI: 10.1016/S0734-9750(00)00034-3

49. Roy A, Saraf S. Limonoids: Overview of Significant Bioactive Triterpenes Distributed in Plants Kingdom. Biological and Pharmaceutical Bulletin. 2006; 29: 191-201. DOI:10.1248/bpb.29.191

50. Manners GD, Jacob RA, Breksa, Schoch TK, Hasegawa S. Bioavailability of Citrus Limonoids in Humans. Journal of Agricultural and Food Chemistry. 2003; 51: 4156-4161. DOI: 10.1021/jf0300691

51. Yu J, Wang L, Walzem RL, Miller EG, Pike LM, Patil BS. Antioxidant Activity of Citrus Limonoids, Flavonoids, and Coumarins. Journal of Agricultural and Food Chemistry. 2005; 53: 2009-2014. DOI:10.1021/jf0484632

52. Zunino SJ, Storms DH, Freytag TL, Adkins YC, Bonnel EL, Woodhouse LR, Breksa AP, Manners GD, Mackey BE, Kelley DS. Dietary supplementation with purified citrus limonin glucoside does not alter ex vivo functions of circulating T lymphocytes or monocytes in overweight/obese human adults. Nutrition Research. 2016; 36: 24-30. DOI: 10.1016/j.nutres.2015.10.011

53. Kim J, Jayaprakasha GK, Patil BS. Obacunone exhibits anti-proliferative and anti-aromatase activity in vitro by inhibiting the p38 MAPK signaling pathway in MCF-7 human breast adenocarcinoma cells. Biochimie. 2014; 105: 36-44. DOI:10.1016/j.biochi.2014.06.002

54. Ohta H, Hasegawa S. Limonoids in Pummelos [Citrus grandis (L.) Osbeck]. Journal of Food Science. 1995; 60: 1284-1285. DOI:10.1111/j.1365-2621.1995.tb04574.x

55. Wattanasiritham L, Taweesuk K, Ratanachinakorn B. Limoninand Naringin in Pummelos (Citrus grandis (L.) Osbeck) In: 31st Congress on science and technology of Thailand, Suranaree University of Technology, Thailand. 2005

56. Fong CH, Hasegawa S, Herman Z, Ou P. Biosynthesis of limonoid glucosides in lemon (Citrus limon). Journal of the Science of Food and Agriculture. 1991; 54: 393-398. DOI: 10.1002/jsfa.2740540310

57. Fong CH, Hasegawa S, Herman Z, Ou P. Limonoid Glucosides in Commercial Citrus Juices. Journal of Food Science. 1989; 54:1505-1506. DOI:10.1111/j.1365-2621.1989.tb05146.x

58. Adetunji VO, Alonge DO, Singh RK, Chen J. Production of wara, a West African soft cheese using lemon juice as a coagulant. LWT - Food Science and Technology. 2008; 41: 331-336. DOI:10.1016/j.lwt.2007.02.012

59. Aleson-Carbonell L, Fernández-López J, Pérez-Alvarez JA, Kuri V. Characteristics of beef burger as influenced by various types of lemon albedo. Innovative Food Science and Emerging Technologies. 2005; 6: 247-255. DOI: 10.1016/j.ifset.2005.01.002

60. Lario Y, Sendra E, García-Pérez J, Fuentes C, Sayas-Barberá E, Fernández-López J, Pérez-Alvarez JA. Preparation of high dietary fiber powder from lemon juice by-products. Innovative Food Science and Emerging Technologies. 2004; 5: 113-117. DOI: 10.1016/j.ifset.2003.08.001

61. Mesbahi G, Jamalian J, Farahnaky A. A comparative study on functional properties of beet and citrus pectins in food systems. Food Hydrocolloids. 2005; 19: 731-738..DOI: 10.1016/j.foodhyd.2004.08.002

62. Chavan P, Singh AK, Kaur G. Recent progress in the utilization of industrial waste and by-products of citrus fruits: A review. Journal of Food Process Engineering. 2018; 41:1-10. DOI: 10.1111/jfpe.12895

63. Kanmani P, Dhivya E, Aravind J, Kumaresan K. Extraction and analysis of pectin from citrus peels: augmenting the yield from citrus limon using statistical experimental design. Iranica Journal of Energy and Environment. 2014; 5: 303-312. DOI: 10.5829/idosi.ijee.2014.05.03.10

64. Moneim A, Sulieman E, Khodari KMY, Salih ZA. Extraction of pectin from lemon and orange fruits peels and its utilization in jam making. International Journal of Food Science and Nutrition Engineering. 2013; 3: 81-84. DOI: 10.5923/j.food.20130305.01

65. Roy MC, Alam M, Saeid A, Das BC, Mia MB, Rahman MA, Eun JB, Ahmed M. Extraction and characterization of pectin from pomelo peel and its impact on nutritional properties of carrot jam during storage. Journal of Food Processing and Preservation. 2018; 42: 1-9. DOI: 10.1111/jfpp.13411

66. Methacanon P, Krongsin J, Gamonpilas C. Pomelo (Citrus maxima) pectin: Effects of extraction parameters andits properties. Food Hydrocolloids. 2014; 35: 383-391. DOI: 10.1016/j.foodhyd.2013.06.018

67. Figueiredo CA, Barroso JG, Pedro LG, Scheffer JJC. Factors affecting secondary metabolite production in plants: volatile components and essential oils. Flavour and Fragrance Journal. 2008; 23:213-226. DOI: 10.1002/ffj.1875

68. Palazzolo E, Laudicina VA, Germanà MA. Current and potential use of citrus essential oils. Current Organic Chemistry. 2013; 17: 3042-3049. DOI: 10.2174/13852728113179990122

69. Sarkic A, Stappen I. Essential oils and their single compounds in cosmetics—A critical review. Cosmetics. 2018; 5:1-11. DOI: 10.3390/cosmetics5010011

70. Mahato N, Sharma K, Koteswararao R, Sinha M, Baral ER, Cho MH. Citrus essential oils: Extraction, authentication and application in food preservation. Critical Reviews in Food Science and Nutrition. 2019; 59: 611-625. DOI: 10.1080/10408398.2017.1384716

71. Klimek-Szczykutowicz M, Szopa A, Ekiert H. Citrus limon (lemon) phenomenon—a review of the chemistry, pharmacological properties, applications in the modern pharmaceutical, food, and cosmetics industries, and biotechnological studies. Plants. 2020; 9: 119; DOI:10.3390/plants9010119

72. Ammad F, Moumen O, Gasem A, Othmane S, Hisashi KN, Zebib B, Merah O. The potency of lemon (Citrus limon L.) essential oil to control some fungal diseases of grapevine wood. Comptes Rendus Biologies, 2018; 341: 97-101. DOI: 10.1016/j.crvi.2018.01.003

73. Gomes MDS, Cardoso MDG, Guimarães ACG, Guerreiro AC, Gago CML, Vilas Boas EVDB, Dias CMB, Manhita ACC, Faleiro ML, Miguel MGC. Effect of edible coatings with essential oils on the quality of red raspberries over shelf-life. Journal of the Science of Food and Agriculture. 2017; 97:929-938. DOI: 10.1002/jsfa.7817

74. Rahmawati D, Chandra M, Santoso S, Puteri MG. Application of lemon peel essential oil with edible coating agent to prolong shelf life of tofu and strawberry. AIP Conference Proceedings. American Institute of Physics. 2017; 0200371-5 DOI: 10.1063/1.4973164

75. Larios G, Garcia JM, Huitrón C. Endo-polygalacturonase production from untreated lemon peel by Aspergillus sp. CH-Y-1043. Biotechnology Letters. 1989; 11:729-734. DOI: 10.1007/BF01044106

76. Martínez-Trujillo A, Arreguín-Range L, García-Rivero M, Aguilar-Osorio G. Use of fruit residues for pectinase produc- tion by Aspergillus flavipes FP-500 and Aspergillus terreus FP- 370. Letters in Applied Microbiology. 2011; 53: 202-209.DOI: 10.1111/j.1472-765X.2011.03096.x

77. Ruiz HA, Rodríguez-Jassoa RM, Rodríguez R, Contreras-Esquivel JC, Aguilar CN. Pectinase production from lemon peel pomace as support and carbon source in solid-state fermentation column-tray bioreactor. Biochemical Engineering Journal. 2012; 65:90-95. DOI: 10.1016/j.bej.2012.03.007

78. Seyis I, Aksoz N. Xylanase production from Trichoderma harzianum 1073 D3 with alternative carbon and nitrogen sources. Food Technology and Biotechnology. 2005; 43:37-40

79. Darah I, Taufiq MMJ, Lim SH. Pomelo citrus grandis (1.) Osbeck peel as an economical alternative substrate for fungal pectinase production. Food Science and Biotechnology. 2013; 22: 1683-1690. DOI: 10.1007/s10068-013-0267-6

80. Maller A, Damásio AR, da Silva TM, Jorge JA, Terenzi HF, Polizeli M. Biotechnological potential of agro-industrial wastes as a carbon source to thermostable polygalacturonase production in aspergillus niveus. Enzyme Research. 2011; 2011: 1-6. DOI: 10.4061/2011/289206

81. Koser S, Anwar Z, Iqbal Z, Anjum A, Aqil T, Mehmood S, Irshad M. Utilization of Aspergillus oryzae to produce pectin lyase from various agro-industrial residues. Journal of Radiation Research and Applied Sciences. 2014; 7: 327-332.DOI: 10.1016/j.jrras.2014.05.001

82. Chen DX, Niu TG, Cai HN. Optimizing culture medium for debittering constitutive enzyme naringinase production by Aspergillus oryzae JMU316. African Journal of Biotechnology. 2010; 9: 4970-4978

83. Hussain K, Wajid A, Babar ME, Anwar Z, Farooqi S, Siddiqa A, Noreen S, Iqbal J. Production and optimization of pectinase from pomelo by Aspergillius niger through solid state fermentation. Annals of Life Sciences. 2019; 6: 21-41

84. Igbonekwu A, Omeje KO, Ezugwu AL, Eze SOO NO, Chilaka CF. Characterization of Nariginase obtained from Aspergillus niger by submerged fermentation using naringin extracted from lemon peels. Research and Development in Material Science. 2018; 4: 420-424. DOI: 10.31031/rdms.2018.04.000599

85. Immanuel G, Singh S. Extraction of antioxidants from fruit peels and its utilization in paneer. Journal of Food Processing and Technology. 2014; 5:1-5. DOI: 10.4172/2157-7110.1000349

86. Gorinstein S, Martín-Belloso O, Park YS, Haruenkit R, Lojek A, Íž M, Caspi A, Libman I, Trakhtenberg S. Comparison of some biochemical characteristics of different citrus fruits. Food Chemistry. 2001; 74: 309-315. DOI: 10.1016/S0308-8146(01)00157-1

87. Shinde M, Sonawane SK, Patil S. Fruit peel utilization in food packaging. Indian Food Industry Mag. 2019; 1:19-24

88. Wu H, Lei Y, Zhu R, Zhao M, Lu J, Xia D, Jiao C, Zhang Z, Shen G, Li S. Preparation and characterization of bioactive edible packaging films based on pomelo peel flours incorporating tea polyphenol. Food Hydrocolloids. 2019; 90: 41-49. DOI: 10.1016/j.foodhyd.2018.12.016

89. Al-Sahlany STG. Production of biodegradable film from soy protein and essential oil of lemon peel and use it as cheese preservative. Basrah Journal of Agricultural Sciences. 2017; 30: 27-35. DOI: 10.37077/25200860.2017.40

90. Dias MV, de Medeiros HS, Nilda de FFS, de Melo NR, Borges SV, João de DSC, Pereira JMT de AK. Development of low-density polyethylene films with lemon aroma. LWT - Food Science and Technology. 2013; 50: 167-171. DOI: 10.1016/j.lwt.2012.06.005

91. Das P, Borah PP, Badwaik LS. Transformation of chicken feather keratin and pomelo peel pectin into biodegradable composite film. Journal of Polymers and the Environment. 2018; 26:2120-2129. DOI: 10.1007/s10924-017-1109-z

Chapter-10

Citrus Essential Oils: A Suite of Insecticidal Compounds

1. Introduction

The genus Citrus has tremendous industrial value all over the globe not only for its nutritive juicy high valued fruits but also for the essential oils present in its different vegetative parts. Thus, both the Citrus fruits and citrus essential oils bear potential to generate livelihood & to boost the country's economy. Citrus essential oils (CEOs) with diverse biologically active compounds of terpene groups with pleasant aroma have already achieved significant positions in flavor, food, cosmetic industries. At the same time, because of their antimicrobial activities as well as anticancer, antioxidant, anti-inflammatory, metabolic disorder alleviating activities etc. these oils and their compounds have been getting importance in pharmaceutical and medical sectors for the last few decades [1]. A good number of studies also reported insecticidal potential of citrus EOs extracted from different citrus sp. and their constituents at different times, a few of which are commercialized to be used by the consumers against insect pests. There are 33 recorded species of citrus worldwide (ThePlantList.org) with many recorded and unexplored varieties present in different parts of the world. The essential oil profile of different citrus species varies although some of the constituent compounds are common but present in different amounts in the total bulk oil. Even the oil profile of different vegetative parts of a single citrus species are not identical. Understanding of essential oil profile of diverse citrus species grown in wild, semi wild and cultivated state across the globe at different seasons is the much-needed task as the quality of the oil, oil yield percentage, consistency of the constituents even varies with seasonal changes, geographical location, harvesting time of the plant parts, soil type etc. however from the existing GC–MS profile of different *Citrus* sp. reported at different times and from different places, it is apparent that two -three dominant compounds are mostly present in most of the Citrus species. The literature revealed that the Citrus EO comprises more than 200 compounds of which %99–85 are volatile and 1–15% nonvolatile compounds. The volatile compounds comprise mostly monoterpenes (predominant limonene), some sesquiterpenes and their oxygenated derivatives [2].

Pest control sector is dominated by synthetic pesticidal products for many decades. At recent times with increasing concern to ecofriendly product, plant essential oils are getting renewed interest as they are not only effective but also comparatively safe and environment friendly in comparison to synthetic counterparts. Essential oils are part of natural plant defense system and many of them are proved effective and some are exploited for integrated management practices of pest and pathogens. As some citrus species are naturally resistant to certain group of pests and or pathogens, it is assumed that certain bioactive compounds may present in the essential oil part of those citrus species. It is already established that citrus essential oils of different citrus species are effective against wide range of pest and pathogens. It is also important to have an insight about the interaction of citrus constituents against its own insect pest and pathogen complex to be used as insecticidal, repellent and bactericidal etc. A few papers highlighted beneficial effects of using citrus essential oil against its own pest and pathogen complex. The added advantage of considering CEO as insecticidal and insect repellent is that the plant is edible therefore safe for residual contamination or toxicity to consumer. At the same time the pleasant aroma offers consumer acceptance.

2. Citrus EO against insect sp

CEO and extracts have been tried against a wide range of insect pests for assessing their insecticidal as well as repellent properties. In some parts of the world citrus plants have been traditionally used to ward off a insect pests. Some recent reports especially of the last two decades of the insecticidal and repellent effects of different citrus sp. are presented below. Most of the works were carried out on dipteran, lepidopteran, hemipteran and coleopteran insect pests.

Topical toxicity of the essential oil of *Citrus hystrix* with LD50 of 26.748 μL/g and antifeedant properties leading to severe growth inhibition has been reported against tobacco armyworm *Spodoptera litura* [3]. The fumigant toxicity and repellent effect of the n-hexane extract of the plant leaf was documented against stored grain pest *Lasioderma serricorne* [4]. Fumigant toxicity of peel oils of lime, orange, mandarin, tangerine, grapefruit and lemon were reported against three store grain pest species *Callosobruchus maculatus, Sitophilus zeamais and Dermestes maculatus* [5].

The peel essential oil of the plant is reported to possess repellent effect against *Callosobruchus maculatus* [6], *Aedes aegypti* and *Anopheles minimus* [7]. Similarly, the insecticidal and repellent activity of *Citrus reticulata, Citrus limon* and *Citrus aurantium* peel oils was demonstrated against *Callosobruchus maculatus* [8]. Insecticidal activity of *Citrus limon* and *Citrus sinensis* against vine mealybug, *Planococcus ficus* [9]. The larvicidal and adulticidal effects of *Citrus limon* and *Citrus sinensis* are mentioned against *Attagenus fasciatus* and *Lasioderma serricorne* [10]. The seed and peel extracts of *Citrus limon* L. was reported to have the highest larvicidal toxicity (LC50 values of 395.59 ppm for seed; 468.69 ppm for peel) after 24 hours over EOs of *Citrus grandis, Citrus sinensis, Citrus paradise, Citrus reticulate* [11]. Essential oil of *Citrus reticulata* and *Citrus sinensis* was reported effective against the fourth instar larvae and adults of *Tribolium castaneum* with higher potency of *Citrus reticulata* [12].

The seed EOs of *Citrus reticulate* var. kinnow, *Citrus reticulate* var. freutrall, *Citrus sinensis* and *Citrus jambhiri* was tested against *Tribolium castaneum* with promising efficacy in terms of LC50 for *Citrus jambhiri* followed by Citrus reticulate and *Citrus sinensis* [13]. Similarly Oboh et al. [14] recorded insecticidal efficacy of *C. sinensis* peel essential oil against Callosobruchus mamulatus, *Tribolium confusum*, *Sitophilus oryzae* with LC50 value of 60 ,38.9 ,21.8 μl/l.

Comparative evaluation of toxicity of EOs of *C. limon, C. aurantifolia, C.sinensis* in filter paper impregnation method showed highest toxicity of *C. limon* (%95 mortality) followed by *C. aurantifolia* (%92.5) and *C. sinensis* (%82.5) against carpenter ant *Camponotus nearcticus* [15]. But, Guerra et al. [16] comparatively lower topical toxicity (15% mortality) of *C. limon* EO against *Camponotus pennsylvanicus* among the eleven different EOs tested. Essential oils and or extracts of *C. maxima* or *C.grandis* have been reported effective against different mosquito species. In our earlier studies we recorded differential biocidal activities of essential oil extracted from peel and leaf part of *Citrus grandis* grown in Assam against different developmental stages of *Aedes aegypti* and *Culex quinquefasciatus* [17, 18]. EO extracted from leaves was found more effective against egg stage while oil from peel was recorded more effective against larval and adult stages of *A. aegypti* [18]. The leaf and peel oil of the plant was recorded highly effective against egg and larval stage with LC50 value of below 50 ppm but did not found much effective against adult stage of *Culex quinquefasciatus* although having repellent properties with good protection time [19]. In a recent study we observed synergistic larvicidal response of *Citrus grandis* leaf oil with *Allium sativum* bulb oil against *C. quinquefasciatus* [19]. Manorenjitha et al. [20] tested hexane, ethyl acetate, methanol, water and essential oil extract of *C. grandis* peel extract for evaluating oviposition deterrent and repellent properties on *Aedes aegypti*. They observed promising oviposition deterrent activity of ethyl acetate fraction (10 ppm concentration) in breeding plates kept within mosquito cage and effective repellency (94.7%) of 20% essential oil fraction of the peel while offering animal bait in modified tunnel test. A study for toxicity assessment on worker termites *Odontotermes feae*, essential oils of *Citrus grandis* with LC50 value of 273.36 ppm was found to show maximum toxicity out of *Citrus paradisi, Cassia fistula, Citrus grandis* EOs [21].

The peel essential oils of *Citrus aurantifolia* has been reported as insecticidal, repellent, and larvicidal against *Aedes aegypti* [22]. In our previous study, we observed the ovicidal, larvicidal and adulticidal effects of leaf and peel essential oil of *Citrus aurantifolia* against *Aedes aegypti* [23].

Promising fumigant toxicity of the peel EO of *Citrus aurantium* and *Citrus sinesis* from the north east Brazil with LC50 value of 5.80 μL/L of air and 3.80 μL/L of air respectively and oviposition deterrent activity at 3.5 and 7.0 μL/L of air against the whitefly *Bemisia tabaci* [24]. The same oil was reported effective against the larval and adult stages of tomato leafminer *Tuta asoluta* (Lepidoptera: Gelechiidae) [25]. The insecticidal activity of *Citrus aurantium* EO against adult housefly *Musca domestica* was reported next to the activity of EO of *Citrus sinensis* [26]. *Citrus aurantium* leaf EO was found as effective fumigant against sawtoothed grain beetle *Oryzaephilus surinamensis*, cigarette beetle *Lasioderma serricorne* and rice weevil *Sitophilus oryzae* with LC50 value of 202.49 ,64.94 and 364.25 μL/L of air respectively [27]. Similarly, Bnina et al. [28] noted fumigant toxicity of peel, leaves and flowers essential oil of *Citrus aurantium* against four stored grain pests namely *Tribolium castaneum, Liposcelis bostrychophila, Sitophilus granarium, Cryptolestes ferrugineus* with LC50 value of %101.50 ,%23.11 ,%64.78 and %20.62 respectively. They also noted repellent property of the oil against these pests. Yazdgerdian et al. [29] tested contact and residual toxicity of eleven essential oils including *Citrus aurantium*, *Citrus sinensis*, *Citrus limon* against wooly beech aphid, *Phyllaphis fagi* (Hemiptera: Aphididae) and rice weevil, *Sitophilus oryzae* (Coleoptera: Curculionidae) and recorded highest residual toxicity (%40) of *C. aurantium* among the citrus sp. tested against the targeted species. Similarly, without affecting the seed viability of stored cowpea and consumer acceptability, peel EOs of *Citrus nobilis* and *Citrus medica* was reported to show significant reduction of egg laying, egg hatching, and adult emergence percentage of pulse beetle *Callosobruchus maculatus*. Both the oils showed dose dependent repellency with higher effect for EO of *C. nobilis* [30]. Like that of EOs, the nonpolar petroleum ether extract of ripe fresh fruit of *C. aurantium* was reported effective against adults of olive fruit fly *Bactrocera oleae* (Diptera:Tephritidae) in petri dish residual exposure test [31]. The same solvent extract also reported to have good toxicity against medfly *Ceratitis capitata* adults (LC50 value of 70.6 and 147.1 lg/cm^2 for male and female respectively at 96 h in Petri dish residual bioassay) [32].

Moravvej et al. [33] tested fumigant toxicity of EOs from four citrus species namely *C. paradisi, C. limonium, C. sinensis* and *C. aurantium* among which *C. paradisi* was the most effective with LC50 value of 125 μl/L and *C.sinensis* is the least effective with LC50 value of 269 μl/L against *Callosobruchus maculatus*. However, fumigant toxicity of *C. sinensis* was reported against *Solenopsis invicta* (Hymenoptera: Formicidae) with %100 mortality at 3 mg/tube after 24 hrs [34]. Ethanolic extract of the same plant was found effective against *Anopheles stephensi, Aedes aegypti* and *Culex quinquefasciatus* with larval and adulticidal LC50 value of below 500 ppm along with more than %50 repellency at 150 ppm concentration till 180 min [35].

Ezeonu et al. [36] reported the insecticidal properties of the volatile peel extracts of *Citrus sinensis* and *Citrus aurantifolia* against mosquito, cockroach and housefly and recorded higher insecticidal potency of the peel extract of *Citrus sinensis* with maximum fumigant effect (%85 at 60 min) against cockroach.

Zewde and Jembere [37] evaluated the solvent extract and essential oil of *Citrus sinensis* against *Zabrotes subfasciatus* (Coleoptera: Bruchidae) for their repellent, fumigant and protectant properties and recorded no progeny emergence on application of oil at low dose (30 mg of EO), prominent percent mortality at high dose (750 mg of essential oil killed %67 of Z. subfaciatus after 96 hours). Fumigant toxicity of *C. sinensis* EO was also reported effective against second instar larvae of *Musca domestica* (Diptera: Muscidae) with LC50 values of 71.2 μl/L as well as percent inhibition of pupae with %46.4 at 40 μl/L exposure concentration [38].

Orange oil extract was also recorded effective against the subterranean termite *Coptotermes formosanus* (Isoptera: Rhinotermitidae). Application of the oil extract at 5 ppm concentration resulted in %96 and %68 mortality respectively in closed container. At the same time termites did not show tunneling behavior on the %0.4 & 0.2 oil treated soil [39].

Majeed et al. [40] reported the insecticidal activity of the acetone, ethanol and aqueous extracts of seeds, leaves and fruit peels and leaves of *Citrus aurantium* and *Citrus sinensis* and two other plants against mealy bug *Drosicha mangiferae* (Hemiptera; Pseudococcidae) of which ethanol extracts of *C. sinensis* seeds and *C. aurantium* leaves were suggested as considerably toxic against the said insect.

C. medica peel essential oil was reported as more effective in filter paper fumigation method against stored grain pest *Tribolium castaneum* (Coleoptera) with LC50 value of 29.5 mg/L air than the leaf EO [41].

Abdel-Kawy et al. [42] showed *Citrus trifoliata* essential oil loaded nano-cubosome significantly enhanced insecticidal property of the essential oil against the second instar larvae of *Spodoptera littoralis*.

While working as biocidal and repellents, plant products including EOs and constituent terpene compounds are reported to act on cholinergic system [43], voltage - gated sodium channel of the nerve membrane, glutamate-gated chloride channel [44], GABA-system [45], Octopaminergic system [46], mitochondrial system [47], endocrine system disrupting the endocrinological balance and respiratory system of insect body. However, not much studies yet conducted on detailed study on the mode of action of EOs and their constituent compounds.

3. GC-MS profile of citrus EO

With the development of GC–MS technique, profiling of essential oil became easier. The composition of different citrus species from different parts of the world have been reported utilizing this technique. Most of the profiling results although detected average 20–50 numbers of compounds, a few compounds mostly occupy the major share of the bulk oil. The dominating compounds in most citrus species is limonene. In some species like *C. hystrix* the major compound is β-citronellal. The other common constituent compounds observe to be present in many GC–MS results of citrus species are linalool, pinene, β-caryophyllene, β-myrcene, terpinene, citral etc. Some of the reported constituent profiles of citrus species are mentioned below.

3.1 *Citrus hystrix*

From the leaf essential oil of *C. hystrix* in Malaysia, 29 compounds were reported, out of which beta-citronellal was the major compound (3] (%66.85]. The other compounds present in more than 1% total compositions were β-citronellol (6.59%), linalool (3.90%), 5,9-dimethyl-1-decanol (4.96%), methyl citronellate (1.90%), geranyl acetate (1.80%), citronellol (1.76%), 3-undecanol (1.04%). The same compound (beta-citronellal) with 86.43% amount was reported in another study from Indonesia along with 11.48% citronellol and 1.65% β-linalool [4].

3.2 Citrus maxima (synonym *Citrus grandis*)

In the leaf EO of *C. maxima*, 42 constituent compounds were reported to be present with citronellol as the major compound (28.26%) of the essential oil. The compounds comprised more than 1% of the oil compositions were β- caryophyllene (16.89%), – spathulenol (9.32%), α-caryophyllene (2.48%), γ-cardinol (3.16%), α-cardinol (2.51%), 2n-hexylcyclopentanone (2.22%), caryophyllene oxide (1.03%). Again, 34 compounds were identified from the peel essential oil of the same plant with limonene comprising bulk of the oil constituents (89.04%) [48]. The major compounds identified in more than 1% of the total composition were β-pinene (2.25%), β-myrcene (2.06%), β- copaene (1.76%). Compounds comprising more than 0.3% amount but below 1% were linalool, β-phellandrene, α-pinene, terpinene-4-ol. Earlier 35 compounds with limonene as the major constituent compound (93.2%) was reported from the fruit peel oil [49]. Myrcene comprised 2.9% and other six compounds namely α-pinene, octyl acetate, germacrene -D, linalool, decanal, geranial comprised above 0.2% but below 1% of the total composition.

3.3 *Citrus aurantium*

Phytochemical profiling of essential oil of *C. aurantium* showed the presence of 25 compounds with limonene occupying %87.52 followed by linalool (%3.365) and β-myrcene (%1.628) as dominant compounds [25]. From Tunisian *Citrus aurantium*, limonene percentage in leaves, flowers and peel EOs were 6.52, 5.03, 73.6% respectively, linalool occupied 37.24, 41.82, 4.8% respectively, linaly acetate occupied 7.87, 13.75, 1.6% respectively and neral share was almost of similar (3.40, 4.80, 3.26% respectively) percentage. β-pinene composition for leaves and flowers were 9.68 and 9.21% respectively and α-thujene composition were 10.65 and 6.15% respectively but both β-pinene and α-thujene present below 0.5% in peel EO of the plant [28].

3.4 *Citrus aurantifolia*

In our recent studies 31 compounds from the leaf oil and 26 compounds from the peel essential oil was recorded from GC–MS analysis of EO of *Citrus aurantifolia* in India. Citral and limonene were noted as the major constituent compound of leaf oil and limonene and palatinol1-C as the major constituent of peel essential oil of the plant [19, 23]. The EO of the plant from Italy reported to comprise limonene (53.8%), γ-terpinene (16.5%), β-pinene (12.6%), β-Bisabolene (1.33%), Geranyl acetate (1.06%), Neryl acetate (1.12%), Geranial (1.84%), sabinene (1.74%), and α-Pinene (1.97%) [50]. Phytochemical analysis of leaf and peel EOs of the plant from Brazil [51] showed limonene as the dominant compound in both leaf (32.7%) and peel (77.5%) part. Other prominent compounds in the leaf EO were linalool (20.1%), citronellal (14.5%), citronellol (14.2%), trans-β-Ocimene (2.7%), geranial (2.6%), neral (2.1%), trans-β-Caryophyllene (2%), myrcene (1.4%) etc. The other major compounds of leaf EO were myrcene (4.4%), linalool (3.5%), citronellal (3.2%), citronellol (2%), β-Bisabolene (1.5%) etc.

3.5 *Citrus sinensis*

Phytochemical analysis of peel essential oil from three varieties of *C. sinensis* from Kenya showed presence of 56 components in Salaustiana variety, 73 in Valencia and 72 in Wshington varirty. Limonene occupied more than %90 in all the essential oil (Salustiana %94.6, Valencia %92.5 and Washington %90.5); alpha terpinene occupied %1.5 in Valencia and Washington and %1.7 in Salustiana [52]. Limonene with 90% share and β-myrcene, γ-terpine, linalool with 1.88%, 1.21% and 0.88% share out of 32 identified compounds in peel EO of *C. sinensis* has been reported from China [53]. GC–MS analysis of the plant EO from Argentina showed limonene (92.47%), linalool (1.43%), and β-myrcene (0.88%) as the major constituent compounds along with terpineol (0.28%) in lesser amount [26]. Almost similar constituents were identified from EO of *C. sinensis* with D-limonene (%80.18–65.28), Linalool (%2.20–0.32) and β- pinene (%5.58 1.71) as major part in another study [34].

3.6 *Citrus nobilis*

Phytochemical study on constituents from the peel of *C. nobilis* from China showed D- limonene (12,601 μg/g) as the major constituent compound followed by β-myrcene (1600 μg/g), β-pinene (82.64 μg/g), p-mentha-1,8-dien-3-one (41.33 μg/g), α-pinene (25.41 μg/g), geranial (21.32 μg/g), sabinene (21.18 μg/g), E-β-ocimene (14.97 μg/g), linalool (10.38 μg/g), α-terpineol (6.36 μg/g). Other compounds are present in lower amounts (below 5 μg/g) [54]. In another study from Sri Lanka, 37 compounds were reported from peel part of which D-limonene (45%) was the major one followed by cyclopentane-2-methyl-1-methylene-3-(1-methylethenyl) (3.94%), p-mentha-4,8-diene (3.73%), α-terpinolene (3.03%), methyl-2-(methylamino) benzoate (2.4%), α-fernesene (1.1%). Other compounds were present in below 1% [30].

3.7 *Citrus limon*

Phytochemical analysis of citrus leaf EO from Iran showed presence of 27 compounds of which the major compound was linalool (30.62%). The other compounds present in significant amount were geraniol (15.91%), α-terpineol (14.52%), linalyl acetate (13.76%), geranyl acetate (6.75%), B-pinene (4.51%), neryl acetate (4.24%), p-Cymene (1.86%), and limonene (1.13%) [55]. Chemical composition of EO of *C. limon* grown in Iraq [56] showed presence of 24 compounds with limonene as principal compound with 29.52% share. Other major compounds were β-Pinene (23.89%), α-Pinene (2.25%), Myrcene (1.31%), (Z)-β-Ocimene (2.09%), Linalool (1.41%), (R)-Citronellal (15.10%), α-Citronellal (3.57%), (+)-α-Terpineol (1.57%), Neral (Z-Citral) (1.19%), Geranial (E-Citral) (1.73%), Thymol (9.79%), Citronellyl acetate (1.87%), Caryophyllene (1.36%), Phytol (1.36%). The analysis of the essential oil of *Citrus limon* from North-East India reported presence of 43 constituent compounds of which limonene (%55.40), neral (%10.39) trans-verbenol (%6.43) and decanal (%3.25) were the major constituent compounds [57].

3.8 Citrus paradisi

Phytochemical analysis result of *C. paradisi* peel EO from Turkey demonstrated presence of 25 constituent compounds of which limonene occupied the highest percentage (%88.6) of the oil. The other major compounds were α-terpinene (%1), and β-pinene ([58] (%1.2].

From Nigeria, fifteen phytochemical constituents of the plant oil were reported. Among the compound limonene (94.2%) occupied the major share [59].

3.9 *Citrus medica*

A total of 19 constituent compounds were identified from leaf essential oil of *Citrus medica* from Bangladesh of which erucylamide (%28.43), limonene (%18.36), citral (%12.95), Mehp (6-,2 ,(%8.96octadien-1ol,3,7-dimethyl-acetate, (Z) (5.23%) were the major compounds. From peel essential oil 43 compounds were reported out of which isolimonene (39.37%), citral (23.12%), limonene (21.78%) were the major constituents. Three other compounds namely β-myrcene, neryl acetate and neryl alcohol were reported to present at around 2% each in total composition and remaining compounds were present in traces amount [60]. In a study carried out by Li et al. [61], all total 28 compounds were reported to present in the fruit essential oil of *Citrus medica* of which limonene (%45.36), γ-terpinene (%21.23), dodecanoic acid (%7.52) were documented as major constituent compounds. Compounds like β-bisabolene, tetradecanoic acid, α-terpineol, terpinene-4-ol, hexadecenoic acid, α-bergamotene, α-pinene, β-pinene comprised between 5–1% range in total composition. In another study fruit peel EO of the plant was reported to comprise limonene (38.7%), γ- terpinene (28%) and o-cymene (15.2%) as major compounds [41].

4. Citrus EO compounds against insect sp

Essential oil composition of different citrus sp. across the globe although may vary but some of the compounds are observed as common in most of the oil profile. The most dominating and commonly present compound is limonene. Other common compounds are citronellal, citronellol, linalool, pinene, myrcene, ocimene, terpinene, caryophyllene etc. The bioactivity of EOs is often related to the activity of major compounds present in the crude oil and some of the studies have already established this fact. Individual assessment for insecticidal property of these common constituent compounds have been performed by different researchers and some of them were found active against insect pest. Limonene and other *Citrus limon*oids are reported as insect repellents, feeding deterrents, growth disruptors, and reproduction inhibitors against a wide range of pest complexes. Insecticidal activity of limonene was reported effective against *Tuta asoluta* (Lepidoptera: Gelechiidae) [25]. Yoon et al. [62] revealed repellent property of different citrus oil and its major compound limonene against different species of cockroaches like *Blatella germanica, Periplaneta americana* and *Periplaneta fuliginosa*. However, Karr and Coats [63] did not get significant insecticidal activity of d-limonene against *Blattella germanica, Musca domestica, Sitophilus oryzae* and *Diabrotica virgifera virgifera*. In contrast they reported enhanced growth of nymph of *Blatella germanica* after oral administration of d-limonene.

In another study against cat flea species *Ctenocphalides felis* (Siphonaptera: Pulicidae), d-limonene (LD 50 against larvae, adults 160 ,226 μg/cm^2 respectively) and d-limonene with piperonyl butoxide (PB) (LD 50 against larvae, adults 49 ,157 μg/cm^2 respectively) were reported effective against all the life stages except the pupal stage of the flea species [64]. Fumigant toxicity of d-limonene, α- terpineol etc. also reported against honey bee *Apis mellifera* and tracheal mite parasite species *Acarapis woodi* [65].

Fouad and da Camara [66] extracted the essential oil from *Citrus aurantiifolia* and *Citrus reticulate* and analyzed the phytochemical constituents using GC–MS and found limonene as the major constituent compound, 38.9% of the *C. aurantiifolia* oil and %80.2 of the *C. reticulata* oil. They analyzed the enantiomers of limonene against the said insect. They found that *Citrus reticulate* was more toxic than *Citrus aurantifolia* towards the said insects. (R)-limonene was shown to have greater toxicity against *S. zeamais* than the (S)-limonene as found in the ingestion bio assay. Repellent bioassay showed (S)-limonene more susceptible to *S. zeamais* than (R)-limonene.

After identifying limonene as major compound in the EOs of *Citrus aurantiifolia* (%38.9) and *Citrus reticulate* (%80.2) Fouad and da Camara [66] tested enantiomers of limonene against *Sitophilus zeamais* and recorded greater toxicity of (R)-limonene than the (S)-limonene in the ingestion bio assay. But in the repellency test they found more susceptibility of *S. zeamais* towards (S)-limonene than (R)-limonene.

In a recent study, Sowler et al. [67] comparatively evaluated the effect of laboratory grade limonene and a commercial limonene-based insecticide against *Haematobia irritans irritans* in terms of deterrence, mortality, and reproduction. They showed that the egg viability was decreased in both the treatment, however, commercial limonene that caused loss of viability at %5.8 concentration was ovicidal in case of laboratory grade limonene. However, in terms of knockdown effect commercial limonene was better. Interestingly, at a concentration of less than %0.1, both the commercial and laboratory grade limonene were acted as attractant.

Giatropoulos et al. [68] tested essential oil of *Citrus sinensis, Citrus limon,* and *Citrus paradise* and their constituents and recorded γ- terpinene as the most toxic compound against *Aedes albopictus* larvae. They also reported that the constituent compound tested for repellency were better mosquito repellent than the parent essential oil. Similarly, Luo et al. [41] analyzed composition of leaf and peel essential oil of *C. medica* and tested both crude oil and major compounds viz. limonene, terpinene, o-cymene, β-caryophyllene against *Tribolium castaneum* and recorded γ-terpinene as the most effective insecticidal compound having LC50 value of 4.1 mg/l air and β- caryophyllene as the effective repellent compound. Limonene was reported to have almost similar fumigant toxicity like that of crude EO of *C. aurantium* against *Tribolium castaneum, Sitophilus granarium* and *Cryptolestes ferrugineus* [28]. In our earlier studies we found higher toxicity of citral, that is the major compound of the essential oil of *Citrus aurantifolia* than the crude oil as mosquito larvicidal, ovicidal and adulticidal against *Aedes aegypti* [23]. Plata-Reuda et al. [69] reported the insecticidal activity of citral and geranyl acetate against peanut beetle *Ulomoides dermestoides.* These compounds affected the survivorship, locomotor activity and reduced the respiration rate of the said species.

Nootketone and carvacrol, a phytochemical constituent present in essential oil of Citrus [70] acts as insecticidal compound against *Aedes aegypti* [71]. Pajaro-Castro et al. [72] recorded the neurotoxic effects of linalool and β-pinene on *Tribolium castaneum.* They observed that at low concentration both the compounds were attractant towards the insect and at higher concentration the compounds were repellent. Individual treatment of limonene and linalool was found to have fumigant toxicity against two ant species namely *Acromyrmex balzani* and *Atta sexdens* with LC50 values of 5.72 μl/L, 5.38 μl/L and 2.40 μl/L, 5.34 μl/L respectively [73]. Similarly individual treatment of β-caryophyllene is reported to have good contact toxicity against these two ant sp. but with low fumigant toxicity [74]. Fumigant toxicity of limonene, linalool and β- pinene were also reported effective against fire ant *Solenopsis invicta* (Hymenoptera: Formicidae) [34].

Linalool, α-terpinene was reported to show 100% fumigant toxicity against adult rice weevil *S. oryzae* at 3.9 mg/L [75].

Muller et al. [76] recorded 85.4%, 71.1%, and 29% repellency of the candles prepared with 5% geraniol, 5% linalool and 5% citronella against mosquitoes on human landing bioassay. They observed similar repellency against sand flies too. 78% repellency of *Culex pipiens pallens* was reported after using %30 citronellal [77]. Progeny deterrent, antifeedant, egg hatching inhibition activity was documented after application of 5–10 μl citronellol against *Callosobruchus analis* [78]. Compounds like limonoids act as synergists to enhance the activities of other biological and or synthetic insecticides [79]. We recorded medium larvicidal and adulticidal potential of limonene against *Aedes aegypti*, but found higher toxicity when combined with diallydisulphide and carvone respectively [80]. So, it is not always the individual compound that act as the most active but appropriate combinations of compounds having synergistic effects would be more fruitful as insecticidal against insect pests.

5. Citrus of North East India

North East India is enriched with Citrus species having documented 23 species and 68 varieties out of the 27 species of Citrus found in India [81, 82]. It is established that some of the citrus species are endemic and some are in endangered status [83]. According to Hore and Barua [84], there are eight citrus species indigenous to this region scattered in the form of semi-wild, wild state and some in cultivated state. Some of the species are naturally tolerant to viral and bacterial diseases and also for drought, cold and rainfall. For instance, *Citrus limon* is reported to resist scab, canker and gummosis, *C. indica* resistant to greening disease [84]. Due to diverse ecogeographical conditions the Citrus species and varieties of this region may bear specific traits including its aroma and essential oil constituents which needs to be investigated. However citrus crop and citrus essential oil-based industry is not yet flourishing in this part of the country. The information regarding Citrus essential oils extracted from the Citrus species grown in this particular area is relatively scanty. As the Citrus plants are rich in secondary metabolites to naturally defend an array of pathogens and pest complexes, it is expected that some of the key compounds for controlling insect pest may lie within the secondary metabolite compounds especially in the diverse aromatic essential oil part of the plants at least in the resistant citrus species.

Here we have attempted to evaluate insecticidal properties of essential oil extracted from the fruit peel of four citrus species namely *Citrus limon, Citrus maxima, Citrus paradisi* and *Citrus medica* grown in North Eastern part of India against one of the household ants *Dolichoderus affinis* (Hymenoptera: Formicidae). Fruits of *Citrus limon, Citrus paradisi* and *Citrus medica* were collected from Udalguri district, Assam (26.7210 °N, 91.9906 °E) and *Citrus maxima* from a daily market at Guwahati, Assam (26.1445 °N, 91.736 °E), in September, 2020. Essential oils from fresh peels were extracted by hydro-distillation using Clevenger's apparatus. After 6 hours, essential oils were collected, anhydrous sodium sulfate was added to absorb traces of moisture and were stored at 4 °C till its use. The worker ants of *Dolichoderus affinis* from naturally existed colony located in the wooden frame of house wall were considered for the assessment. Fumigant toxicity of these oils were assessed following the method described by Hu et al. [34]. Six different concentrations viz. 0.25 μL, 0.5 μL, 1 μL, 2.5 μL, 5 μL and 7.5 μL. of each EO was individually loaded in 1.5 ml centrifuged tubes, evaporation of EOs was allowed by making five small holes of 1–1.2 mm diameter and placed into 500 ml properly cleaned borosilicate conical flask. Twenty worker ants were taken per flask in a replication and the flask was covered by aluminum foil and bound tightly by rubber band to prevent the loss of volatile compounds. For each concentration three replications were made. Equal number of controls set without oil were placed against each treatment. The environmental temperature range was 21–30°C and relative humidity range 56–99 during the experimental period. The Percent mortality [percent mortality = (Total no. of dead ants/ Total no. of treated ants) × 100] data was recorded after 12 h and 24 h of treatment. The ants were considered to be dead if touched with a needle but did not show any movement. Based on the results sublethal concentration was determined using probit analysis with the help of SPSS and Minitab software. As shown in the figure, after treatment with *C. limon*, maximum 71.66% mortality was recorded at 12 h and 91.66% was recorded after 24 h at 7.5 μL treatment. The calculated LC50 for the oil was 2.66 μl / 500 air volume. For the EO of *C.paradisi*, maximum 15% mortality was recorded at 12 h and 66.66% mortality was recorded after 24 h of treatment. LC50 for the oil at 24 h was 7.32 μl / 500 air volume. For the EO of *C. maxima*, not more than 10% mortality was recorded even after 24 h at the highest dose applied and LC50 could not be computed for the oil. While for the EO of *C. medica*, maximum 56.6% mortality was recorded at 5 μl concentration at 12 h and maximum 88.33% mortality was recorded after 24 h at 7.5 μl. LC50 for *C. medica* at 24 h was recorded as 2.09 μl/500 air volume (Figure 1, Table 1). Highest toxic effect was recorded for *C. medica* followed by *C. limon*. Earlier Adusei-Mensah et al. [15] evaluated insecticidal properties of three citrus species viz. *Citrus aurantifolia*, *Citrus sinensis* and *Citrus limon* against *Camponotus nearcticus* (Formicidae) and recorded highest performance from *C. limon* with %95 mortality. But Guerra et al. [16] recorded only 15% mortality of *Camponotus pennsylvanicus* (Hymenoptera: Formicidae) on topical application of *C. limon*, the efficacy of which was comparatively lower than other eight EOs tested against the ant species. Not much studies on insecticidal activities of citrus EO against ants have been found to be reported. The findings showed the prospect of using *C. medica* and *C. limon* oil for controlling household ants.

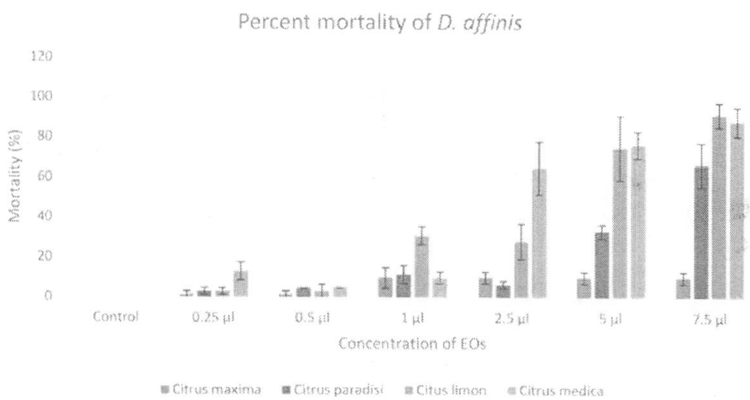

Figure 1. Relation between concentration of EOs and respective percent mortality.

Table 1. LC50 values of the individual citrus oils against *Dolichoderus affinis* at 24 h.

Essential oils	Time	LC50 value μl/500 ml air	95% confidence level		Regression equation	Chi-square value
			Lower limit	Upper limit		
Citrus paradisi	24 h	7.32	1.143	1.902	Y = 3.67118 + 1.53688X	31.750
Citrus limon	24 h	2.66	1.739	2.486	Y = 4.03677 + 2.26321X	49.452
Citrus medica	24 h	2.09	1.609	2.266	Y = 4.37759 + 1.94055X	52.747

6. Conclusion

With the increasing awareness of consumers for ecofriendly products and at the same time increasing resistance of insect pests against insecticides, the demand for novel, safe and effective products is increasing. As discussed above, the existing literature revealed presence of a good number of terpene compounds in different *Citrus* species which are present in different ratios although in most cases limonene is the predominant constituent. Both the crude oil as well as individual compounds possess good insecticidal and repellent properties against diverse insect pests, both indoor and outdoor. Our study also showed promising potential against *Dolichoderus affinis* while using four Citrus essential oils with higher efficacy of *Citrus medica* and *Citrus limon* essential oils. It is expected that in near future Citrus plant essential oils with their pleasant aroma and array of chemical compounds shall take leading space in development of insecticidal and repellent products to be used in both indoor and outdoor pest management practices against insect pests.

References

1. Jing L, Lei Z, Li L, Xie R, Xi W, Guan Y, Sumner LW, Zhou Z. Antifungal activity of citrus essential oils. J Agric Food Chem. 2014; 62(14): 3011-33
2. Tranchida PQ, Bonaccorsi I, Dugo P, Mondello L, Dugo G. Analysis of Citrus essential oils: state of the art and future perspectives. A review. Flavour Fragr J. 2012; 27(2): 98-123
3. Loh FS, Awang RM, Omar D, Rahmani M. Insecticidal properties of Citrus hystrix DC leaves essential oil against Spodoptera litura fabricius. J Med Plant Res. 2011; 5(16): 3739-44
4. Ikawati S, Dhuha MS, Himawan T. Bioactivity of Citrus hystrix DC Leaf Extract against Cigarette Beetle Lasioderma serricorne (F.). J Trop Life Sci. 2017; 7(3): 189-96
5. Don-Pedro KN. Fumigant toxicity of citruspeel oils against adult and immature stages of storage insect pests. Pestic Sci. 1996; 47(3): 213-23
6. Harshani HS, Karunaratne MM. Volatile profiling and bio-efficacy of Citrus hystrix fruit peel as a seed protectant against Callosobruchus maculatus. J Entomol Zool Stud. 2018; 6(4): 27-31
7. Nararak J, Sathantriphop S, Kongmee M, Bangs MJ, Chareonviriyaphap T. Excito-repellency of Citrus hystrix DC leaf and peel essential oils against Aedes aegypti and Anopheles minimus (Diptera: Culicidae), vectors of human pathogens. J Med Entomol. 2017; 54(1): 178-86
8. Saeidi M, Moharramipour S, Sefidkon F, Aghajanzadeh S. Insecticidal and repellent activities of Citrus reticulata, Citrus limon and Citrus aurantium essential oils on Callosobruchus maculatus. Integrated Protection of Stored Products IOBC/WPRS Bulletine. 2011; 69: 289-93
9. Karamaouna F, Kimbaris A, Michaelakis A, Papachristos D, Polissiou M, Papatsakona P, Tsora E. Insecticidal activity of plant essential oils against the vine mealybug, Planococcus ficus. J Insect Sci. 2013; 13(1): 142

10. Reda FA, Abdel Fattah HM, Salim NM, Atiya NH. Insecticidal activity of four volatile oils on two museum insects pests. Egypt Acad J Biolog Sci, F. Toxicology & Pest Control. 2010; 2(2): 57-66
11. Din, S. U., Akram, W., KhaN, H. A. A., Hussain, A., Hafeez, F. Citrus Waste-Derived Essential Oils: Alternative Larvicides for Dengue Fever Mosquito, Aedes albopictus (Skuse) (Culicidae: Diptera). Pak J Zool. 2011; 43(2): 367-372
12. Saleem MU, Hussain DI, Rashid RS, Saleem HM, Ghouse GH, Abbas MU. Insecticidal activities of two citrus oils against Tribolium castaneum (herbst). Am J Res Commun. 2013; 1(6): 67-74
13. Bilal H, Akram W, Hassan SA, Zia A, Bhatti AR, Mastoi MI, Aslam S. Insecticidal and repellent potential of citrus essential oils against Tribolium castaneum Herbst (Coleoptera: Tenebrionidae). Pak J Zool. 2015; 47(4)
14. Oboh G, Ademosun AO, Olumuyiwa TA, Olasehinde TA, Ademiluyi AO, Adeyemo AC. Insecticidal activity of essential oil from orange peels (Citrus sinensis) against Tribolium confusum, Callosobruchus maculatus and Sitophilus oryzae and its inhibitory effects on acetylcholinesterase and Na+/K+-ATPase activities. Phytoparasitica. 2017; 45(4): 501-8
15. Adusei-Mensah F, Inkum E, Mawuli Agbale C, Eric A. Comparative Evaluation of the Insecticidal and Insect Repellent Properties of the Volatile Oils of Citrus aurantifolia (Lime), Citrus sinensis (Sweet Orange) and Citrus limon (Lemon) on Camponotus nearcticus (Carpenter Ants). International Journal of Novel Research in Interdisciplinary Studies. 2014; 1:19-25
16. Guerra MD, Suiter DR, Scocco CM. Topical toxicity of nine essential oils to Camponotus pennsylvanicus (Hymenoptera: Formicidae). Sociobiology. 2011; 58(2): 419
17. Mahanta S, Khanikor B, Sarma R. Potentiality of essential oil from Citrus grandis (Sapindales: Rutaceae) against Culex quinquefasciatus Say (Diptera: Culicidae). J Entomol Zool Stud. 2017; 5(3): 803-9
18. Sarma R, Khanikor B, Mahanta S. Essential oil from Citrus grandis (Sapindales: Rutaceae) as insecticide against Aedes aegypti (L) (Diptera: Culicidae). Int J Mosq Res. 2017; 4(3): 88-92
19. Mahanta S, Khanikor B, Sarma R. Allium sativum (Liliales: Asparagales) essential oil-based combinations–a potential larvicide for Culex quinquefasciatus (Diptera: Culicidae). Int J Trop Insect Sci. 2020
20. Manorenjitha Malar S, Jamil M, Hashim N, Kiong LS, Jaal Z. Repellency effect of white flesh Citrus grandis osbeck fruit peel extracts against Aedes aegypti (Linn.) Mosquitoes. Int J Mosq Res. 2017; 4(4): 88-94
21. Khanikor B, Barman J, Sarma R, Mahanta S, Adhikari K. Evaluation of Efficacy of Three Essential Oils against Odontotermes feae (Isoptera: Termitidae). Int J Environ Res Public Health. 2018; 6(2): 68-76
22. Roekmi-ati K, Murad S, Imron SS, Asnari H. A new formulation of insecticide, repellent, and larvacide against mosquitoes from the waste product of Jeruk Nipis (Citrus aurantifolia) syrup industry. 2008. Corpus ID: 67759427

23. Sarma R, Adhikari K, Mahanta S, Khanikor B. Insecticidal activities of Citrus aurantifolia essential oil against Aedes aegypti (Diptera: Culicidae). Toxicology Reports. 2019a; 6: 1091-6

24. Ribeiro ND, da Camara CA, Born FD, de Siqueira HA. Insecticidal activity against Bemisia tabaci biotype B of peel essential oil of Citrus sinensis var. pear and Citrus aurantium cultivated in northeast Brazil. Nat Prod Commun. 2010; 5(11): 1819-22

25. Zarrad K, Chaieb I, Ben H, Bouslama T, Laarif A. Chemical composition and insecticidal effects of Citrus aurantium of essential oil and its powdery formulation against Tuta absoluta. Tunis J Plant Prot. 2017; 12(Special Issue):83-94

26. Palacios SM, Bertoni A, Rossi Y, Santander R, Urzúa A. Efficacy of essential oils from edible plants as insecticides against the house fly, Musca domestica L. Molecules. 2009; 14(5): 1938-47

27. Abad MK, Besheli BA. Insecticidal potential of essential oil from the leaves of Citrus aurantium L. against Oryzaephilus surinamensis (F.), Lasioderma serricorne (L.) and Sitophilus oryzae (L.). J Entomol Zool Stud. 2016; 4: 865-9

28. Bnina EB, Hajlaoui H, Chaieb I, Said MB, Jannet HB. Chemical composition, antimicrobial and insecticidal activities of the tunisian Citrus aurantium essential oils. Czech J Food Sci. 2019; 37(2):81-92

29. Yazdgerdian AR, Akhtar Y, Isman MB. Insecticidal effects of essential oils against woolly beech aphid, Phyllaphis fagi (Hemiptera: Aphididae) and rice weevil, Sitophilus oryzae (Coleoptera: Curculionidae). J Entomol Zool Stud. 2015; 3(3): 265-71

30. Harshani HS, Karunaratne S. Chemical composition and insecticidal effect of fruit peel powders of two citrus species against Callosobruchus maculatus (F.)(Coleoptera: Bruchidae) in stored cowpea (Vigna unguiculata). Int J Pest Manage. 2019: 1-8

31. Siskos EP, Konstantopoulou MA, Mazomenos BE, Jervis M. Insecticidal activity of Citrus aurantium fruit, leaf, and shoot extracts against adult olive fruit flies (Diptera: Tephritidae). J Econ Entomol. 2007; 100(4): 1215-20

32. Siskos EP, Konstantopoulou MA, Mazomenos BE. Insecticidal activity of Citrus aurantium peel extract against Bactrocera oleae and Ceratitis capitata adults (Diptera: Tephritidae). J Appl Entomol. 2009; 133(2): 108-16

33. Moravvej G, Abbar S. Fumigant toxicity of citrus oils against cowpea seed beetle Callosobruchus maculatus (F.)(Coleoptera: Bruchidae). Pak J Biol Sc. 2008; 11(1): 48-54

34. Hu W, Zhang N, Chen H, Zhong B, Yang A, Kuang F, Ouyang Z, Chun J. Fumigant activity of sweet orange essential oil fractions against red imported fire ants (Hymenoptera: Formicidae). J Econ Entomol. 2017; 110(4): 1556-62

35. Murugan K, Kumar PM, Kovendan K, Amerasan D, Subrmaniam J, Hwang JS. Larvicidal, pupicidal, repellent and adulticidal activity of Citrus sinensis orange peel extract against Anopheles stephensi, Aedes aegypti and Culex quinquefasciatus (Diptera: Culicidae). Parasitology research. 2012; 111(4): 1757-69

36. Ezeonu FC, Chidume GI, Udedi SC. Insecticidal properties of volatile extracts of orange peels. Bioresour technol. 2001; 76(3): 273-4
37. Zewde DK, Jembere B. Evaluation of orange peel Citrus sinensis (L) as a source of repellent, toxicant and protectant against Zabrotes subfasciatus (Coleoptera: bruchidae). Momona Ethiop J Sci. 2010; 2(1)
38. Kumar P, Mishra S, Malik A, Satya S. Insecticidal evaluation of essential oils of Citrus sinensis L.(Myrtales: Myrtaceae) against housefly, Musca domestica L.(Diptera: Muscidae). Parasitol Res. 2012; 110(5): 1929-36
39. Raina A, Bland J, Doolittle M, Lax A, Boopathy R, Folkins M. Effect of orange oil extract on the Formosan subterranean termite (Isoptera: Rhinotermitidae). J Econ Entomol. 2007; 100(3): 880-5
40. Majeed MZ, Nawaz MI, Khan RR, Farooq U, Ma CS. Insecticidal effects of acetone, ethanol and aqueous extracts of Azadirachta indica (A. Juss), Citrus aurantium (L.), Citrus sinensis (L.) and Eucalyptus camaldulensis (Dehnh.) against mealybugs (Hemiptera: Pseudococcidae). Trop Subbtrop Agroecosystems. 2018; 21(3)
41. Luo C, Li D, Wang Y, Guo S, Zhang D, Du S. Chemical Composition and Insecticide Efficacy of Essential Oils from Citrus medica L. var. sarcodactylis Swingle Against Tribolium castaneum Herbst in Stored Medicinal Materials. J Essent Oil-Bear Plant. 2019; 22(5): 1182-94
42. Abdel-Kawy MA, Michel CG, Kirollos FN, Hussien RA, Al-Mahallawi AM, Sedeek MS. Chemical composition and potentiation of insecticidal and fungicidal activities of Citrus trifoliata L. fruits essential oil against Spodoptera littoralis, Fusarium oxysporum and Fusarium solani via nano-cubosomes. Nat Prod Res. 2019: 1-6
43. Ryan, M.F., Byrne, O., 1988. Plant-insect coevolution and inhibition of acetylcholineesterase. J Chem Ecol. 1988; 14: 1965e1975
44. Gershenzon J, Dudareva N. The function of terpene natural products in the natural world. Nat Chem Biol. 2007; 3(7): 408-14
45. 45.Bloomquist JR, Boina DR, Chow E, Carlier PR, Reina M, Gonzalez-Coloma A. Mode of action of the plant-derived silphinenes on insect and mammalian GABAA receptor/chloride channel complex. Pestic Biochem Physiol. 2008; 91(1): 17-23
46. 46.Enan EE. Molecular and pharmacological analysis of an octopamine receptor from American cockroach and fruit fly in response to plant essential oils. Arch Insect Biochem Physiol: Published in Collaboration with the Entomological Society of America. 2005; 59(3): 161-71
47. s47.Khambay BP, Batty D, Jewess PJ, Bateman GL, Hollomon DW. Mode of action and pesticidal activity of the natural product dunnione and of some analogues. Pest Manage Sci. 2003; 59(2): 174-82
48. 48.Prasad DA, Prasad BR, Prasad DK, Shetty P, Kumar KS. GC-MS compositional analysis of essential oil of leaf and fruit rind of Citrus maxima (Burm.) Merr. from Coastal Karnataka, India. J Appl Pharm Sci. 2016; 6(5): 68-72

49. 49.Bordoloi AK, Pathak MG, Sperkova J, Leclercq PA. Volatile Constituents of the Fruit Peel Oil of Citrus maxima (J. Burman) Merrill, from Northeast India. J Essent Oil Res. 1999; 11(5):629-32
50. 50.Rosta R, Bisignano C, Filocamo A, Grasso E, Occhiuto F, Spadaro F. Antimicrobial activity and chemical composition of Citrus aurantifolia (Christm.) Swingle essential oil from Italian organic crops. J Essent Oil Res. 2014; 26(6): 400-8
51. Lemes RS, Alves CC, Estevam EB, Santiago MB, Martins CH, SANTOS TC, Crotti AE, Miranda ML. Chemical composition and antibacterial activity of essential oils from Citrus aurantifolia leaves and fruit peel against oral pathogenic bacteria. Anais da Academia Brasileira de Ciências. 2018; 90(2): 1285-92
52. Njoroge SM, Koaze H, Karanja PN, Sawamura M. Essential oil constituents of three varieties of Kenyan sweet oranges (Citrus sinensis). Flavour Fragr J. 2005; 20(1):80-5
53. Qiao Y, Xie BJ, Zhang Y, Zhang Y, Fan G, Yao XL, Pan SY. Characterization of aroma active compounds in fruit juice and peel oil of Jinchen sweet orange fruit (Citrus sinensis (L.) Osbeck) by GC-MS and GC-O. Molecules. 2008; 13(6): 1333-44
54. Liu C, Cheng Y, Zhang H, Deng X, Chen F, Xu J. Volatile constituents of wild citrus Mangshanyegan (Citrus nobilis Lauriro) peel oil. J Agric Food Chem. 2012; 60(10): 2617-28
55. Hojjati M, Barzegar H. Chemical composition and biological activities of lemon (Citrus limon) leaf essential oil. Nutrition and Food Sciences Research. 2017 Sep 10;4(4):15-24
56. Kaskoos RA. Essential oil analysis by GC-MS and analgesic activity of Lippia citriodora and Citrus limon. J Essent Oil-Bear Plant. 2019; 22(1): 273-81
57. Paw M, Begum T, Gogoi R, Pandey SK, Lal M. Chemical Composition of Citrus limon L. Burmf Peel Essential Oil from North East India. J Essent Oil-Bear Plants. 2020; 23(2): 337-44
58. Uysal B, Sozmen F, Aktas O, Oksal BS, Kose EO. Essential oil composition and antibacterial activity of the grapefruit (Citrus paradisi. L) peel essential oils obtained by solvent-free microwave extraction: comparison with hydrodistillation. Int J of Food Sci Technol. 2011; 46(7): 1455-61
59. Karioti A, Skaltsa H, Gbolade AA. Constituents of the distilled essential oils of Citrus reticulata and C. paradisi from Nigeria. J Essent Oil Res. 2007; 19(6): 520-2
60. Bhuiyan MN, Begum J, Sardar PK, Rahman MS. Constituents of peel and leaf essential oils of Citrus medica L. J Sci Res. 2009; 1(2): 387-92
61. Li ZH, Cai M, Liu YS, Sun PL, Luo SL. Antibacterial Activity and Mechanisms of Essential Oil from Citrus medica L. var. sarcodactylis. Molecules. 2019; 24(8): 1577
62. Yoon C, Kang SH, Yang JO, Noh DJ, Indiragandhi P, Kim GH. Repellent activity of citrus oils against the cockroaches Blattella germanica, Periplaneta americana and P. fuliginosa. J Pestic Sci. 2009; 34(2): 77-88
63. Karr LL, Coats JR. Insecticidal properties of d-limonene. J Pestic Sci. 1988; 13(2): 287-90

64. Hink WF, Feel BJ. Toxicity of D-limonene, the major component of citrus peel oil, to all life stages of the cat flea, Ctenocephalides felis (Siphonaptera: Pulicidae). J Med Entomol. 1986; 23(4): 400-4

65. Ellis MD, Baxendale FP. Toxicity of seven monoterpenoids to tracheal mites (Acari: Tarsonemidae) and their honey bee (Hymenoptera: Apidae) hosts when applied as fumigants. J Econ Entoml. 1997; 90(5): 1087-91

66. Fouad HA, da Camara CA. Chemical composition and bioactivity of peel oils from Citrus aurantiifolia and Citrus reticulata and enantiomers of their major constituent against Sitophilus zeamais (Coleoptera: Curculionidae). J Stored Prod Res. 2017; 73: 30-6

67. Showler AT, Harlien JL, Perez de Léon AA. Effects of Laboratory Grade Limonene and a Commercial Limonene-Based Insecticide on Haematobia irritans irritans (Muscidae: Diptera): Deterrence, Mortality, and Reproduction. J Med Entomol. 2019; 56(4): 1064-70

68. Giatropoulos A, Papachristos DP, Kimbaris A, Koliopoulos G, Polissiou MG, Emmanouel N, Michaelakis A. Evaluation of bioefficacy of three Citrus essential oils against the dengue vector Aedes albopictus (Diptera: Culicidae) in correlation to their components enantiomeric distribution. Parasitology research. 2012; 111(6): 2253-63

69. Plata-Rueda A, Martínez LC, da Silva Rolim G, Coelho RP, Santos MH, de Souza Tavares W, Zanuncio JC, Serrão JE. Insecticidal and repellent activities of Cymbopogon citratus (Poaceae) essential oil and its terpenoids (citral and geranyl acetate) against Ulomoides dermestoides. Crop Protection. 2020; 137: 105299

70. Houda AK, Ali B, Ahmed A, Salah O, Yazid FC. Chemical composition, antimicrobial and insecticidal activities of citrus paradisi peel essential oil from Algeria. J Microbiol Biotechnol Food Sci. 2020; 10(1): 1093-8

71. Anderson JA, Coats JR. Acetylcholinesterase inhibition by nootkatone and carvacrol in arthropods. Pestic Biochem Physiol. 2012; 102(2): 124-8

72. Pajaro-Castro N, Caballero-Gallardo K, Olivero-Verbel J. Neurotoxic effects of linalool and β-pinene on Tribolium castaneum Herbst. Molecules. 2017; 22(12): 2052

73. De Oliveira BM, Melo CR, Alves PB, Santos AA, Santos AC, Santana AD, Araújo AP, Nascimento PE, Blank AF, Bacci L. Essential oil of Aristolochia trilobata: synthesis, routes of exposure, acute toxicity, binary mixtures and behavioral effects on leaf-cutting ants. Molecules. 2017; 22(3): 335

74. Feitosa-Alcantara RB, Bacci L, Blank AF, Alves PB, Silva IM, Soares CA, Sampaio TS, Nogueira PC, Arrigoni-Blank MD. Essential oils of Hyptis pectinata chemotypes: isolation, binary mixtures and acute toxicity on leaf-cutting ants. Molecules. 2017; 22(4): 621

75. Kim SW, Lee HR, Jang MJ, Jung CS, Park IK. Fumigant toxicity of Lamiaceae plant essential oils and blends of their constituents against adult rice weevil Sitophilus oryzae. Molecules. 2016; 21(3): 361

76. Muller GC, Junnila A, Kravchenko VD, Revay EE, Butler J, Schlein Y. Indoor protection against mosquito and sand fly bites: a comparison between citronella, linalool, and geraniol candles. J Am Mosq Control Assoc. 2008; 24(1): 150-3

77. Kim, JK, Kang, CS., Lee, JK., Kim, YR., Han, HY. and Yun, HK. Evaluation of repellency effect of two natural aroma mosquito repellent compounds, citronella and citronellal. Entomological Research. 2005; 35(2): 117-120

78. Brari J, Thakur DR. Insecticidal potential properties of citronellol derived ionic liquid against two major stored grain insect pests. J Entomol Zool Stud. 2016; 4(3): 365-70

79. Alford AR, Murray KD. Prospects for citrus limonoids in insect pest management. 2000: 201-211

80. Sarma R, Adhikari K, Mahanta S, Khanikor B. Combinations of plant essential oil based terpene compounds as larvicidal and adulticidal agent against Aedes aegypti (Diptera: Culicidae). Scientific reports. 2019b; 9(1): 1-2

81. Sharma BD, Hore DK, Gupta SG. Genetic resources of Citrus of north-eastern India and their potential use. Genet Resour Crop Evol. 2004; 51(4): 411-8

82. Sanabam R, Singh NS, Sahoo D, Devi HS. Genetic relationship of rough lemon landraces and under-utilised citrus genotypes from North-East India revealed by SSR and RAPD markers. Trees. 2018; 32(4): 1043-59

83. Borah D, Kafley P, Tangjang S, Das AP. Population structure and conservation of endangered Citrus indica Yu. Tanaka (Rutaceae) in Behali Reserve Forest of Assam, India. 2018: 181-186

84. Hore DK, Barua U. Status of citriculture in North Eastern region of India–A review. Agricultural Reviews. 2004;25(1):1-5

Chapter-11

The Orange Peel: An Outstanding Source of Chemical Resources

1. Introduction

Citrus sinensis (CS) (L.) Osbeck is a perennial species growing in warm climate areas of the world and largely employed as food in form of fresh fruit, with a global production of ca. $6.7X10^7$ tons. per year (TPY) in 2016 [1], or as a processed derivative (ca. $1.85x10^7$ TPY) such as juice, marmalade, flavor, fragrance and coloring additive, pectin.

CS is an evergreen tree, 3 to 9 mts. high with sparingly barbed branches, alternate leaves with toothed blades differently shaped, oval or elliptical, connected to the stem by winged-petioles. Axillary flowers are present singly or in whorls of 6 and possess 5 white petals and up to about 25 yellow colored stamens. The pericarp of CS has a spherical or oval shape of 6–10 cm diameter with the color changing from green to yellow-orange during the ripening; the endocarp containing juice sac glands is enclosed within a wrinkled epicarp or exocarp or flavedo containing a great number of essential oil glands protected by a waxy epidermis. Below the flavedo is the albedo, also called mesocarp, a white filamentary tissue composed of tubular-like cells.

The principal industrial application of CS is the production of frozen concentrated juice. The procedure of juice extraction eventually accompanied by the extraction of the essential oil, implies the generation of a major "by-product" constituted by a pomace, mainly containing peels, accounting for up to around 60% w/w of the original fruit mass processed [2]. This huge amount of biomass does pose serious environmental concerns because of its high level of total organic carbon (TOC) and biological oxygen demand (BOD) that make disposal procedures rather complex and demanding from both the legal and industrial points of view. This is because there is an increasing trend to modify the way of approaching this problem by reconsidering the post-production orange pomace more like a by-product rather than a waste. In the last years, many producers have subjected this material to processings involving partial acidic fermentation, drying, and packaging to biologically and chemically stabilize the biomass before its application as animal feed in zootechnics, soil conditioners in agriculture, or the manufacturing of compost and biogas [2].

Beyond the standard workup of the *Citrus sinensis* peel (CSP) waste, new perspectives have been being opened in the context of the high chemical added value of the CSP [3, 4, 5] also by the complete knowledge of the rich metabolomics profile of this species. The use of CS peel has been proposed for a variety of purposes that include the production of antioxidant-enriched dietary supplements in veterinary [6], the preparation of human dietary supplements, and nutraceuticals such as citric acid [7] and flavonoids [8, 9]. The extract of CS peel is the source of a huge variety of phytochemicals and has been investigated on several applications including its chemotherapeutic and chemopreventive potential for several relevant human pathologies, such as cancer [10, 11] and obesity [12]. The extraction procedures vary in function of the main components that have to be obtained: from the simple cold pressing of pomace and the extraction with water to obtain highly hydroxylated compounds to the employment of mixtures organic protic solvent/water and finally low polar organic solvents such as Chloroform and Ethyl acetate to obtain polymethoxylated phenols (PMF, see below). New extraction technologies such as ultrasounds and microwaves may help to obtain better extraction yields.

In the following sections, the chemical structures and the biological effects of these compounds will be discussed.

2. The chemistry of *Citrus sinensis* peel

2.1 Essential oils

The essential oil (EO) is mainly obtained from the CS peel as a major by-product of the juice production process by a cold-pressing method that can provide the intact blend of compounds without losing the lighter, more volatile, components of the complex mixture that can be lost in the standard EO extraction procedure that is the hydrodistillation. The last one is mainly used in small scale applications, for example in research laboratories.

The chemical composition of CSP EO [13, 14, 15] is reported in Table 1. As it can be seen, the major component is D-Limonene, accompanied by several minor components belonging to the classes of monoterpene alkenes, oxygenated monoterpenes including alcohol aldehydes and esters, sesquiterpenes as well as linear alkanes and aldehydes. This rather complex blend accounts for the numerous deal of biological activities reported for the CSP EO [14, 15, 16], which include anthelmintic, anti-aflatoxigenic [17], antibacterial [18, 19, 20], anticarcinogenic, antifungal [21], antioxidant [17], anti-tumor [22], anxiolytic [23], food preservative [24], hepatocarcinogenesis suppressant, insecticidal and larvicidal [25], pain relief and relaxant [26]. It can be argued that the main effects are due to the presence of the major component Limonene that showed several bioactivities when tested as pure compound [27]. However, it is possible that synergistic effects due to the combination of Limonene with other minor components may be speculated and should have to be demonstrated.

Table 1. Composition of *C. sinensis* essential oil obtained from peels.

Comp.	Comp. name	%	Compound.	Comp. name	%
1	Aromadendrene	0.01	**21**	β-Linalool	0.4–5.6
2	δ-Amorphene	0.05	**22**	β-Myrcene	1.3–3.3
3	D-Cadinene	0.01–0.03	**23**	Neral	0.1–1.3
4	δ-3-Carene	0.18	**24**	Neryl acetate	0.02
5	β-Citral	0.12–0.15	**25**	Nonanal	0–0.1
6	L-(+)-Citronellal	0.01–0.1	**26**	Nootkatone	0.01
7	Citronellyl acetate	0.01	**27**	*cis*-β-Ocimene	0.03–0.26
8	α-Copaene	0.04	**28**	Octanal	0.02–0.8
9	α-Cubebene	0.02–0.26	**29**	Perillaldehyde	0.03
10	β-Cubebene	0.03	**30**	α-Phellandrene	0.02–0.07
11	Decanal	0.04–0.4	**31**	α-Pinene	0.49–0.59
12	n-Dodecanal	0.06	**32**	(+)-Sabinene	0.2–1.0
13	β-Elemene	0.01–0.02	**33**	γ-Terpinene	0–1.21
14	Geranial	0–1.8	**34**	γ-Terpineol	0.04–008
15	Germacrene-D	0.02–0.08	**35**	α-Terpineol	0.07–0.42
16	β-Gurjurene	0.01	**36**	Terpinolene	0–0.08
17	Hexadecanol	0.04	**37**	α-Thujene	0.04
18	D-Limonene	Ca. 95			
19	L-Limonene	0.02			
20	*trans*-Limonene oxide	0.01			

2.2 Polyphenols

2.2.1 Flavanoids

Polyphenols extracted from the CS peel belongs to the general structural categories of flavanones (Figure 1a), flavones (Figure 1b), flavonols (Figure 1b), with and without sugar moieties attached to one or more of the hydroxyl groups [28]. It is worthy of particular mention the rarely occurring class of C-glycolflavones (Figure 1b, compounds **63–65, 85, 86**).

These compounds are produced *in vivo* from the biogenetic mixed pathway of the Acetate and Shikimate that implies the enantiospecific formation of the basic aromatic bicyclic framework of the flavanone, from which a huge number of flavonoids originate employing selective enzymatic hydroxylations, methylations, and glycosylation steps. As can be seen from the structures shown in Figure 1, most of the chemical entities found in the peel extract contain several methoxy fragments that decorate the carbon skeleton. This characteristic makes those molecules to get a rather apolar character that explains their presence in the hydrophobic environment of the waxy peel. On the contrary, compounds containing a major number of hydroxyl groups are less present in the peel and are instead more significantly concentrated in the juice of the pericarp. However, some glycosylated compounds are present in the peel. In these molecules, the aglicone bears a monosaccharide unit (mainly glucose) or a disaccharide, in most of the cases being Rutinose (**91**) – Rhamnosyl (α1 → 6) glucose – or Neohesperidose (**92**)- Rhamnosyl (α1 → 2) glucose (Figure 2).

a)

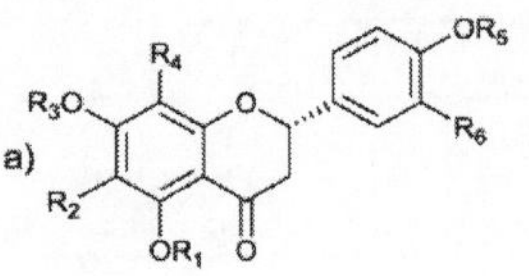

Cmp.	R1	R2	R3	R4	R5	R6	Name
38	H	H	H	H	Me	OH	Hesperetin
39	H	H	Rut	H	Glu	H	Narirutin-4'-glucoside
40	H	H	Rut	H	Me	OH	Hesperidin
41	H	H	Neohesp	H	Me	OH	Neohesperidin
42	H	H	Rut	H	H	H	Narirutin
43	H	H	Rut	H	Me	H	Didymin
44	H	H	Glu-Glu	H	H	H	Naringenin-7-O-bglucoside
45	Me	OMe	Me	H	Me	H	5,6,7,4'-tetramethoxyflavanone
46	H	OMe	Me	OMe	Me	OMe	5-hydroxy-6,7,8,3',4'-pentamethoxyflavanone
47	H	H	Neohesp	H	H	H	Naringin

b)

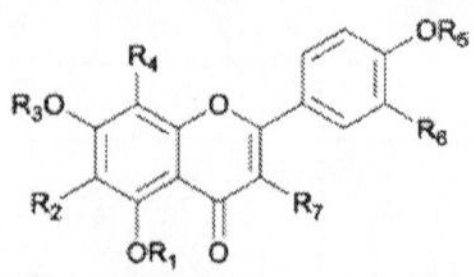

Cmp.	R1	R2	R3	R4	R5	R6	R7	Name
48	H	H	H	H	H	H	H	Apigenin
49	H	H	H	H	Me	H	H	Acacetin
50	Me	OMe	Me	H	Me	H	H	Tetra-O-methylscutellarein
51	Me	OMe	Me	H	Me	OMe	H	Sinensetin
52	Me	OMe	Me	OMe	Me	H	H	Tangeretin
53	Me	OMe	Me	OMe	Me	OMe	H	Nobiletin
54	Me	OMe	Me	H	Me	OMe	OMe	Hexa-O-methylquercetagetin
55	Me	OMe	Me	OMe	Me	OMe	OMe	3',4',3,5,6,7,8-Heptamethoxyflavone
56	H	H	H	H	Me	H	OH	Kaempferide
57	H	H	H	H	H	H	OH	Kaempferol
58	H	H	H	H	H	OH	H	Luteolin
59	H	H	H	H	H	OH	OH	Quercetin
60	Me	OMe	Me	OMe	Me	OMe	OH	Natsudaidain
61	Me	OMe	Me	OMe	Me	H	OH	3-hydroxy-5,6,7,8,4'-pentamethoxyflavone
62	Me	OMe	Me	H	Me	H	OH	3-hydroxy-5,6,7,4'-tetramethoxyflavone
63	H	H	H	C-Glu	H	H	H	Vitexin
64	H	C-Glu	H	C-Glu	H	H	H	6,8-di-C-Glucosylapigenin
65	H	C-Glu	H	C-Glu	Me	OH	H	6,8-di-C-Glucosyldiosmetin

Figure 1. Chemical structures of flavonoids from C. sinensis peels.

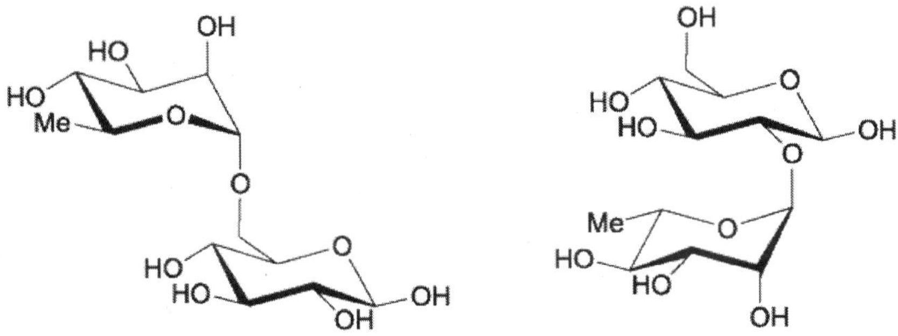

Rutinose **91** Neohesperidose **92**

Figure 2. Chemical structures of the disaccharides most commonly bound to flavonoids of C. sinensis peel.

The composition of the peel extracts described in the literature may slightly vary depending on the cultivar and the region of harvesting but some general points are common, that is the presence of the high amount of bioactive polymethoxyflavonoids [29, 30](PMF) some of which are rather ubiquitous, e.g. Nobiletin **53**, Sinensetin **51**, 3′,4′,3,5,6,7,8-Heptamethoxyflavone **55**; some other compounds containing one to six methoxy groups in place of the hydroxyl groups are present at variable amounts. The presence of one or more residual hydroxy groups in the molecule may result in a higher bioavailability and in other general differences in their mechanism of biological and therapeutic actions [30, 31].

The biological role of these secondary metabolites in the plant is still matter of debate. It has been proposed their involvement in the mechanism of defense of the fruits exposed to the attack of phytopathogens, such as *Phytophthora citrophthora* [32].

Further, the composition of the PMF blend can be employed for the chemiotaxonomic characterization of the *Citrus* genus [33].

However, it needs to be stressed that in many cases the reported compounds were recognized by mass spectrometry and electronic spectroscopy. It is not always a matter of simplicity to discern the exact structure of a given PMF and to discriminate between different regioisomers, normally quite similar in terms of mass and electronic spectra, if an isolation procedure is not conducted and followed by a complete bi-dimensional NMR characterization. Significant differences in the extract composition do arise also in consequence of the extraction method; non-polar solvents such as Methanol, Chloroform Ethyl acetate let to obtain PMFs-rich extracts while, on the other hand, hydroalcoholic and aqueous extracts do contain a low concentration of PMFs and a higher concentration of un-methylated polyphenols as well as glycosylated compounds.

The biological activities disclosed for the flavonoids extracted from CSP are variegated. They include antioxidant [9, 34, 35, 36, 37, 38, 39], anti-inflammatory [40, 41], antimicrobial [39, 42, 43, 44], antimalarial [45], antitrypanosomal [46], cardio-protective [47], anti-osteoporosis [48], anti-ulcer [49], vascular protective [50], anti-diabetes [51, 52], hepatoprotective [53, 54], neurotrophic [55], anti-adipogenesis and anti-obesity [56, 57, 58], anti-hypertensive [59], cataract prevention [60], sun protection [61], metabolic syndrome control [62]. Further, it has been demonstrated [63] that while both flavonoid set **40**, **42**, **43** and the PMFs **51–53** were able to inhibit the anion transportin polypeptide OATP2B1 in HEK293 cells, only the PMF group displayed this inhibitory activity also for the OATP1B1 and OATP1B3 carriers.

The most abundant PMF occurring in CSP, Nobiletin **53**, was proven to possess sevral bioactivities, such as antioxidant, anti-inflammatory, cancer preventive [64] and also a significant protective effect *in vivo* against the endotoxic shock [65] and ethanol-induced acute gastric lesions [66] in mice. Further, compound **53** demonstrated the capacity to induce autophagy in human keratinocyte HaCaT cells [67], vasodilatator effect in the rat aorta [68] and to protect the intestinal barrier from the demages induced by dextran sulfate sodium [69].

PMFs can be considered as especially promising lead compounds for cancer therapy as asignificant cytotoxic activity has been demonstrated toward a number of cancer cells [70, 71] with several mechanisms of action [72, 73]; the cytotypes investigated include MCF-7 [73, 74, 75, 76], Hs578T triple-negative breast cancer [73, 77]; colon cancer cells CaCo-2 [19], LoVo [78], HTC-116 [79, 80] and HT-29 [79, 81]; lung cancer cells A549 [80, 82], H460 [82, 83], H1299 [82, 83]; gastric cancer cell lines AGS, BGC-823, and SGC-7901 [84]; leukemia cells HL-60 [85]. However, data regarding a possible antitumor activity *in vivo* are still rather uncommon. An interesting example is the case of the significant reduction of the intestinal tumor mass in ApcMin/+ mice treated with a CSP extract containing various PMF [86]. Further, CSP extract and pure Naringin **47** were tested for their efficacy against a YM1 esophageal cancer in an animal model [87].

Given the development of pharmacological applications of CSP extract components, further investigations are needed to better understand the bioavailability, safety, and efficacy of these compounds in humans. Most of the data reported concern *in vitro* experimentations or animal model tests. For example, the toxicity of Hesperidin **40** was evaluated [88] in Sprague Dawley rats showing a 50% lethal dose (LD50) of about 5 g/Kg body weight (BW) and a lowest-observed-adverse-effect level (LOAEL) of ca. 1 g/Kg BW.

In general, it should be emphasized as the body of evidence concerning the actual efficacy of sweet orange-derived compounds in human health is still far to be exhaustive. For example, while this work is under typewriting, a severe acute respiratory syndrome pandemic due to a COVID-19 virus is in act and a big deal of research has been being directed toward antiviral remedies and therapies. Research on nutraceuticals is not an exception and in particular some authors have shown by computational and molecular docking methods how Hesperidin **40**, the most abundant polyphenol obtained from *C. sinensis*, would be able to bind the spyke protein of this virus thus inhibiting its activity [89]. Despite their undoubted interest, these results need to be further investigated with different experimental approaches.

The pharmacological potential of pure Hesperidin **40** was also investigated for several relevant human morbidity, such as cancer, hypertension, and ulcer [90].

2.2.2 Hydroxy-acids

Several hydroxylated carboxylic acids belonging to several structural sub-classes are present foremostly in the extract obtained with mixed hydro-organic solvents, such as

MeOH/water and EtOH / water [37, 38, 51, 78]; these include the aliphatic Ascorbic, Citric, Kojic, Lactic, and L-Malic acids; the aromatic 4-Hydroxybenzoic, Protocatechulic, and Gallic acids. Further, the cinnamyl compounds (Figure 3) Cinnamic (**93**), p-Cumaric (**94**), Caffeic (**95**), Ferulic (**96**), Sinapinic (**97**) acids, and Artepillin (**98**) were identified in some CSP extracts that showed relevant biological activities, such as antioxidant [34, 37, 38] and antidiabetes [51].

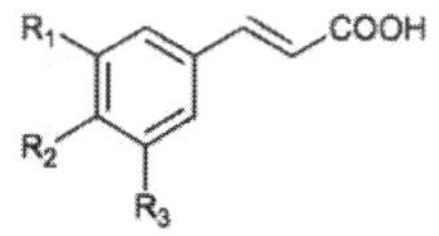

Compound	R1	R2	R3	Name
93	H	H	H	Cinnamic Acid
94	H	OH	H	p-Cumaric acid
95	OH	OH	H	Caffeic acid
96	MeO	OH	H	Ferulic Acid
97	MeO	OH	MeO	Sinapinic acid
98	3'-Methylbut-2-enyl	OH	3'-Methylbut-2-enyl	Artepillin C

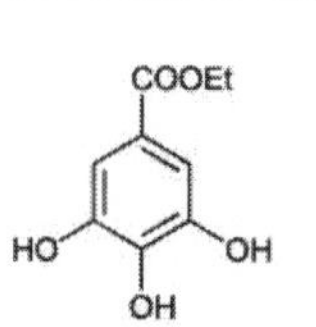

Ethyl gallate **99**

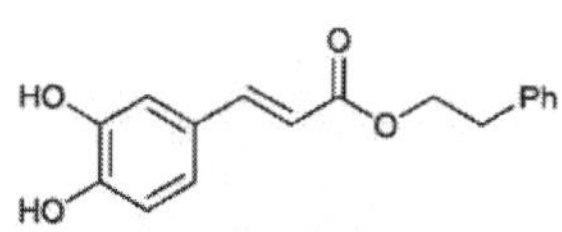

Phenylethyl caffeate **100**

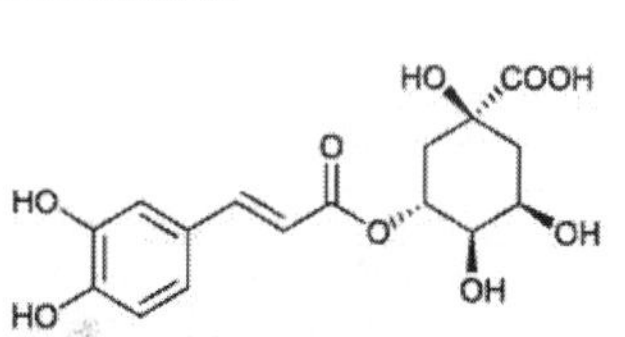

Chlorogenic acid **101**

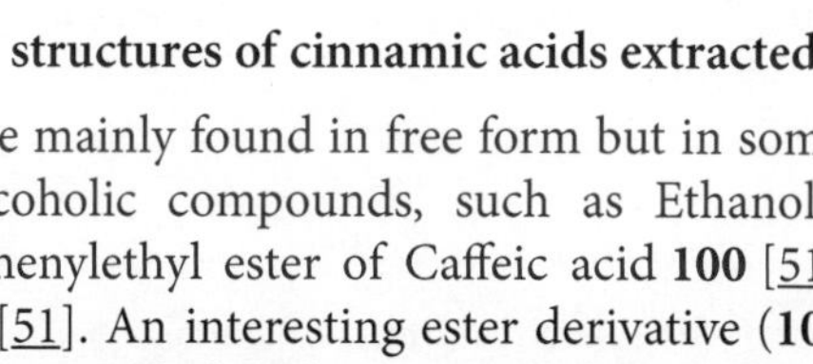

102

Figure 3. Chemical structures of cinnamic acids extracted from C. sinensis peels.

These organic acids are mainly found in free form but in some cases, they are esterified with a variety of alcoholic compounds, such as Ethanol in Ethyl gallate **99** [51], 2-Phenylethanol in Phenylethyl ester of Caffeic acid **100** [51] and (−)-Quinic acid in Chlorogenic acid **101** [51]. An interesting ester derivative (**102**) in which the anomeric hydroxyl of Glucose is esterified with a O-Caffeylsinapoyl acid unit was found in the methanolic extract of a Greek cultivar of *C. sinensis* [34].

It was shown [38] that the antioxidant properties of a CSP extract better correlated with the total phenols content (TPC) of the sample rather than with its total flavonoid content (TFC), as it can be expected from the known relevant antioxidant character of hydroxycynamic derivatives.

2.2.3 Coumarins

Coumarins are aromatic compounds biogenetically related to the o-hydroxysubstituted cynamic acids from which originate by the intramolecular condensation between the carboxylic and the o-hydroxy groups. These compounds are most commonly encountered in other species of *Citrus* taxa [91], such as *C. aurantium* (bitter orange), *C. limon*, (lemon), *C. limetta* (lime), *C. paradisi* (grapefruit) and only a few molecules of this class were Isolated from extracts of CSP endowed with activity against osteoporosis [48] and antioxidant [92]; these compounds are shown in Figure 4. As coumarins are relatively less common in *C. sinensis* cultivars compared to other species of the *Citrus* taxa, their rarity can be considered as a chemotaxonomic marker for *C. sinensis*.

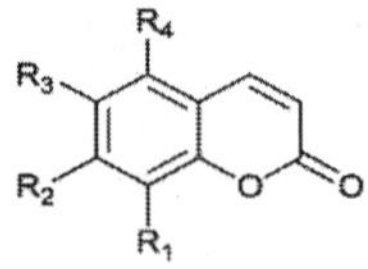

Compound	R_1	R_2	R_3	R_4	Name
103	H	OH	H	H	Umbelliferone
104	3'-methyl but-2'-enyl	OMe	H	H	Osthol
105	H	OMe	OMe	H	Scoparone
106	H	OMe	H	OMe	Limettin

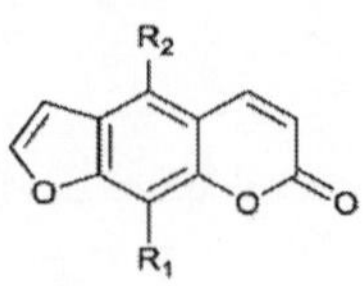

Compound	R_1	R_2	Name
107	H	H	Psoralen
108	H	OH	Bergaptol
109	OMe	H	Xanthotoxin
110	H	OMe	Bergapten
111	3'-methyl but-2'-enyl	H	Imperatorin
112	H	3'-methyl but-2'-enyl	Isoimperatorin

Figure 4. Chemical structure of coumarins extracted from C. sinensis peels.

2.2.4 Catechins

The NADPH dependent bioreduction of flavanols is the biogenetic origin of this class of compounds, present as minor constituents in CSP extract possessing significant antioxidant activity [38]; they are the two enantiomeric forms Catechin **113** and Epicatechin **114**, together with Epigallocatechin **115** (Figure 5).

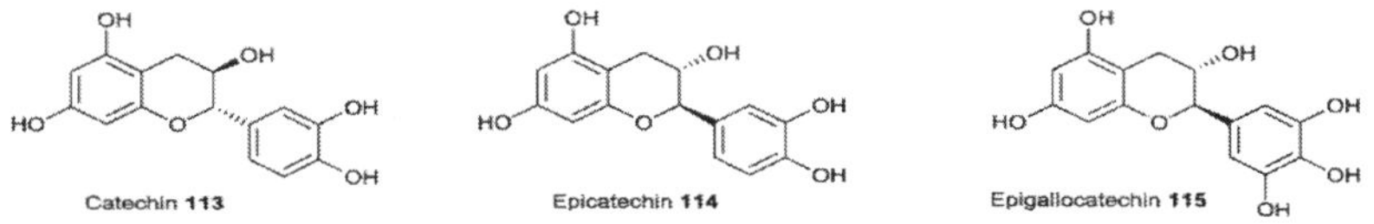

Figure 5. Chemical structure of catechins from C. sinensis peels.

2.3 Pectins

Pectins [93] are chemically definable as complex mixtures of polyglyconic acids in which a linear polymeric backbone is structured by a series of α (1 → 4) linkages (Figure 6). The main sugar monomer is always Galacturonic acid with the presence of possible heterogeneous domains of other sugars such as Xylogalacturonan and Rhamnogalacturonan-I. A variable amount of the free carboxy functions may be esterified with methyl groups, while the hydroxy groups at C-2 and C-3 positions of the sugar monomers may be acetylated. Even though the primary structure of the main chain is linear, a possible degree of ramification, depending on the pectin source, may also be found. The differences in the pectins composition and structures, depending on their natural source, do confer them different physio-chemical properties, such as water solubility, sol–gel concentrations, etc. On the ground of the degree of methylation of the acid moieties, pectins are classified as "low methoxyl" (LMP, -COOMe/-COOH <50% mol.) or as "high methoxyl" (> 50% mol). A simplified representation of pectin structure is given in Figure 6.

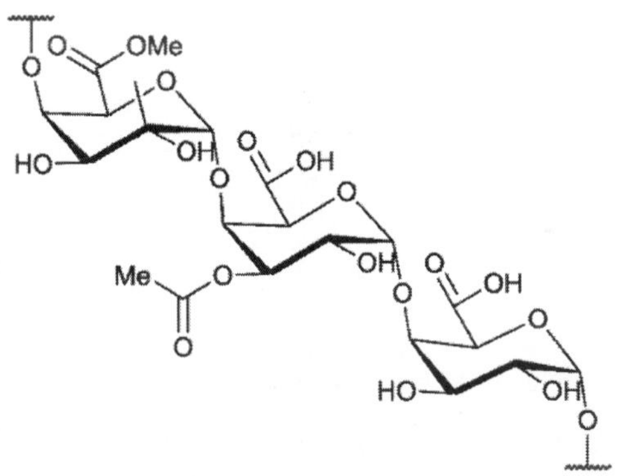

Figure 6. Minimal representation of a Homopolygalacturonic acid domain of the linear primary pectin structure with a 1/3 Mol. /Mol. Esterification degree.

Pectins find many applications in the food and drug industry as a thickening and gelling agents, excipients, and colloidal stabilizers [93].

As it has been already mentioned, the extraction method does affect the structure and the properties of the final product; the traditional acidic water extraction implies a certain degree of hydrolytic deterioration, so that new extraction technologies have been being investigated to improve the quality of the final pectins, that is microwave-assisted extraction (MAE) [94] and ultrasounds assisted extraction (USAE) [35, 95].

2.4 Enzymes

As it can be easily argued, the CSP cellular system, whose genomic profile has been fully characterized [96], is the site of a complex network of enzymatic activity. Some of the enzymes of CSP have been characterized and employed in many applications.

The acetylesterase (international enzymatic classification: EC 3.1.1.6) from CSP is known since 1947 [97] and was isolated and characterized [98]. The acetylesterase activity of the partially purified enzyme was used for the removal of the acetyl group at the 3 positions of β-lactamic antibiotics **116** [98] (Figure 7a). Further, the whole CSP, as well as pomace from the industrial waste of the orange juice production, was successfully employed to catalyze several relevant biotransformations [99] such as the conversion of Geranyl acetate **118** to Geraniol **119** (Figure 7b) and the di-acetoxynaphtalene derivative **120** to the vitamin k1 precursor **121** (Figure 7c).

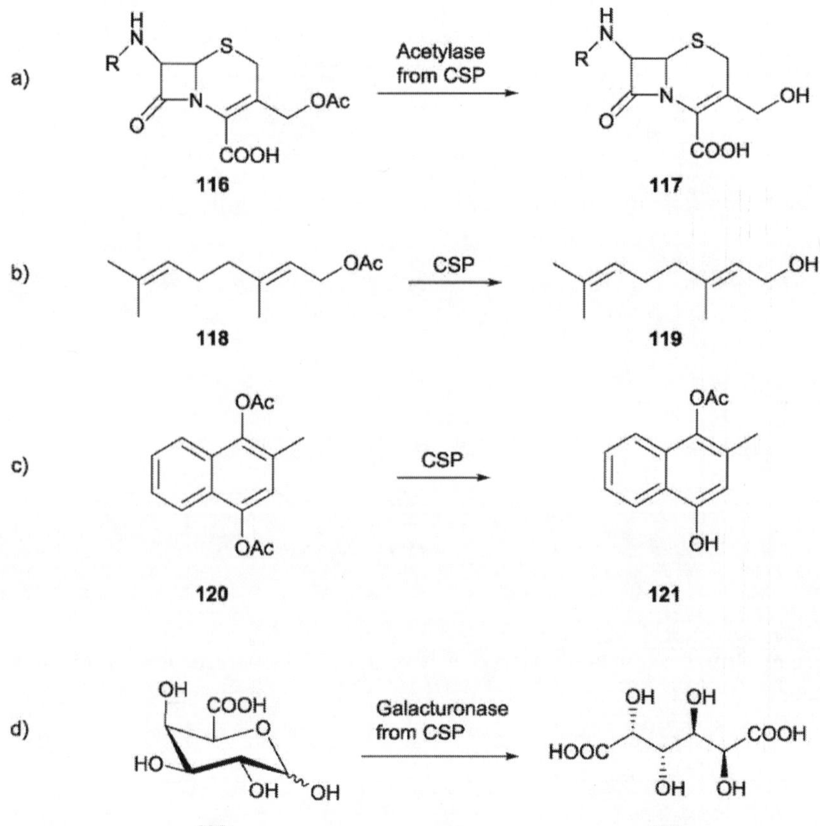

Figure 7. Chemical reactions biocatalysed by enzymes from C. sinensis peels.

Recently, partial purification and functional characterization of a Uronic acid oxidase from CSP was accomplished [100]; this enzyme promotes the oxidation by O_2 of Galacturonic acid **122** to Galactaric acid **123** (Figure 7d).

2.5 Miscellaneous

2.5.1 Highly lipophilic compounds

The waxy environment of flavedo in CSP does contain several long-chain saturated and unsaturated compounds: alkanes, fatty acids, waxes, higher terpenoids.

Tetracosane, Tetratriacontanoic acid, and Ethyl pentacosanoate were identified in CSP of a Pineapple variety [101]. Further, some carotenoids were identified in the CSP extract obtained with a solvent mixture composed of Ethanol, Ethyl acetate, Petroleum ether 1: 1:1 [102]. This complex blend of carotenoids includes α- and β-Carotene, Phytoene, Phytofluene, (all-E)- and (9Z)-Violoxanthin, (all-E)-Neoxanthin, (13Z)-, (13Z')- and (all-E)-Lutein, (9Z)-Zeaxanthin, (all-E)-Zeaxanthin; the mono and di-esters of violaxanthin, antheroxanthins, Xanthophyll, β-Citraurin with various fatty acids, including Lauraic, Myristic, Oleic, Palmitic, Stearic. The composition of the blend has been correlated with the maturity stage of the fruit.

2.5.2 Peptides

Three cyclic peptides have been isolated from the hot water extract of CSP and were structurally characterized by FAB-MS and 2D-NMR techniques [103]. Their amino-acidic sequences, including a mostly lipophlic heptapeptide **124**, a di-hydroxylated heptapeptide **125,** and a Glutamate-rich octapeptide **126**, are reported in Figure 8.

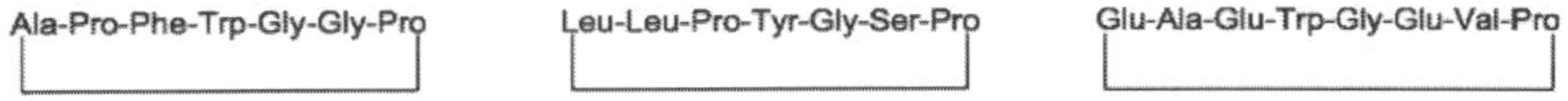

124 **125** **126**

Figure 8. Primary structure of cyclic peptide isolated from the C. sinensis peels.

3. Conclusions

The chemical richness of the primary and secondary metabolome of *C. sinesnis* species is undoubtedly impressive. Thousands of different compounds belonging to dozens of structural classes have been isolated and described. The most deeply investigated are sure, on one hand, the mixtures of volatile compounds composing the blend of the essential oil and, on the other hand, polyphenols, especially flavonoids.

The chemical composition of the extract from the exocarp of *C. sinensis* does differ from the composition of juice, or leaf extracts for some aspects [104]: the presence of a higher amount of more lipophilic compounds such as polymethoxy-flavonoids, r carotenoids, higher alkanes; a lesser extent of lighter terpenoids, a lower content of glycosylated compounds, the absence of cyanidins and sterols.

It is also a matter of fact that several interesting bioactivities were disclosed in the last years for the *C. sinensis* extracts that have been variously associated with the well-recognized beneficial effects that regular sweet oranges consumption may have on human health. However, a great deal of research work is still needed to clarify the molecular basis and the mechanism of these chemopreventive effects and to relate them with precise chemical entities that can be recognized as valuable nutraceuticals, as it is already the case for the well-established antioxidants Ascorbic acid, Hesperidin, Hesperetin, Quercetin, etc.

References

1. Food and Agriculture Organization of the United nations: Citrus fruit fresh and processed - Statistical Bulletin 2016 [Internet]. 2017. Available from: http://www. http://www.fao.org/3/a-i8092e.pdf [Accessed: 2020-12-16]

2. Tamburino V, Zama DA. I sottoprodotti dell'industria di trasformazione: il pastazzo di agrumi. In: Vacante V, editor. Citrus - Trattato di agrumicoltura. Il Sole 24 Ore-Edagricole; 2009. p. 459–470

3. Putnik P, Kovačević DB, Jambrak AR, Barba FJ, Cravotto G, Binello A, Lorenzo JM, Shpigelman A. Innovative "Green" and Novel Strategies for the Extraction of Bioactive Added Value Compounds from CitrusWastes—A Review. Molecules. 2017:22:680–704. DOI: 10.3390/molecules22050680

4. Senit JJ, Velasco D, Gomez Manrique A, Sanchez-Barba M, Toledo LM, Santos VE, Garcia-Ochoa F, Yustos P, Ladero M. Orange peel waste upstream integrated processing to terpenes, phenolics, pectin and monosaccharides: Optimization approaches. Ind. Crops & Prod. 2019:134:370–381. DOI: 10.1016/j.indcrop.2019.03.060

5. Lamine M, Gargouri M, Rahali FZ, Mliki A. Recovering and Characterizing Phenolic Compounds From Citrus By-Product: A Way Towards Agriculture of Subsistence and Sustainable Bioeconomy. Waste Biomass Valor. 2020. DOI: 10.1007/s1264 9-020-01306 -9

6. Williams CA. Specialized dietary supplements. In: Geor RJ, Harris PA, Coenen M, editors. Equine Applied and Clinical Nutrition. Saunders; 2013. p. 351–366. DOI: 10.1016/C2009-0-39370-8

7. Patel S, Shukla S. Fermentation of Food Wastes for Generation of Nutraceuticals and Supplements. In: Frias J, Martinez-Villaluenga C, Peñas E, editors. Fermented Foods in Health and Disease Prevention. Academic Press; 2016. p. 707–734. DOI: 10.1016/C2014-0-01734-0

8. Khan N, Mongas M, Urpi-sarda M, Llorach R, Andres-La Cueva C. Contribution of Bioactive Foods and Their Emerging Role in Immunomodulation, Inflammation, and Arthritis. In: Watson RR. Preedy VR, editors. Bioactive Food as Dietary Interventions for Arthritis and Related Inflammatory Diseases. Academic Press; 2013. p. 43–65. DOI: 10.1016/C2011-0-07467-7

9. Rehman Z. Citrus peel extract – A natural source of antioxidant. Food Chem. 2006;99:450–454. DOI: 10.1016/j.foodchem.2005.07.054

10. Abe S, Fan K, Ho CT, Ghai G, Yang K. Chemopreventive Effects of Orange Peel Extract (OPE) II. OPE Inhibits Atypical Hyperplastic Lesions in Rodent Mammary Gland. J. Med. Food 2007;10:18–24; DOI: 10.1089/jmf.2006.0215

11. Lai CS, Li S, Miyauchi Y, Suzawa M, Hob CT, Pan MH. Potent anti-cancer effects of citrus peel flavonoids in human prostate xenograft tumors. Food Funct. 2013;4:944–949; DOI: 10.1039/c3fo60037h

12. Tung YC, Chang WT, Li S, Wu JC, Badmeav V, Ho CT, Pan MH. Citrus peel extracts attenuated obesity and modulated gut 1 microbiota in a high-fat diet-induced obesity mice. Food Funct. 2018;9:3363–3373; 10.1039/C7FO02066J

13. Qiao Y, Jun Xie B, Zhang Y, Zhang Y, Fan G, Yao XL, Pan SY. Characterization of Aroma Active Compounds in Fruit Juice and Peel Oil of Jinchen Sweet Orange Fruit (Citrus sinensis (L.) Osbeck) by GC-MS and GC-O. Molecules. 2008;13:1333–1344; DOI: 10.3390/molecules13061333

14. Dosoky NS, Setzer WN. Biological Activities and Safety of Citrus spp. Essential Oils. Int. J. Mol. Sci. 2018;19:1966; DOI: 10.3390/ijms19071966

15. Ettoumi KY, Zouambia Y, Moulai-Mostefa N. Chemical composition, antimicrobial and antioxidant activities of Algerian Citrus sinensis essential oil extracted by hydrodistillation assisted by electromagnetic induction heating. J Food Sci Technol DOI: 10.1007/s13197-020-04808-5

16. Leherbauer I, Stappen I. Selected essential oils and their mechanisms for therapeutic use against public health disorders. An overview. Z. Naturforsch. 2020;75: 205–223

17. Singh P, Shukla R, Prakash B, Kumar A, Singh S, Mishra PK, Dubey NK. Chemical profile, antifungal, antiaflatoxigenic and antioxidant activity of Citrus maxima Burm. and Citrus sinensis (L.) Osbeck essential oils and their cyclic monoterpene, DL-limonene. Food Chem. Toxicol. 2010;48:1734–1740; DOI:10.1016/j.fct.2010.04.001

18. O'Bryan CA, Crandall PG, Chalova VI, Ricke SC. Orange Essential Oils Antimicrobial Activities against Salmonella spp. J. Food Sc. 2008;73:M264-M267; DOI: 10.1111/j.1750-3841.2008.00790.x

19. Fisher K, Phillips CA. The effect of lemon, orange and bergamot essential oils and their components on the survival of Campylobacter jejuni, Escherichia coli O157, Listeria monocytogenes, Bacillus cereus and Staphylococcus aureus in vitro and in food systems. J. Appl. Microbiol. 2006;101:1232–1240; DOI:10.1111/j.1365-2672.2006.03035.x

20. Muthaiyan A, Martin EM, Natesan S, Crandall PG, Wilkinson BJ, Ricke SC. Antimicrobial effect and mode of action of terpeneless cold-pressed Valencia orange essential oil on methicillin-resistant Staphylococcus aureus. J. Appl. Microbiol. 2012;112:1020–1033; DOI:10.1111/j.1365-2672.2012.05270.x

21. Hung PV, Chi PTL, Phi NTL. Comparison of antifungal activities of Vietnamese citrus essential oils. Nat. Prod. Res. 2013;27:506–508; DOI: 10.1080/14786419.2012.706293

22. Najar B, Shortrede JE, Pistelli L, Buhagiar J. Chemical Composition and in Vitro Cytotoxic Screening of Sixteen Commercial Essential Oils on Five Cancer Cell Lines. Chem. Biodiversity 2020;17:e1900478; DOI: 10.1002/cbdv.201900478

23. Faturi CB, Leite JR, Alves PB, Canton AC, Teixeira-Silva F. Anxiolytic-like effect of sweet orange aroma in Wistar rats. Prog. Neuro-Psychoph. 2010;34:605–609; DOI:10.1016/j.pnpbp.2010.02.020

24. Uçar Y. Antioxidant Effect of Nanoemulsions Based on Citrus Peel Essential Oils: Prevention of Lipid Oxidation in Trout. Eur. J. Lipid Sci. Technol. 2020;122:1900405–1900419; DOI: 10.1002/ejlt.201900405

25. Ciriminna R, Meneguzzo F, Pagliaro M. Orange Oil. In: Nollet LML, Rathore HS editors. Green Pesticides Handbook – Essential Oils for Pest Control. CRC Press; 2017. p. 215–265

26. Lehrner J, Eckersberger C, Walla P, Pötsch G, Deecke L. Ambient odor of orange in a dental office reduces anxiety and improves mood in female patients. Physiology & Behavior 2000;71:83–86

27. Anandakumar P, Kamaraj S, Vanitha MK. D-limonene: A multifunctional compound with potent therapeutic effects. J. Food Biochem. 2020;45:e13566; DOI: 10.1111/jfbc.13566

28. Li S, Lo CY, Ho CT. Hydroxylated Polymethoxyflavones and Methylated Flavonoids in Sweet Orange (Citrus sinensis) Peel. J. Agric. Food Chem. 2006;54:4176–4185. DOI: 0.1021/jf060234n

29. Gao Z, Gao W, Zeng SL, Li P, Liu EH. Chemical structures, bioactivities and molecular mechanisms of citrus polymethoxyflavones. J. Funct. Foods 2018;40:498–509; DOI: 10.1016/j.jff.2017.11.036

30. Owis AL. Citrus Polymethoxyflavones: Biofunctional Molecules of Therapeutic Interest. Studies in Nat. Prod. Chem. 2019;59: 509–530. DOI: 10.1016/B978-0-444-64179-3.00015-3

31. Lai CS, Wu JC, Ho CT, Pan MH. Disease chemopreventive effects and molecular mechanisms of hydroxylated polymethoxyflavones. Biofactors 2015;41:301–313; DOI 10.1002/biof.1236

32. Del Río JA, Gómez P, Báidez AG, Arcas MC, Botía JM, Ortuño A. Changes in the Levels of Polymethoxyflavones and Flavanones as Part of the Defense Mechanism of Citrus sinensis (Cv. Valencia Late) Fruits against Phytophthora citrophthora. J. Agric. Food Chem. 2004;52:1913–1917; DOI: 10.1021/jf035038k

33. Mizuno M, Iinuma M, Ohara M, Tanaka T, Iwamasa M. Chemotaxonomy of the Genus Citrus Based on Polymethoxyflavones. Chem. Pharm. Bull. 1991;39:945–949; DOI: 10.1248/cpb.39.942

34. Kanaze FI, Termentzi A, Gabrieli C, Niopas I, Georgarakis M, Kokkaloua E. The phytochemical analysis and antioxidant activity assessment of orange peel (Citrus sinensis) cultivated in Greece–Crete indicates a new commercial source of hesperidin. Biomed. Chromatogr. 2009;23:239–249. DOI: 10.1002/bmc.1090

35. Montero-Calderon A, Cortes C, Zulueta A, Frigola A, Esteve MJ. Green solvents and Ultrasound- Assisted Extraction of bioactive orange (Citrus sinensis) peel compounds. Scientific Reports. 2019;9:16120–16128. DOI: 10.1038/s41598-019-52717-1

36. Manthey JA. Fractionation of Orange Peel Phenols in Ultrafiltered Molasses and Mass Balance Studies of Their Antioxidant Levels. J. Agric. Food Chem. 2004;52:7586–7592. DOI: 0.1021/jf049083j

37. Anagnostopoulou MA, Kefalas P, Kokkalou E,. Assimopoulou AN, Papageorgiou VP. Analysis of antioxidant compounds in sweet orange peel by HPLC–diode array detection–electrospray ionization mass spectrometry. Biomed. Chromatogr. 2005;19:138–148. DOI: 10.1002/bmc.430

38. Liew SS, Ho WY, Yeap SK, Bin Sharifudin SA. Phytochemical composition and in vitro antioxidant activities of Citrus sinensis peel extracts. PeerJ. 2018;6:e5331. DOI: 10.7717/peerj.5331

39. Guo C, Shan Y, Yang Z, Zhang L, Ling W, Liang Y, Ouyang Z, Zhong B, Zhang J. Chemical composition, antioxidant, antibacterial, and tyrosinase inhibition activity of extracts from Newhall navel orange (Citrus sinensis Osbeck cv. Newhall) peel. J Sci Food Agric 2020;100:2664–2674. DOI: 10.1002/jsfa.10297

40. Gosslau A, Chen KY, Ho CT, Li S. Anti-inflammatory effects of characterized orange peel extracts enrichedwith bioactive polymethoxyflavones. Food Sci. Human Well. 2014;3:26–35; DOI:10.1016/j.fshw.2014.02.002

41. Hagenlocher Y, Feilhauer K, Schäffer M, Bischoff SC, Lorentz A. Citrus peel polymethoxyflavones nobiletin and tangeretin suppress LPS- and IgE-mediated activation of human intestinal mast cells. Eur J Nutr 2017;56:1609–1620; DOI 10.1007/s00394-016-1207-z

42. Liu L, Xu X, Cheng D, Yao X, Pan S. Structure–Activity Relationship of Citrus Polymethoxylated Flavones and Their Inhibitory Effects on Aspergillus niger. J. Agric. Food Chem. 2012;60:4336–4341. DOI: 10.1021/jf3012163

43. Shetty SB, Mahin-Syed-Ismail P, Varghese S, Thomas-George B, Kandathil-Thajuraj P, Baby O, Haleem S, Sreedhar S, Devang-Divakar D. Antimicrobial effects of Citrus sinensis peel extracts against dental caries bacteria: An in vitro study. J Clin Exp Dent. 2016;8:e70–7. DOI:10.4317/jced.52493

44. Ortuño A, Báidez A, Gómez P, Arcas MC, Porras I, García-Lidón A, Del Río JA. Citrus paradisi and Citrus sinensis flavonoids: Their influence in the defence mechanism against Penicillium digitatum. Food Chem. 2006;98:351–358; DOI:10.1016/j.foodchem.2005.06.017

45. Bagavan A, Rahuman AA, Kamaraj C, Kaushik NK, Mohanakrishnan D, Sahal D. Antiplasmodial activity of botanical extracts against Plasmodium falciparum. Parasitol Res 2011;108:1099–1109. DOI: 10.1007/s00436-010-2151-0

46. Nakanishi M, Hino M, Yoshimura M, Amakura Y, Nomoto H. Identification of sinensetin and nobiletin as major antitrypanosomal factors in a citrus cultivar. Exp. Parasitol. 2019;200:24–29; DOI: 10.1016/j.exppara.2019.03.008

47. Lara Testai L, Calderone V. Nutraceutical Value of Citrus Flavanones and Their Implications in Cardiovascular Disease. Nutrients. 2017;9:502–515. DOI:10.3390/nu9050502

48. Shalaby NMM, Abd-Alla HI, Ahmed HH, Basoudan N. Protective effect of Citrus sinensis and Citrus aurantifolia against osteoporosis and their phytochemical constituents. J. Med. Plant. Res. 2011;5:579–588

49. Selmi S, Rtibi K, Grami D, Sebai H, Marzouki L. Protective effects of orange (Citrus sinensis L.) peel aqueous extract and hesperidin on oxidative stress and peptic ulcer induced by alcohol in rat. Lipids in Health and Disease. 2017;16:152–164. DOI: 10.1186/s12944-017-0546-y

50. Chen PY, Li S, Koh YC, Wu JC, Yang MJ, Ho CT, Pan MH. Oolong Tea Extract and Citrus Peel Polymethoxyflavones Reduce Transformation of L-Carnitine to Trimethylamine-N-Oxide and Decrease Vascular Inflammation in L-Carnitine Feeding Mice. J. Agric. Food Chem. 2019;67:7869–7879. DOI: 10.1021/acs.jafc.9b03092

51. Sathiyabama RG, Gandhia GR, Denadaib M, Sridharanc G, Jothic G, Sasikumard P, Siqueira Quintanse JdS, Narain N, Cuevas LE, Melo Coutinho HD, Barbosa Ramose AG, Quintans-Júniore LJ, Queiroz Gurgela R. Evidence of insulin-dependent signalling mechanisms produced by Citrus sinensis (L.) Osbeck fruit peel in an insulin resistant diabetic animal model. Food Chem. Toxicol. 2018;116:86–99. DOI: 10.1016/j.fct.2018.03.050

52. Ahmeda OM, Hassanb MA, Abdel-Twabc SM, Abdel Azeem MN. Navel orange peel hydroethanolic extract, naringin and naringenin have anti-diabetic potentials in type 2 diabetic rats. Biomed. & Pharmacotherapy 2017;94:197–205. DOI: 10.1016/j.biopha.2017.07.094

53. Ahmed OM, Fahim H, Ahmed HY, Al-Muzafar HM, Ahmed RR , Amin KA, El-Nahass ES, Abdelazeem WH. The Preventive Effects and the Mechanisms of Action of Navel Orange Peel Hydroethanolic Extract, Naringin, and Naringenin in N-Acetyl-p-aminophenol-Induced Liver Injury in Wistar Rats. Oxidative Medicine and Cellular Longevity. 2019; DOI: 10.1155/2019/2745352

54. Kim TW, Lee DR, Choi BK, Kang HK, Jung JY, Lim SW, Hwan Yang S, Suh JW. Hepatoprotective effects of polymethoxyflavones against acute and chronic carbon tetrachloride intoxication. Food Chem. Toxicol. 2016;91: 91–99; DOI: 10.1016/j.fct.2016.03.004

55. Chiu SP, Wu MJ, Chen PY, Ho YR, Tai MH, Ho CT, Yen JH. Neurotrophic action of 5-hydroxylated polymethoxyflavones: 5-demethylnobiletin and gardenin A stimulate neuritogenesis in PC12 cells. J. Agric. Food Chem. 2013;61:9453–9463; DOI: 10.1021/jf4024678

56. Lai CS, Ho MH, Tsai ML, Li S, Badmaev V, Ho CT, Pan MH. Suppression of Adipogenesis and Obesity in High-Fat Induced Mouse Model by Hydroxylated Polymethoxyflavones. J. Agric. Food Chem. 2013;61:10320–10328; DOI: 10.1021/jf402257t

57. Sergeev IN, Li S, Ho CT, Rawson NE, Dushenkov S. Polymethoxyflavones Activate Ca2+-Dependent Apoptotic Targets in Adipocytes. J. Agric. Food Chem. 2009;57:5771–5776; DOI:10.1021/jf901071k

58. Yu Wang Y, Lee PS, Chen YF, Ho CT, Pan MH. Suppression of Adipogenesis by 5-Hydroxy-3,6,7,8,30,40-Hexamethoxyflavone from Orange Peel in 3T3-L1 Cells. J. Med. Food 2016;19:830–835; DOI: 10.1089/jmf.2016.0060

59. Li GJ, Wang J, Cheng YJ, Tan X, Zhai YL, Wang Q, Gao FJ, Liu GL, Xin Zhao X, Hua Wang H. Prophylactic Effects of Polymethoxyflavone-Rich Orange Peel Oil on N^{ω}-Nitro-L-Arginine-Induced Hypertensive Rats. Appl. Sci. 2018;8:752–768; DOI:10.3390/app8050752

60. Miyata Y, Oshitari T, Okuyama Y, Shimada A, Takahashi H, Natsugari H, Kosano H. Polymethoxyflavones as agents that prevent formation of cataract: Nobiletin congeners show potent growth inhibitory effects in human lens epithelial cells. Bioorg. Med. Chem. Lett. 2013;23:183–187; DOI: 10.1016/j.bmcl.2012.10.133

61. Li G, Tan F, Zhang Q, Tan A, Cheng Y, Zhou Q, Liu M, Tan X, Huang L, Rouseff R, Wu H, Zhao X, Liang G, Zhao X. Protective effects of polymethoxyflavone-rich cold-

pressed orange peel oil against ultraviolet B-induced photoaging on mouse skin. J. Funct. Foods 2020; 67:103834–103844; DOI: 10.1016/j.jff.2020.103834

62. Zeng SL, Li SZ, Xiao PT, Cai YY, Chu C, Chen BZ, Li P, Li J, Liu EH. Citrus polymethoxyflavones attenuate metabolic syndrome by regulating gut microbiome and amino acid metabolism. Sci. Adv. 2020;6: eaax6208; DOI: 10.1126/sciadv.aax6208

63. Bajraktari-Sylejmani G, Weiss J. Potential Risk of Food-Drug Interactions: Citrus Polymethoxyflavones and Flavanones as Inhibitors of the Organic Anion Transporting Polypeptides (OATP) 1B1, 1B3, and 2B1. Eur. J. Drug Metab. Pharm. 2020;45:809–815; DOI: 10.1007/s13318-020-00634-4

64. Li S, Wang H, Guo L, Zhao H, Ho CT. Chemistry and bioactivity of nobiletin and its metabolites. J. Funct. Food 2014;6;2–10. DOI: 10.1016/j.jff.2013.12.011

65. Li W, Wang X, Niu X, Zhang H, He Z, Wang Y,1 Zhi W, Liu F. Protective Effects of Nobiletin Against Endotoxic Shock in Mice Through Inhibiting TNF-α, IL-6, and HMGB1 and Regulating NF-κB Pathway. Inflammation 2016;39:786–797. DOI: 10.1007/s10753-016-0307-5

66. Li W, Wang X, Zhi W, Zhang H, He Z, Wang Y, Liu F, Niu X, Zhang X. The gastroprotective effect of nobiletin against ethanol-induced acute gastric lesions in mice: impact on oxidative stress and inflammation. Immunopharm. Immunot. 2017;39:354–363; DOI: /10.1080/08923973.2017.1379088

67. Abe S, Hirose S, Nishitani M, Yoshida I, Tsukayama M, Tsuji A, Yuasa K. Citrus peel polymethoxyflavones, sudachitin and nobiletin, induce distinct cellular responses in human keratinocyte HaCaT cells. Biosci. Biotechnol. Biochem. 2018;82:2064–2071; DOI: 10.1080/09168451.2018.1514246

68. Kaneda H, Otomo R, Sasaki N, Omi T, Sato T, Kaneda T. Endothelium-independent vasodilator effects of nobiletin in rat aorta. J. Pharmacol. Sci. 2019;140:48–53; DOI: 10.1016/j.jphs.2019.04.004

69. Wen X, Zhao H, Wang L, Wang L, Du G, Guan W, Liu J, Cao X, Jiang X, Tian J, Wang M, Ho CT, Li S. Nobiletin Attenuates DSS-Induced Intestinal Barrier Damage through the HNF4α-Claudin-7 Signaling Pathway. J. Agric. Food Chem. 2020;68:4641–4649; DOI: 10.1021/acs.jafc.0c01217

70. Koolaji N, Shammugasamy B, Schindeler A, Dong Q, Dehghani F, Valtchev P. citrus Peel Flavonoids as Potential Cancer Prevention Agents. Current Developments In Nutrition. 2020;4:ID nzaa025;DOI: 10.1093/cdn/nzaa025

71. Tung, YC, Chou YC, Hung WL, Cheng AC, Yu RC, Ho CT, Pan MH. Polymethoxyflavones: Chemistry and Molecular Mechanisms for Cancer Prevention and Treatment. Curr. Pharmacol. Rep. 2019;5:98–113; DOI: 10.1007/s40495-019-00170-z

72. Wang L, Wang J, Fang L, Zheng Z, Zhi D, Wang S, Li S, Ho CT, Zhao H. Anticancer Activities of Citrus Peel Polymethoxyflavones Related to Angiogenesis and Others. Biomed. Res. Int. 2014; Article ID 453972; DOI: 10.1155/2014/453972

73. Chan EWC, Soo OYM, Tan YH, Wong SK, Chan HT. Nobiletin and tangeretin (citrus polymethoxyflavones): an overview on their chemistry, pharmacology and cytotoxic

activities against breast cancer. J. Chin. Pharm. Sci. 2020;29:443–454; DOI: 10.5246/jcps.2020.07.042

74. Sergeev IN, Ho CT, Li S, Colby J, Dushenkov S. Apoptosis-inducing activity of hydroxylated polymethoxyflavones and polymethoxyflavones from orange peel in human breast cancer cells. Mol. Nutr. Food Res. 2007;51:1478–1484. DOI: 10.1002/mnfr.200700136

75. Rahideh ST, Keramatipour M, Nourbakhsh M, Koohdani F, Hoseini M, Talebi S, Shidfar F. Comparison of the effects of Nobiletin and Letrozole on the activity and expression of aromatase in MCF-7 breast cancer cell line. Biochem. Cell. Biol. 2017;95:468–473;DOI: 10.1139/bcb-2016-0206

76. Sergeev IN, Li S, Colby J, Ho CT, Dushenkov S. Polymethoxylated flavones induce Ca2+–mediated apoptosis in breast cancer cells. Life Sci. 2006;80:245–253; DOI: 10.1016/j.lfs.2006.09.006

77. Borah N, Gunawardana S, Torres H, McDonnell S, Van slambrouck S. 5,6,7,3′,4′,5'-Hexamethoxyflavone inhibits growth of triple-negative breast cancer cells via suppression of MAPK and Akt signaling pathways and arresting cell cycle. Int. J. Oncol. 2017;51:1685–1693; DOI: 10.3892/ijo.2017.4157

78. Ademosun AO, Oboh G, Passamonti S, Tramer F, Ziberna L, Augusti Boligon A, Athayde ML. Inhibition of metalloproteinase and proteasome activities in colon cancer cells by citrus peel extracts. J Basic Clin Physiol Pharmacol 2015;26: 471–477. DOI: 10.1515/jbcpp-2013-0127

79. Kaur J, Kaur G. An insight into the role of citrus bioactives in modulation of colon cancer. J. Funct. Foods. 2015;13:239–261;DOI: 10.1016/j.jff.2014.12.043

80. 80.Qiu P, Cui Y, Xiao H, Han Z, Ma H, Tang Y, Xu H, Zhang L. 5-Hydroxy polymethoxyflavones inhibit glycosaminoglycan biosynthesis in lung and colon cancer cells. J. Funct. Foods. 2017;30:39–47. DOI: 10.1016/j.jff.2017.01.008

81. Silva I, Estrada MF, Pereira CV, Bento da Silva A, Bronze MR, Alves PM,. Duarte CMM, Brito C, Serra AT. Polymethoxylated Flavones from Orange Peels Inhibit Cell Proliferation in a 3D Cell Model of Human Colorectal Cancer. Nutrition and Cancer. 2018;70:257–266. DOI: 10.1080/01635581.2018.1412473

82. Charoensinphon N, Qiu P, Dong P, Zheng J, Ngauv P, Cao Y, Li S, Ho CT, Xiao H. 5-Demethyltangeretin inhibits human nonsmall cell lung cancer cell growth by inducing G2/M cell cycle arrest and apoptosis. Mol. Nutr. Food Res. 2013;57:2103–2111; DOI: 10.1002/mnfr.201300136

83. Xiao H, Yang CS, Li S, Jin H, Ho CT, Patel T. Monodemethylated polymethoxyflavones from sweet orange (Citrus sinensis) peel inhibit growth of human lung cancer cells by apoptosis. Mol. Nutr. Food Res. 2009;53:398–406; DOI 10.1002/mnfr.200800057

84. Wang Y, Chen Y, Zhang H, Chen J, Cao J, Chen Q, Li X, Sun C. Polymethoxyflavones from citrus inhibited gastric cancer cell proliferation through inducing apoptosis by upregulating RARβ, both in vitro and in vivo. Food Chem. Toxicol. 2020;146:111811–111821; DOI: 10.1016/j.fct.2020.111811

85. Li S, Pan MH, Lai CS, Lo CY, Dushenkovc S, Ho CT. Isolation and syntheses of polymethoxyflavones and hydroxylated polymethoxyflavones as inhibitors of HL-60 cell lines. Bioorg. Med. Chem. 2007;15:3381–338. DOI: 10.1016/j.bmc.2007.03.021

86. Fan K, Kurihara N, Abe S, Ho CT, Ghai G, Yang K. Chemopreventive Effects of Orange Peel Extract (OPE) I. OPE Inhibits Intestinal Tumor Growth in ApcMin+ Mice. Med. Food 2007;10:11–17. DOI: 10.1089/jmf.2006.0214

87. Tajaldinia M, Samadia F, Khosravib A, Ghasemnejadd A, Asadie J. Polymethoxylated Protective and anticancer effects of orange peel extract and naringin in doxorubicin treated esophageal cancer stem cell xenograft tumor mouse model. Biomed. Pharm. 2020;121:e109594. DOI: 10.1016/j.biopha.2019.109594

88. Lia Y, Kandhare AD, Mukherjee AA, Bodhankar SL. Acute and sub-chronic oral toxicity studies of hesperidin isolated from orange peel extract in Sprague Dawley rats. Regul. Toxicol Pharm. 2019;105:77–85. DOI: 10.1016/j.yrtph.2019.04.001

89. Bellavite P, Donzelli A. Hesperidin and SARS-CoV-2: New Light on the Healthy Function of Citrus Fruits. Antioxidants. 2020;9:742–760. DOI: 10.3390/antiox9080742

90. Ganeshpurkar A, Saluja A. The pharmacological potential of hesperidin. Indian J. Biochem. Biophys. 2019;56:287–300

91. Murray RDH. Naturally Occurring Plant Cumarins. In: Herz W, Grisebach H, Kirby GW, editors. Progress in the Chemistry of Organic Natural Products. Springer-Verlag. New York; 1978. p. 199–400

92. Mohamed TK. Chemical constituents and antioxidant activity of Citrus paradisi (star-ruby red grapefruit) and Citrus sinensis (blood sweet orange) Egyptian cultivars. Asian J. Chem. 2004:887815

93. Cunha AG, Gandini A. Turning polysaccharides into hydrophobic materials: a critical review. Part 2. Hemicelluloses, chitin/chitosan, starch, pectin and alginates. Cellulose. 2010;17:1045–1065. DOI: 10.1007/s10570-010-9435-5

94. Sua DL, Li PJ, Quekc SY, Huang ZQ, Yuan YJ, Li GY, Shan Y. Efficient extraction and characterization of pectin from orange peel by a combined surfactant and microwave assisted process. Food Chem. 2019;286:1–7. DOI: 10.1016/j.foodchem.2019.01.200

95. Hosseini SS, Khodaiyan F, Kazemi M, Najari Z. Optimization and characterization of pectin extracted from sour orange peel by ultrasound assisted method. Int. J. Biol. Macromol. 2019;125:621–629. DOI: 10.1016/j.ijbiomac.2018.12.096

96. Jiao WB, Huang D, Xing F, Hu Y, Deng XX, Xu Q, Chen LL. Genome-wide characterization and expression analysis of genetic variants in sweet oranged. The Plant Journal 2013;75:954–964. DOI: 10.1111/tpj.12254

97. 97.Jansen EF, Jang R, Mac Donnel LR. Citrus acetylesterase. Arch Biochem. 1947;15:415–431

98. 98.Pasta P, Verga R, Zambianchi F, Daminati M. Acetyl Esterase from Mediterranean Oranges: Partial Purification, Immobilisation and Biotransformations. Biocatal. & Biotrasf. 2004;22:221–224. DOI: 10.1080/10242420410001697089

99. Fontana G, Bruno M, Maggio A, Rosselli S. Functional investigation and applications of the acetylesterase activity of the Citrus sinensis (L.) Osbeck peel. Nat. Prod. Res. 2020; DOI: 10.1080/14786419.2020.1737055

100. Wei Y, Tan YL, Ang FL, Zhao H. Identification and Characterization of Citrus Peel Uronic Acid Oxidase. ChemBioChem 2020;21:797–800. DOI: 10.1002/cbic.201900546

101. 101.Rani G, Yadav L, Kalidhar SB. Chemical Examination of Citrus sinensis Flavedo Variety Pineapple. Indian J. Pharm. Sci. 2009;71:677–679

102. Lux PE, Carle R, Zacarías L, Rodrigo MJ, Schweiggert RM, Steingass CB. Genuine Carotenoid Profiles in Sweet Orange [Citrus sinensis (L.) Osbeck cv. Navel] Peel and Pulp at Different Maturity Stages. J. Agric. Food Chem. 2019; 67:13164–13175. DOI: 10.1021/acs.jafc.9b06098

103. Matsubara Y, Yusa T, Sawabe A, Itzuka Y, Takekuma S, Yoshida Y. Structures of New Cyclic Peptides in Young Unshiu (Citrus unshiu Marcov.), Orange (Citrus sinensis Osbeck.) and Amanatzu (Citrus natzudaidai) Peelings. Agric. Biol. Chem. 1991;55:2923–2929

104. Favela-Hernández JMJ, González-Santiago O, Ramírez-Cabrera MA. Esquivel-Ferriño PC, Camacho-Corona MdR. Chemistry and Pharmacology of Citrus sinensis. Molecules. 2016;21, 247–271. DOI: 10.3390/molecules21020247

Chapte-12

Physiological Functions Mediated By Yuzu (*Citrus Junos*) Seed-Derived Nutrients

1. Introduction

In 1997, the World Cancer Research Fund published 14 articles concerning dietary recommendations in addition to smoking cessation for the prevention of cancer in *Food, Nutrition and the Prevention of Cancer: a Global Perspective* (2007 revised edition) to promote international awareness of the relationship between nutrition, diet, and cancer. Articles 4 ,1, and 5 strongly recommend the consumption of foods of plant origin, and especially emphasized the importance of fruits and vegetables for the prevention of many types of cancer [1].

There are more than 1000 species of citrus and various varieties account a major part of all fruit production worldwide. In particular, citrus species native to Asia are believed to have originated in the Assam area of India around 30–40 million years ago and were propagated in China, Thailand, Malaysia, Indonesia, and Taiwan before being brought to Japan [2]. The tachibana orange, which is the oldest variety of mandarin orange in Japan, was introduced in Japan from Taiwan via the Korean Peninsula from mainland China, and is listed in the *Manyoshu*, the oldest extant collection of classical Japanese poetry compiled sometime after 759 AD during the Nara period, as the only citrus fruit that existed in the wild. After that, it is estimated that the daidai, an Asian variety of bitter orange, and other small oranges arrived in Japan around 2 to 300AD. Yuzu (*Citrus junos Sieb. ex Tanaka*) originated in China and was introduced to Japan and other countries around the 4th to 8th centuries, as this fruit is mentioned in the *Shyoku-Nihongi*, an imperially-commissioned Japanese history text completed in 797 AD.

The traditional Japanese meal *washoku* was recognized as a UNESCO Intangible Cultural Heritage of Humanity in 2013.The Japanese have the highest life expectancy of any other ethnicity. Therefore, washoku has attracted attention as a healthy diet. Especially, yuzu is an essential ingredient of the Japanese diet in the winter months. A traditional 0sechi dish, including yuzu, to be eaten on New Year's Day is shown in the photo in Figure 1.

Figure 1. A typical Osechi package for New Year's day in Japan.

Yuzu is a commercially important fruit, as compared to other sour citrus fruits, and has become very popular in Japan. Although rarely eaten as a fruit, yuzu is a common ingredient in Japanese cuisine, where the aromatic zest (outer rind) as well as juice are used much in the same way as lemons in other cuisines. The yuzu fruit and juice are traditionally used in making vinegar and seasoning yuzu peel and juice, and along with sudachi, daidai, and other similar citrus fruits, are integral ingredients in the citrus-based sauce ponzu. In addition, yuzu is often used as an ingredient in alcoholic drinks, such as the yuzu sour. Recently, yuzu kosho "yuzu and pepper" has become a very popular spicy Japanese sauce made from the peel (zest) of green or yellow yuzu, combined with green or red chili peppers and salt. Yuzu is also well-known because of its pleasant aroma and essential oil of the outer rind. In fact, in Japan, it has been customary since ancient times to take a bath with yuzu in hot water during the winter solstice. The yuzu peel is particularly high in aromatic compounds and pectin; therefore, the waste peel from juice extraction is sometimes used to produce essential oils and flavorings as well as for medicinal purposes. Similarly, yuzu is industrially used in the production of sweetened beverages, cosmetics, and perfumes, as well as oils for aromatherapy [3]. Only a small portion of produce is used for natural medicine, while satsuma mandarins, oranges, and grapefruits are commonly used for the production of fruit juices.

There is a reason why yuzu is not often eaten as a fruit in Japan because it contain large seeds that convey a bitter taste to the juice. The well-known constituents of citrus fruits include essential oil components, flavonoid glycosides, and other basic substances with biological activities, including limonene, a major component of essential oils found in the juice and rind, polymethoxyflavones, coumarins, carotenoids, which are pigments, vitamins, and terpenoids [4, 5, 6, 7]. In the past, the bitterness of citrus juices, skins, and seeds hindered the demand for citrus fruits. Much research has been conducted to produce bitter-free citrus fruits. At the time when there was very little demand for bitter fruit and juice, Hasegawa et al. [8, 9, 10] reported that the high physiological activity of limonoids was responsible for the bitter taste in citrus juice. Limonoids are a group of triterpene derivatives found in plants of the Rutaceae and Meliaceae families. So far, more than 300 types of limonoids have been reported, and about 100 types have been isolated from the neem and sendan plants of the family Meliaceae.

Limonoids are characterized by a furan ring at C-17, a lactone ring at C-3 or C-6, and an epoxide between C-14 and C-15 (Figure 2). Of the four basic structures of limonoids, reversible opening and closing occurs in A, D lactone rings. For example, in the case of limonin, one of the major limonoids, a closed D-ring creates a bitter taste, while the open D-ring form (limonate A-ring lactone) has no bitter taste. Furthermore, the limonate A-ring lactone at C-17 is converted into the limonoid glycoside 17-β-D-glucopyranoside, which is D-glucose bound with β-glucoside.

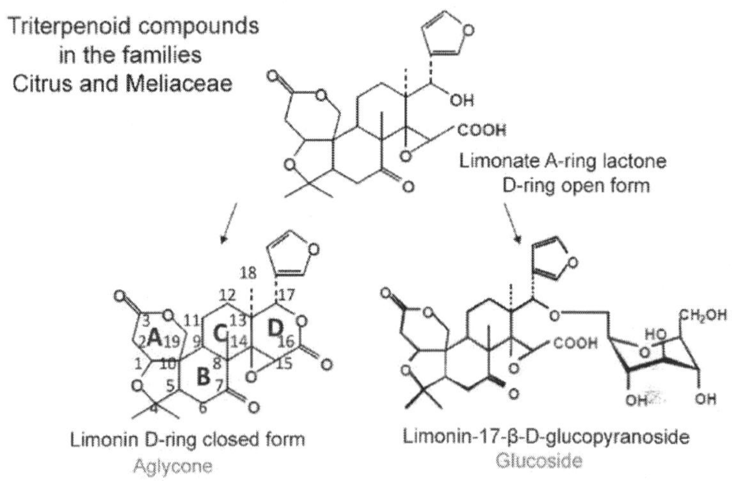

Figure 2. D lactone ring structures of limonin.

Until recently, not much had been known about the metabolism of limonoids found in fruit. In 1991, Hasegawa et al. [11] discovered that limonate A-ring lactone, an open D-ring form of a limonoid aglycone, was metabolized to a glucoside derivative in the late stages of fruit growth and maturation, and suggested that this occurred independently in both the seeds and fruit. It is known that the limonoid aglycone and glycosides accumulate in the seeds [12].

To date, 36 types of aglycones and 17 types of glycosides have been identified mainly in citrus fruits of the family Rutaceae, which is composed of 160 genera and about 2,070 species [13]. The first study of the physiological effects of *Citrus limon*oids reported inhibitory effects on the eating behaviors of armyworms and predatory insects [14]. Strong inhibitory effects on eating were subsequently observed in termites. It has been reported that the ligand activity of the bile acid receptor TGR5 increases the inhibition of tumor formation and the activity of glutathione S-transferase, which is a detoxification enzyme that assists with the excretion of toxic substances by the liver and digestive organs, as well as increasing anti-obesity effects via insulin and increased heat production [15]. The seeds of citrus fruits contain particularly potent limonoids. The metabolic pathway of limonoid biosynthesis in citrus fruits has been nearly elucidated by Hasegawa et al. [16] with the use of 14C-labeled radioisotopes as tracers (Figure 3). Within the phloem of the stem, nomilin is synthesized through the metabolism of acetic acid, mevalonic acid, and furnesole phosphate. Especially in the stems of seedlings, nomilin synthesis becomes very active [17].

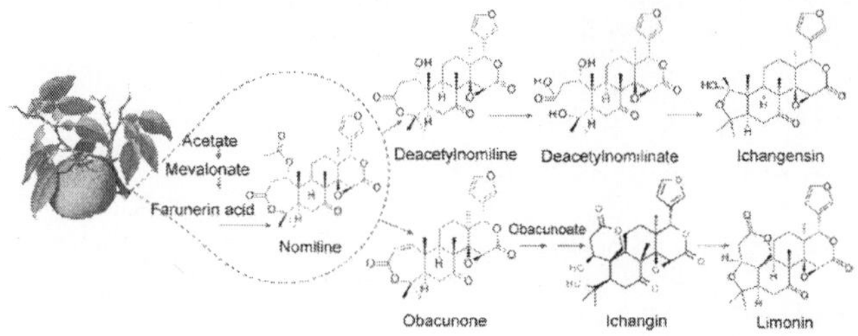

Figure 3. The limonoid biosynthesis pathway of yuzu seeds.

Nomilin is synthesized only in the stem and then transferred to the leaves, fruits, and seeds where it is metabolized into other limonoids. Since metabolism proceeds with the D-ring open form, limonoids exist in the D-ring open form in the stems, fruits, and leaves, and mostly form glycosides. In seeds, metabolism to limonoid aglycone with a closed D-ring also proceeds at the same time, so that aglycone and glycoside are accumulated simultaneously. Therefore, the aglycon in the fruit tissue decreases during maturation, but continues to accumulate in the seeds.

2. Bioactive substances of yuzu seeds

Minamisawa et al. [18, 19] has been searching for new antioxidants to maximize the original functions of living organisms with the use of waste resources derived from natural products, including yuzu seeds, that can be regenerated as many times as possible in the human life span. In 2011, the oldest original species of yuzu in Japan was discovered in the village of Mizuo, which is located in the north-west of Kyoto city [2]. This yuzu is characterized by "seedlings" grown from seeds in the land associated with Emperor Seiwa (9*th* century) and Emperor Hanazono (15*th* century), and it takes about 20 years to harvest. Most of the citrons cultivated in Japan today originate from the yuzu of Mizuo, which is cultivated mainly by grafting, and the growth is faster than that of seedlings. Since the yuzu of Mizuo is considered to be the finest quality, the fruit is highly demanded by high-end restaurants serving Japanese cuisine in Kyoto. However, yuzu seeds, which are closer to the ancestral citrus, account for %30–20 of the fruit weight, but are discarded as waste after the juice extraction process.

Hence, our team chose to evaluate the this development of active natural resources would encompass the application in nutrition and environmental attributes of yuzu seeds as natural resources with bioactivities [20]. In 2014, we reported the development of a relatively simple technique to simultaneously extract secondary metabolites of yuzu seeds, including expensive limonoids and yuzu seed oil with high total antioxidant capability, from the waste of fully ripe fruits [2]. Yuzu seeds contain higher amounts of fat-soluble limonoid aglycones, water-soluble limonoid glycosides, and oil than other citrus fruits (Figure 4).

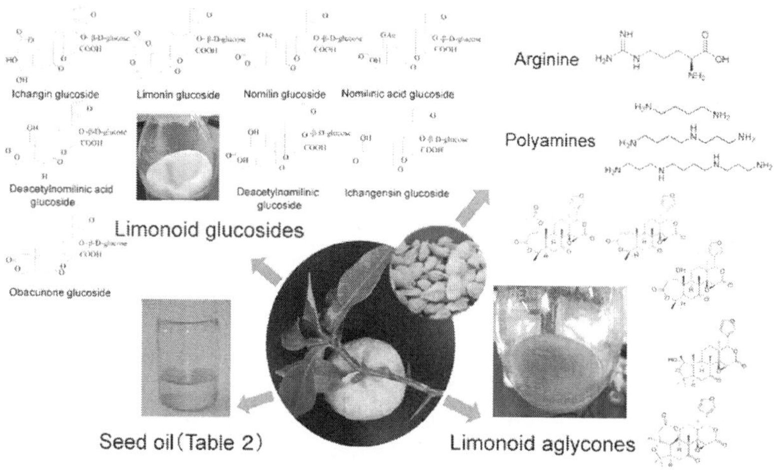

Figure 4. The extracted components from Yuzu seed.

Analysis of the components of limonoids from yuzu seeds by high-performance liquid chromatography–mass spectrometry identified five limonoid aglycones (deacetylnomilin, limonin, nomilin, obacunone, and ichangensin) and eight limonoid glycosides (limonin glucoside, ichangin glucoside, deacetyl nomilinic acid glucoside, deacetylnomilin glucoside, nomilin glucoside, nomilinic acid glucoside, ichangensin glucoside, and obacunone glucoside) (Figure 4, Table 1). Yuzu seed oil extracts (Table 2) contain large amounts of oleic and linoleic acids ([2], in preparation). The contents of limonoids extracted from yuzu seeds compared with the results of previous studies are shown in Table 1 [21, 22, 23].

Table 1. Limonoids in various citrus seeds.*

	Content (mg/g of dry seeds)							
Seeds	**Limonoids aglycones**					**Total**		
	Nomilin	**Deacetyl-nomilin**	**Limonin**	**Obacu-none**	**Ichan-gensin**	**Aglycones**	**Glucosides**	**Aglycones/ glucosides**
Yuzu (*Kyoto, Japan*)	114	106	95.0	16.7	2.10	334	452	0.74
Yuzu (*Tokushima, Japan*)	58.4	48.3	54.0	5.20	1.60	167	192	0.87
Yuzu* [21] (California)	0.25	0.22	0.47	—	0.23	0.94		
Hanayu [22]	0.12	0.86	0.54	0.07	—	1.59	15.9	0.10
Shiikuwasha [22]	0.96	—	1.87	0.45	—	3.28	12.7	0.30
Iyo [22]	2.53	0.72	4.57	0.91	—	8.73	4.46	2.00
Hyuganatu [23]	3.73	0.35	4.68	0.28	—	9.04	3.12	2.90

Seeds	Content (mg/g of dry seeds)							
	Limonoids aglycones					Total		
	Nomilin	Deacetyl-nomilin	Limonin	Obacu-none	Ichan-gensin	Aglycones	Glucosides	Aglycones/glucosides
Grapefruit [23]	1.84	1.10	19.1	1.86	—	23.9	6.98	3.40
Lemon [23]	3.03	—	8.95	0.58	—	12.6	6.37	2.00
Valencia [23]	2.30	1.24	10.0	0.08	—	13.6	8.71	1.60

Units are mg/g of fresh weight.

Table 2. The FA content was determined by gas chromatography with the use of a GC-14 gas chromatograph (Shimadzu corporation, Kyoto, Japan) equipped with a DB-1 column (30 m, 0.25 mm, Agilent Technologies, Inc., Santa Clara, CA, USA), which was maintained at a constant temperature of 300°C. The yield of yuzu seed oil was 100 mg/g of dry seeds.

Fatty acid	16:0	16:1	18:0	18:1	18:2n-6	18:3n-3	20:0	20:1	24:0
%	19.7	0.5	3.8	37.9	34.7	1.6	0.3	0.2	0.1

As compared to other citrus seeds, the concentrations of limonoid aglycones extracted from the seeds of yuzu fruit from Kyoto were two- or three-fold greater than in fruits from Tokushima and California (334 vs. 167 and 0.94 mg/g, respectively). According to Nogata [22], the iyokan fruit (*C. iyo*), Valencia orange (*C. sinensis Osbeck*), and hyuganatsu (*C. tamurana Hort. ex Tanaka*) belong to the same family as the *daidai* (*C. aurantium group V*). Hence, the limonoid compositions of these varieties are similar (Table 1). Although the amount of nomilin in the Valencia orange is similar to that in the *iyokan* and *hyuganatsu* varieties, the amount of limonin is approximately two-fold greater, while the amount of deacetylnomilin is higher and that of obacunone is significantly lower.

The yuzu and hanayu (*C. hanaju*) varieties are classified to yuzu group VI. However, both the compositions and amounts of the limonoid aglycones differed markedly between these two species in the present study, which may be attributed to differences in the metabolism of the seeds and fruits [24, 25]. For this reason, the ratio of aglycone to glycosides in mature fruit tissues is mostly due to glycosides, whereas the glycoside content in seeds may be the same or lower than that of aglycones (Table 1). These findings indicate that limonoids are biosynthesized completely independently of fruit tissues and seeds.

Nogata et al. [22] pointed out that the high amounts of glycosides in seeds of the iyo and shiikuwasha fruits could be due to the high activity of uridine diphosphate-D-glucose transferase, and perhaps in the yuzu seeds as well. The high limonoid content in the seeds of yuzu fruit grown in Kyoto is thought to be related to the seedling cultivation method. Similar to yuzu seeds, the glycosides deacetyl nomilinic acid glucoside and deacetylnomilin glucoside, but not ichangensin glucoside, accumulate in hanayu seeds. Ichangensin is reportedly metabolized from nomilin through the intermediaries deacetylnomilin and deacetyl nomilinic acid [26, 27]. The hanayu are, therefore, different from other citrus varieties. Although both belong to the same yuzu group, there are differences in characteristics, such as aglycones contents.

Several *in vitro* studies have shown that limonoid components mediate the antioxidant properties of citrus. Reactive oxygen is believed to be a factor in diseases with underlying cellular disorders [28]. Modern lifestyles and diets involve a number of factors that can increase the production of active oxygen species, which can overwhelm the body's self-regulating defense mechanisms [29, 30]. One way to protect oneself from active oxygen species is to consume foods containing antioxidants. One of the main reasons for our interest in antioxidants is the link between active oxygen species and aging.

Limonoids components of *C. junos* are known to possess a vast range of biological activities, including antioxidant functions, protective effects on vascular endothelial cells [31], and anti-carcinogenic activities [15, 20, 32, 33, 34].

Study also evaluated the *in vitro* antioxidant activities of yuzu seed aglycones, glycosides, and oil extracts. Notably, the yuzu seed oil, the potential extracts had high antioxidant activities due to the presence of lipophilic aglycones (Figure 5, new unpublished data). Yuzu seed oil is a semi-drying oil that contains large amounts of unsaturated fatty acids (FAs), mainly oleic acid and linoleic acid, in addition to a lot of palmitic acid.

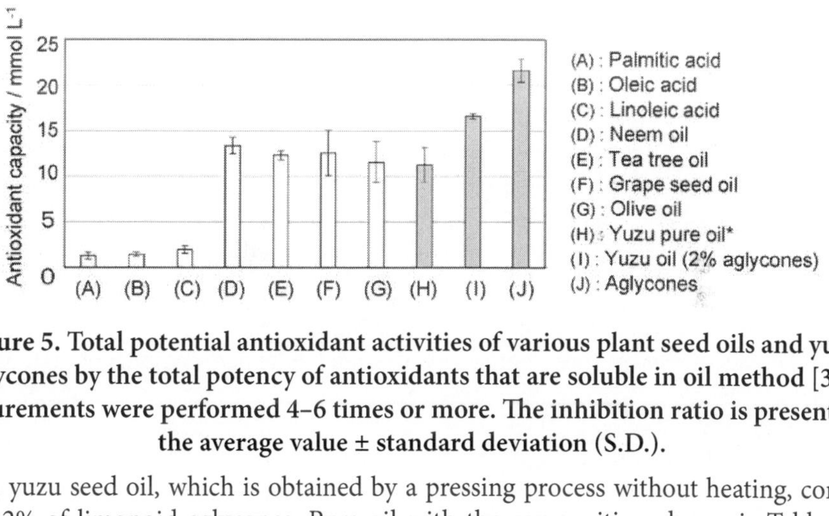

Figure 5. Total potential antioxidant activities of various plant seed oils and yuzu aglycones by the total potency of antioxidants that are soluble in oil method [35]. Measurements were performed 4–6 times or more. The inhibition ratio is presented as the average value ± standard deviation (S.D.).

Virgin yuzu seed oil, which is obtained by a pressing process without heating, contains about 2% of limonoid aglycones. Pure oil with the composition shown in Table 2 can be obtained by heating and drying roasted yuzu seeds, followed by extraction with an organic solvent, such as hexane, and purification.

Lipophilic limonoid aglycones, which were extracted from the residual extracts of yuzu seed oil [2], were composed of the following concentrations of limonoids per gram of dry seeds: deacetylnomilin, 105 mg; limonin, 95 mg; nomilin, 115 mg; obacunone, 17 mg; and ichangensin, 2.1 mg.

The total potential antioxidant capacity of yuzu seed oil and lipophilic limonoid aglycones was measured by utilizing the reduction reaction of copper (Cu^{++}/Cu^{+}) [35]. Many other plant oils, including olive oil [36], tea tree oil, grape seed oil [37], and neem seed oil [38], which have strong antioxidant activities, were measured at the same time for comparisons.

Among all of the tested plant seed oils, limonoid aglycones extracted from yuzu seeds had the highest antioxidant capacity, followed by yuzu seed oil, neem seed, grape seed, tea tree, olive oil, and pure yuzu seed oil. The antioxidant capacity of pure yuzu seed oil was approximately 6–9-fold greater than that of palmitic acid, oleic acid, and linoleic acid.

While water-soluble antioxidants are rapidly excreted through the urine if an excessive amount is ingested, fat-soluble antioxidants are adsorbed onto lipoproteins and cell membrane lipids, and are therefore considered to exhibit a higher activity in the body. For this reason, fat-soluble antioxidants are expected to be beneficial in preventing diseases caused by oxidative stress. Vitamin E, oryzanol, and carotenoids are well-known examples of fat-soluble antioxidants. Neem (*Azadirachta indica*) seed oil, which has the same total antioxidant capacity as yuzu seed oil, contains the triterpene derivative azadirachtin, which is similar to the triterpene limonoids of yuzu seed oil, which is a potent insect repellent [39]. Press-extracted virgin olive oil contains oleocanthal that has a potent anti-inflammatory effect strikingly similar to that of ibuprofen. Both of these molecules inhibit the same cyclooxygenase enzymes in the prostaglandin-biosynthesis pathway [40].

The result in Figure 5 suggest the presence of other types of fat-soluble antioxidants. Limonoid aglycones also contribute to the high antioxidant capacity of yuzu seed oil.

3. Yuzu seeds contain arginine and polyamines (PAs)

Atherosclerosis has become a serious health concern worldwide, as one-third of the global population is at risk for diseases associated with arteriosclerosis, which accounts for about half of deaths in developed countries. In particular, cardiovascular disease (CVD), which is a consequence of atherosclerosis, is the leading cause of death in industrialized nations. Besides lifestyle habits, body weight, socio-economic factors, and certain pre-existing conditions, a number of foods seem to play a role in the incidence of CVD [41, 42]. In addition, many studies have suggested the importance of inflammation in atherosclerosis and CVD [43, 44].

Some food components with anti-inflammatory properties can decrease the risk of CVD [44, 45]. Many foods contain wide-ranging concentrations of natural PAs, such as spermidine (Spd) and spermine (Spm), which suppress the synthesis of pro-inflammatory cytokines [46, 47]. In particular, an epidemiological survey of Westerners found that "people who eat cheese or yogurt every day are less likely to have myocardial infarction." [48]. The Japanese consume a lot of traditional fermented foods, mainly soybeans, which are thought to suppress arteriosclerosis [49]. PAs concentrations are relatively high in yogurt, cheese, and traditional Japanese foods. PAs are aliphatic amines that are essential for the growth of all living cells [50]. PAs exist primarily in association with RNA and are involved in promoting the synthesis of specific proteins and overall protein synthesis via the ribosome activation. As shown in Figure 6, the PAs comprising Put ($NH_2(CH_2)_4NH_2$) → Spd ($NH_2(CH_2)_3NH(CH_2)_4NH_2$) → Spm ($NH_2(CH_2)_3NH(CH_2)_4NH(CH_2)_3NH_2$) are produced from arginine via ornithine or agmatine [51].

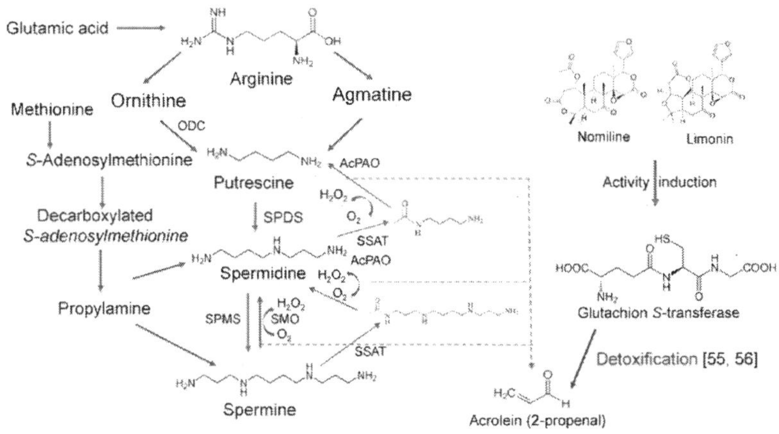

Figure 6. PA synthesis and degradation [50, 51]. AcPAO, acethylpolyamine oxidase; ODC, ornithine decarboxylase; SAMDC, S-adenosylmethionine decarboxylase; SMO, spermine oxidase; SPDS, spermidine synthase; SPMS, spermine synthase; SSAT, spermidine/spermine N1-acetyltransferase.

PAs have been implicated in the regulation of several growth and development processes in plants, including cell division, morphogenesis, flower initiation, pollen tube growth, and senescence [52]. Analyses of the PA contents of various fruits have mainly been conducted in Europe [53, 54].

Recent studies have indicated that citrus limonoids have antitumor, detoxification, and anti-obesity effects [55, 56], which may indirectly contribute to the suppression of acrolein production, which is a side reaction product of PA metabolism (Figure 6). Hence, we measured the PA and arginine contents of yuzu, which produces very high concentrations of limonoids, as well as lemons produced in Japan for comparison. The PA contents, as determined by high-performance liquid chromatography, as well as the arginine and free arginine contents, as determined by automated amino acid analysis, of various citrus fruits are shown in Table 3.

Table 3. PAs and arginine contents in several citrus (nmol/g).

Citrus	PAs				Arginine
	Putrescine	Spermidine	Spermine	Total PAs	
Yuzu (*Kyoto, Japan*) [58]	294.8	117	34.6	446.4	36625
Only seed	79.4	117	34.6	231	30540
Yuzu (*Tokushima, Japan*) [58]	466	144.6	24.7	635.3	20207
Only seed	125	89.5	24.7	635	14007
Lemons (*Kamogawa, Japan*)	320.0	56.2	3.8	380.0	37933
Only seed	71.1	56.2	3.8	371	33950
Grapefruit [54]	436	19.5	1.5	457	
Lemons, Limes [54]	390.5	19	4.5	414	
Oranges, Mandarins [54]	432	15.5	0	448	

Bean	Putrescine	Spermidine	Spermine	Total PAs	
Soybean [54]	297	909	235.5	1442	
Beans [57]	147	644.8	299.7	1092	
Sesame seed [54]	29	126	22	177	

As compared with the juice and peel of yuzu and lemons, the seeds contain higher quantities of PAs and arginine. The Put contents are high in all citrus fruits, but the quantities of Spd and Spm in yuzu seeds were 5–23-fold greater than the reference values. The PA contents of yuzu and lemon fruit are not high as compared to legumes [57], but when the limonoid and arginine contents are also considered, these fruits have high levels of functional constituents.

As mentioned earlier, limonoids and PAs have various bioactivities and reportedly have strong anti-inflammatory capabilities. Hence, the potential antioxidant activities (i.e., H_2O_2-scavenging activity, 2,2-diphenyl-1-picrylhydrazyl radical-scavenging activity, and inhibition of superoxide dismutase [SOD] and antioxidants with SOD-like activities) of PAs (Put, Spd, and Spm) and arginine were investigated (in preparation). The results showed that these compounds have no antioxidant activities or only weak (less than 10%) inhibitory potential. As reported in many studies, the anti-inflammatory activities of PAs and arginine are due to factors other than antioxidant capacity.

4. Yuzu seed limonoids or Spm increased survival of mice with Sandhoff disease

In our previous study, we investigated the life-extending effect of limonoids (lipophilic limonoid aglycones) and Spm as an exogenous anti-inflammatory component in a mouse model of Sandhoff disease (SD), which is a lysosomal disease [58]. Lysosomal storage disorders are caused by functional defects of proteins that are essential for normal lysosome function, such as enzymes that play critical roles in the intracellular digestion of glycoproteins, glycolipids, glycosaminoglycans, and other macromolecules [59]. SD is an autosomal recessive hereditary disease [60]. The gangliosides GM2 and GA2 accumulate in the nervous system, resulting in severe developmental and neurological disorders, and death, which usually occurs during infancy because of the lack of effective treatment methods. Neurological dysfunction is the major clinical manifestation of GM2 gangliosidosis [61, 62, 63, 64].

SD mice present with trembling, startled responses, and decreased motor activities from 11 to 15 weeks of age (105 days) due to damage caused by microglial activation, macrophage infiltration, and oxidation associated with the accumulation of glycolipids. It has been suggested that inflammation may be fatal [51]. The therapeutic effects of enzyme replacement therapy and anti-inflammatory drugs have been reported [65]. Inflammation due to the accumulation of lipids is inhibited by antioxidant and anti-inflammatory treatments, which can delay disease progression, but no cure exists at present.

We consider the degeneration of the nervous system might be rooted in oxidative stress and inflammation. Given that dietary interventions can moderate these phenomena, consuming foods with antioxidants and anti-inflammatory components, such as limonoid aglycons (limonoids) and Spm, could effectively combat or minimize neurological damage. Therefore, the inhibition of SD pathologies could be promoted by factors other than suppressing the storage of gangliosides.

Preventing inflammation appears to be one of the most effective approaches for increasing longevity [66, 67]. To test this hypothesis, the life spans of SD mice treated with limonoids or Spm were assessed. The prognostic outcomes of SD mice, a typical model of abnormal glycolipid metabolism in humans, were observed after administration of limonoids extracted from yuzu seeds and Spm. The treated mice lived significantly longer than untreated littermates (9–10%, $p < 0.01$) and had a slower rate of disease progression ($p < 0.01$) [58]. When limonoid treatment was combined with Spm therapy, synergy resulted in a maximum improvement of 12% in survival ($p < 0.001$) (in preparation). The hematoxylin and eosin (H&E) staining results of thalamus sections of SD mice following administration of limonoids or Spm are shown in Figure 7.

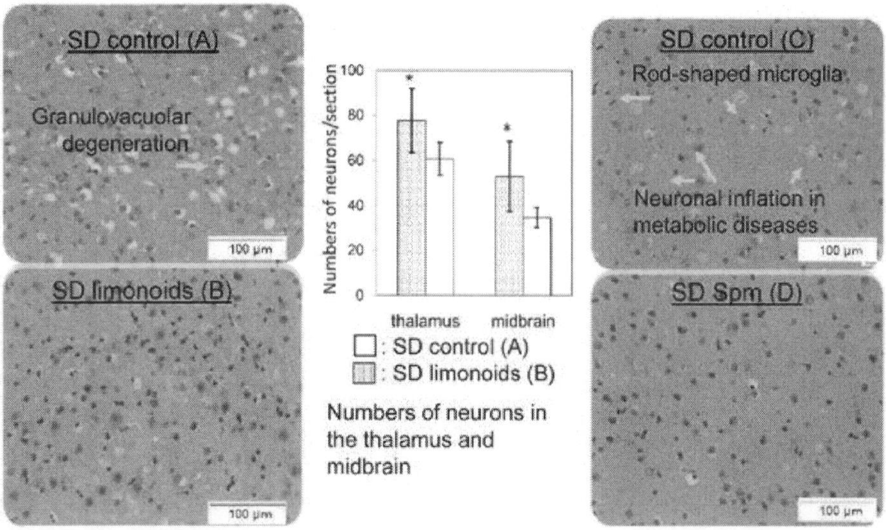

Figure 7. H&E-stained thalamus sections of control SD mice (A and C) and SD mice administered yuzu limonoids (B) and Spm (D). Enlarged cells with ganglioside storage are indicated. The numbers of neurons in the thalamus and midbrain of SD mice administered limonoids (B) and control SD mice (A). Data are presented as the mean ± S.D. *p ≤ 0.05 (Student's t-test).

H&E staining results of the neural tissues of the SD control mice (A) and(C) correspond to SD mice treated with limonoids (B) and Spm (D), respectively.

Gangliosidosis and inflammatory/autoimmune diseases are characterized by degeneration and the accumulation of fat, granulovacuolar degeneration, rod-shaped microglia, and neuronal inflammation in metabolic diseases, as determined by analyses of pathological tissues (Tokyo Metropolitan Institute for Medical Science). The characteristic degeneration was clearly decreased in SD mice treated with limonoids or Spm. The numbers of neurons in the thalamus and midbrains of SD mice treated with limonoids were higher than those in the control SD mice. These results demonstrate that inflammation contributes to disease progression and the anti-inflammatory effects of Spm and limonoid therapies as a potential adjunctive approach to slow the clinical course of inflammatory diseases.

5. Bacterial flora analysis of SD mouse feces by the 16S ribosomal DNA (16S rDNA) terminal restriction fragment length polymorphism (T-RFLP) method

PAs possess anti-inflammatory activities by inhibiting the synthesis of inflammatory cytokines by macrophages and the regulation of nuclear factor-κB activation, which are closely associated with maintaining the intestinal mucosal barrier function [68]. Bilateral signals between the intestine and brain are involved in the control of nerve, hormone, and immune activities, as well as prolonging longevity [69]. Recent studies have shown that bilateral signals between the brain and intestine are important for maintaining homeostasis and extending the life span [70]. In particular, the functions mediated by PAs may be involved in metabolism by indigenous intestinal bacteria and the health of the host [71].

At 12 weeks of age for which the survival period was extended by limonoids or Spm, T-RFLP analysis of 16S rRNA was performed to classify the intestinal microbiota at the order level for each mouse group (Figure 8). The results showed that the taxonomic groups of the bacterial flora in feces after administration of limonoids or Spm were completely different from those of the SD control mice.

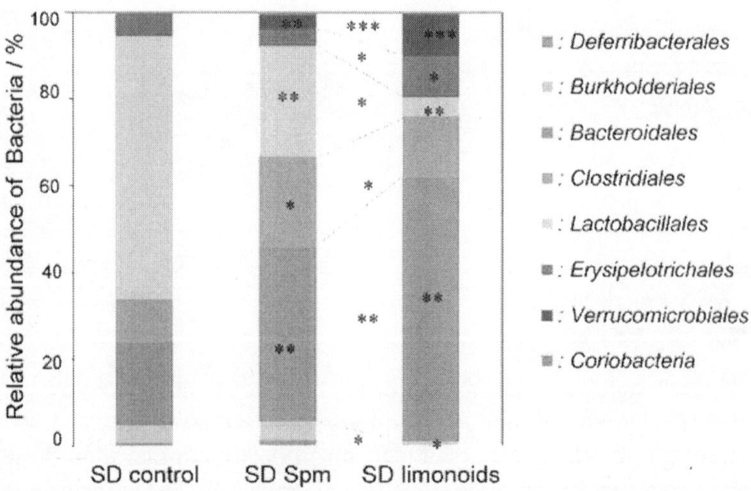

Figure 8. Estimated ratios (%) of the taxonomic categories of the bacterial flora at the order level were identified by T-RFLP analysis of 16S rRNA in the feces of SD mice at 12 weeks of age in the control, Spm, and limonoids groups (n = 9/group). Values are presented as the mean ± S.D. data of the treatment groups are plotted against those of the control group. *p ≤ 0.05, **p ≤ 0.001, and ***p ≤ 0.0001 vs. the untreated control SD mice or each (dashed line----). All experiments were performed at least three times.

The bacterial flora in feces after administration of limonoids or Spm had increased proportions of *Bacteroidales* and *Clostridiales*. However, *Lactobacillus* was remarkably prevalent in feces of the SD control mice. The abundance of *Clostridiales* were significantly increased in the feces of SD mice treated with Spm, whereas *Bacteroidales* were increased in the feces of SD mice treated with limonoids. The administration of Spm or limonoids slightly increased the proportion of *Erysipelotrichales*.

It is generally known that the abundance of *Erysipelotrichaceae* is increased due to fat accumulation in mice [72]. In this case, it was possible that the bacterial flora of the SD control mice caused dysbiosis [73]. In the SD control mice, dysbiosis may have been due to suppressed absorption of dietary fats and other nutrients. Even more interesting was the significant appearance of *Verrucomicrobiaceae* in feces after the administration of limonoids or Spm, which were not found in feces of the control SD mice. *Verrucomicrobiaceae* include mucin-degrading bacteria that are also present in the human intestine, and especially *Akkermansia*, which promote the suppression of obesity, diabetes, and inflammation [74, 75].

It will be necessary to investigate the specific bacteria involved in more detail. Unfortunately, the T-RFLP method made it difficult to analyze the bacterial flora in more detail, and it was not possible to identify particular species. We are currently preparing a report of the findings of next-generation sequencing that allowed for more detailed classification.

6. Short-chain fatty acid (SCFA) production in SD mouse feces

There have been many reports of the relationships between chronic inflammatory diseases and the intestinal bacterial flora that have helped to clarify the balance between the intestinal ecosystem and diseases related to the intestinal tract. For example, genetic abnormalities and the breakdown of the intestinal ecosystem have been detected in inflammatory bowel disease [76]. In particular, members of the genus *Clostridium* promote the production of butyric acid, induce an immune response in the intestinal mucosa, and promote the differentiation of regulatory T cells (Tregs) that contribute to suppression. Thus, changes in intestinal *Clostridium* are considered to be closely related to the onset of inflammatory bowel disease [77, 78]. It has been reported that the SCFAs produced by intestinal bacteria may function as bio-modifying factors. Hence, the SCFA composition of feces from the same mice at 12 weeks were determined (Figure 9).

The production levels of SCFAs comprising acetic acid, propionic acid, and butyric acid were increased in mice administered limonoids or Spm as compared to SD control mice. In particular, the production levels of all SCFAs were higher in SD mice following administration of limonoids or Spm. The experimental results demonstrated differences between the fecal microflora composition and these metabolites after administration of limonoids or Spm. Butyric acid is a SCFA that is produced by clostridia [78].

As shown by the results presented in Figures 8 and 9, the addition of Spm to the diet clearly increased the proportion of *Clostridiales* and butyric acid in feces. Previous metabolomic analyses have shown that butyric acid contributes to the induction of Treg differentiation in the colonic mucosa. Thus, butyric acid functions as a histone deacetylase inhibitor and as an immunomodulator responsible for inducing Treg differentiation in the colonic mucosa, as well as the activation of dendritic cells [79]. Acetic acid produced by intestinal bacteria suppressed colitis in a mouse model by promoting apoptosis via the GPR43 receptor expressed by neutrophils and plays a central role in the inflammatory reaction [80, 81]. Furthermore, the addition of limonoids seems to contribute to the production of acetic acid and propionic acid as well as butyric acid.

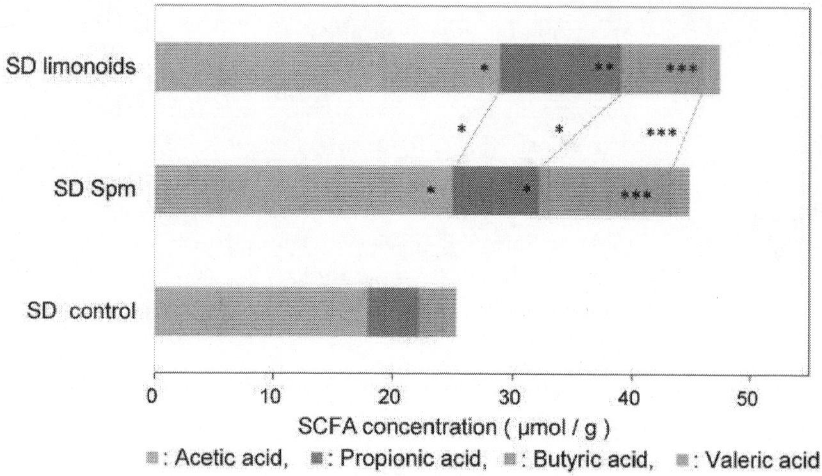

Figure 9. SCFA contents (µmol/g) in feces of SD mice in the control, Spm, and limonoid groups. SD mice were treated at 12 weeks of age (n = 9/group). Values are presented as the mean ± S.D. data of the treatment groups are plotted against those of the control group. *p ≤ 0.05, **p ≤ 0.001, and ***p ≤ 0.0001 vs. the untreated control SD mice or each (dashed line----). All experiments were performed at least three times.

Acetate, butyrate, and propionate are produced by members of the intestinal microbial community through fermentation of dietary fibers and starches, which are unable to be broken down by host metabolism [82]. In turn, these metabolites are sensed by host cells through various G-protein coupled receptors, known as free fatty acid receptors, and intracellular peroxisome proliferator-activated receptor gamma. Furthermore, SCFAs can also regulate cellular responses through inhibition of histone deacetylases. Examples of the effects of SCFAs on the host include differentiation of Tregs and macrophages, and downregulation of pro-inflammatory mediators. These effects underline the fine balance that SCFAs help to maintain between intestinal immunity and inflammation [82, 83, 84].

These results suggest that yuzu conveys anti-inflammatory and lipid metabolism-promoting activities in mice following administration of limonoid aglycones and Spm. Thus, the metabolites of intestinal bacteria may be indirectly involved in suppressing the inflammatory mechanism to directly enhance the health of the host. Furthermore, administration of limonoids or Spm improved the proportions of beneficial bacterial in the intestinal flora and associated metabolites. In the healthy intestinal tract, the microbiota and gut-associated immune system are assumed to be at a dynamic homeostatic equilibrium [85], but the inflammation process may undermine this balance. We consider that the human lifespan can be extended by inhibiting inflammation via control of the intestinal microbiota.

However, it was not possible to elucidate the mechanisms underlying the effects of limonoid aglycones and Spm on the extended life span of SD mice. Thus, in order to clarify the anti-inflammatory effects of yuzu seed extract, limonoids, and Spm, as well as to widely apply yuzu to promote health and enhance longevity, it will be necessary to determine the composition of the bacterial flora based on detailed metagenomic analyses of 16S rRNA. Furthermore, it will be necessary to analyze the anti-inflammatory effects of limonoids and Spm in yuzu seed extracts at the gene level.

PAs quantities have reference values. The reference values for the PAs contents of legumes are also shown in Table 3. The values for Japanese produced yuzu and lemons are shown, as well as the reference values for other citrus fruits. No previous studies reported the quantities of arginine in citrus fruits, so only the compared PA quantities are indicated by reference values. The reference values for the PA contents of legumes are also shown.

7. Conclusions

Yuzu is a natural and renewable resource of limonoids, arginine, and PAs. The results of the present study suggest that yuzu limonoids and Spm improved the proportions of beneficial bacteria and their metabolites in the intestinal flora. Thus, the ingestion of fruits that contain high concentrations of specific ingredients may be a simple method to suppress inflammation, thereby enhancing immune function, improving intestinal health, and increasing lifespan. In other words, our this study demonstrated the possibility that bilateral signals between the brain and intestine are important for maintaining homeostasis and extending lifespan. However, it was not possible to examine the physiological effects of limonoids and Spm. Thus, future studies are needed to evaluate the effects of limonoids and Spm on metabolism and the immune response, and to explore the potential of these molecules as natural antioxidants/antibiotics for lysosomal diseases, such as SD.

References

1. World Cancer Research Fund/American Institute for Cancer Research, 1997
2. Minamisawa M, Yoshida S, Uzawa A. The functional evaluation of waste yuzu (Citrus junos) seeds.Food and Function. 2014;5:330-336. DOI: 10.1039/c3fo60440c
3. Sawamura M. citrus JUNOS SIEB. Ex TANAKA (YUZU), FURUIT. In: Ramadane, ed. Fuits:Growth, nutrition, and quality. Helsinki: WFL Publisher; 2005. p. 1-24
4. Sawamura M, Bando A, Nobuyuki O, Kusunose H. Japan food industry. 1986:33:566
5. Sawamura M, Kuroiwa N, Kuriyama T, Kusunose H. Japan food industry 1986;33:773
6. Lan Phi NT, Sawamura M. Food Science and Technology Research. 2008;14:359-366. DOI: 10.3136/fstr.14.359
7. Lan Phi NT, Nishiyama C, Choi HS, Sawamura M. Evaluation of characteristic aroma compounds of Citrus natsudaidai Hayata (Natsudaidai) cold-pressed peel oil. Bioscience, Biotechnology, and Biochemistry. 2006;70:1832-1838. DOI: 10.1271/bbb.50705
8. Hasegawa S. Quality factors of fruits and vegetables. In: Jen JJ, ACS, ed. Symp. American Chemical Society; 1990. p. 84
9. Hasegawa S,, Pelton VAW. American Chemical Society. Journal of Agricultural and Food Chemistry. Jan/Feb 1983;31:178-180
10. Hsu WJ, Berhow M, Robertson GH, Hasegawa S. Limonoids and flavonoids in juices of Oroblanco and Melogold grapefruit hybrids. Journal of Food Science. 1998;63:57-60. DOI: 10.1111/j.1365-2621.1998.tb15675.x

11. Hasegawa S, Ou P, Fong CH, Herman Z, Coggins CW, Atkin DR. Changes in the limonoate A-ring lactone and limonin 17-β-D-glucopyranoside content of navel oranges during fruit growth and maturation.Journal of Agricultural and Food Chemistry. 1991;39:262-265. DOI: 10.1021/jf00002a008

12. Hasegawa S, Bennett RD, Herman Z, Fong CH, Ou P. Limonoid glucosides in citrus. Phytochemistry. 1989;28:1717-1720. DOI: 10.1016/S0031-9422(00)97831-2

13. Hasegawa S, Citrus Limonoids. American Chemical Society; 2000. 9-30

14. asegawa S, Ifuku Y. Biochemistry of citrus limonoids. Metabolism and Biological Functions. Nippon Shokuhin Kogyo Gakkaishi. 1994;41:372-380. DOI: 10.3136/nskkk1962.41.372

15. Ono E, Inoue J, Hashidume T, Shimizu M, Sato R. Anti-obesity and anti-hyperglycemic effects of the dietary citrus limonoid nomilin in mice fed a high-fat diet. Biochemical and Biophysical Research Communications. 2011;410:677-681. DOI: 10.1016/j.bbrc.2011.06.055

16. Hasegawa S Jen JJ, editors.Quality factors of Fruit and Vegetables Chemistry and Technology . Washington DC: American Chemical Society; 1989. p. 84

17. Ou P, Hasegawa S, Herman Z, Fong CH. Phytochemistry. 1988;27:115-118. DOI: 10.1016/0031-9422(88)80600-9

18. Minamisawa M, Yoshida S, Uzawa A. The pyrolytic synthesis of calcium phosphate compounds from carbonized fowl droppings. Powder Technology. 2012;230:20-28. DOI: 10.1016/j.powtec.2012.06.034

19. Minamisawa M, Minamisawa H, Yoshida S, Takai N. Adsorption behavior of heavy metals on biomaterials. Journal of Agricultural and Food Chemistry. 2004;52:5606-5611. DOI: 10.1021/jf0496402

20. Roy A, Saraf S. Limonoids: overview of significant bioactive triterpenes distributed in plants kingdom. Biological and Pharmaceutical Bulletin. 2006;29:191-201. DOI: 10.1248/bpb.29.191

21. Hashinaga F, Herman Z, Hashegawa S. Limonoids in seeds of yuza (Citrus junos. Sieb. ex Tanaka), Nippon Shokuhin Kogyo Gakkaishi. 1990; 37: 380-382

22. Nogata. National Agricultural Research Organization (NAR) Western Region Agricultural Research Center report. 2005;5:19-84

23. Ozaki Y, Fong CH, Herman Z, Maeda H, Miyake M, Ifuku Y, Hasegawa S. Limonoid glucosides in citrus seeds. Agricultural and Biological Chemistry. 1991;55:137-141. DOI: 10.3390/molecules21111530

24. Ozaki Y, Miyake M, Maeda H, Ifuku Y, Bennett RD, Herman Z, Fong CH, Hasegawa S. Ichangensin glucoside in Citrus junos, Citrus sudachi and Citrus sphaerocarpa. Phytochemistry. 1991;30:2659-2661. DOI: 10.1016/0031-9422(91)85118-J

25. Herman Z, Hasegawa S, Fong CH, Ou P. Limonoids in Citrus ichangensis. Journal of Agricultural and Food Chemistry. 1989;37:850-851. DOI: 10.1021/jf00088a003

26. Fong CH, Hasegawa S, Herman Z, Ou P. Biosynthesis of limonoid glucosides in lemon (Citrus limon). Journal of the Science of Food and Agriculture. 1991;54:393-398. DOI: 10.1002/jsfa.2740540310
27. Hasegawa S, Herman Z, Orme E, Ou P. Biosynthesis of limonoids in Citrus: sites and translocation. Phytochemistry. 1986;25:2783-2785. DOI: 10.1016/S0031-9422(00)83741-3
28. Guthrie N, Hasegawa S, Manners G, D, Carroll K, K. Inhibition of human breast cancer cells by citrus limonoids. In: Berhow MA, Hasegawa S, Manners GD, (Eds.), Citrus limonoids: functional chemicals in agriculture and foods. Washington, DC: American Chemical Society; 2000. p. 164-174
29. Straface E, Matarrese P, Gambardella L, Vona R, Sgadari A, Silveri MC, Malorni W. Oxidative imbalance and cathepsin D changes as peripheral blood biomarkers of Alzheimer disease: A pilot study. FEBS Letters. 2005;579:2759-2766. DOI: 10.1016/j.febslet.2005.03.094
30. Pregel P, Bollo E, Cannizzo FT, Biolatti B, Contato E, Biolatti PG. Antioxidant capacity as a reliable marker of stress in dairy calves transported by road. Veterinary Record. 2005;156:53-54. DOI: 10.1136/vr.156.2.53
31. Yoon J, Park M, Lee Jh, Min BS, Ryoo S. Endothelial nitric oxide synthase activation through obacunone-dependent arginase inhibition restored impaired endothelial function in ApoE-null mice. Vascular Pharmacology. 2014;60:102-109. DOI: 10.1016/j.vph.2014.01.006
32. Manners GD. Citrus limonoids: analysis, bioactivity, and biomedical prospects. Journal of Agricultural and Food Chemistry. 2007;55:8285-8294. DOI: 10.1021/jf071797h
33. Chidambara Murthy KN, Jayaprakasha GK, Patil BS. Obacunone and obacunone glucoside inhibit human colon cancer (SW480) cells by the induction of apoptosis. Food and Chemical Toxicology . 2011;49:1616-1625. DOI: 10.1016/j.fct.2011.04.014
34. Poulose SM, Harris ED, Patil BS. Antiproliferative effects of citrus limonoids against human neuroblastoma and colonic adenocarcinoma cells. Nutrition and Cancer. 2006;56:103-112. DOI: 10.1207/s15327914nc5601_14
35. Advanced protocol oxidative stress II; 594 Sakai K, Kino S, Takeuchi M, Ochi T, Da Cruz G, Tomita I, Armstrong D, ed.; 2010. p. 241-250
36. Reboredo-Rodríguez P, Figueiredo-González M, González-Barreiro C, Simal-Gándara J, Salvador MD, Cancho-Grande B, Fregapane G. State of the art on functional virgin olive oils enriched with bioactive compounds and their properties. International Journal of Molecular Sciences. 2017;18:121-126. -DOI: https://doi.org/10.1016/j.foodchem.2008.03.045
37. Sahin NO, Berköz M, Eker ED, Pomierny B, Przejczowska K. Cytotoxic and antioxidant effects of grape seed oil on the treatment of leukemia with methotrexate. European Journal of Chemistry. 2012;3:147-151. DOI: 10.5155/eurjchem.3.2.147-151.569
38. Biswas K, Chattopadhyay I, Banerjee RK, Bandyopadhyay U. Biological activities and medicinal properties of neem (Azadirachta indica). Current Science. 2002;82:1336-1345

39. Kraus W. The neem tree: source of unique natural products for integrated pest management, medicine, industry and other. In: Purposes, (ed. Schmutterer, H.). VCH, Wenheim; 1995:1-696.; 1995. p. 35-88

40. Beauchamp GK, Keast RSJ, Morel D, Lin J, Pika J, Han Q, Lee C-H, Smith AB, Breslin PAS. Phytochemistry: Ibuprofen-like activity in extra-virgin olive oil. Nature. 2005;437:45-46. DOI: 10.1038/437045a

41. ML Daviglus ML, Stamler J, Greenland P. Letter to the editor. New England Journal of Medicine. 1997;337:498-499

42. Louis J Ignarro LJ, Balestrieri ML, Napoli C. Nutrition, physical activity, and cardiovascular disease. Cardiovascular Research. 2007;15:326-340

43. Dariush Mozaffarian D, Katan MB, Ascherio A, Stampfer MJ, Willett WC. Trans fatty acids and cardiovascular disease. New England Journal of Medicine. 2006;354:1601-1613. DOI: 10.1056/NEJMra054035

44. Jennifer A Nettleton JA, Steffen LM, Mayer-Davis EJ, Jenny NS, Jiang R, Herrington DM, Jacobs DR. Dietary patterns are associated with biochemical markers of inflammation and endothelial activation in the Multi-Ethnic Study of Atherosclerosis(MESA). American Journal of Clinical Nutrition. 2006;83:1369-1379. DOI: 10.1093/ajcn/83.6.1369

45. Mozaffarian D. Trans fatty acids - effects on systemic inflammation and endothelial function. Atherosclerosis. Supplements. 2006;7:29-32. DOI: 10.1016/j.atherosclerosissup.2006.04.007

46. Nettleton JA, Steffen LM, Mayer-Davis EJ, Jenny NS, Jiang R, Herrington DM, Jacobs DR Jr. Dietary patterns are associated with biochemical markers of inflammation and endothelial activation in the Multi-Ethnic Study of Atherosclerosis (MESA). American Journal of Clinical Nutrition. 2006;83:1369-1379. DOI: 10.1093/ajcn/83.6.1369

47. Zhang M, Caragine T, Wang H, Cohen PS, Botchkina G, Soda K, Bianchi M, Ulrich P, Cerami A, Sherry B, Tracey KJ. Spermine inhibits proinflammatory cytokine synthesis in human mononuclear cells: a counterregulatory mechanism that restrains the immune response. Journal of Experimental Medicine. 1997;185:1759-1768. DOI: 10.1084/jem.185.10.1759

48. Soda K, Kano Y, Chiba F. Food polyamine and cardiovascular disease-An epidemiological study-global journal of health science. Global Journal of Health Science. 2012;4:170-178. DOI: 10.5539/gjhs.v4n6p170

49. Clarkson TB. Soy, soy phytoestrogens and cardiovascular disease. Journal of Nutrition. 2002;132:566S–569S. DOI: 10.1093/jn/132.3.566S

50. Igarashi K, Kashiwagi K. Modulation of cellular function by polyamines. The International Journal of Biochemistry and Cell Biology. 2010;42:39-51. DOI: 10.1016/j.biocel.2009.07.009

51. Liu T, Kim DW, Niitsu M, Maeda S, Watanabe M, Kamio Y, Berberich T, Kusano T. Oryza sativa polyamine oxidase 7 catalyzes terminal catabolism reaction and specifically expressed in anthers. Plant and Cell Physiology. 2014;55:1110-1122

52. Bagni N, Tassoni A. Biosynthesis, oxidation and conjugation of aliphatic polyamines in higher plants. Amino Acids. 2001;20:301-317. DOI: 10.1007/s007260170046

53. Chiancone B, Tassoni A, Bagni N, Germanà MA. Effect of polyamines on in vitro anther culture of Citrus clementina Hort. Ex Tan. Plant Cell, Tissue and Organ Culture. 2006;87:145-153. DOI: 10.1007/s11240-006-9149-4

54. Atiya Ali M, Poortvliet E, Strömberg R, Yngve A. Polyamines in foods:development of a food database. Food and Nutrition Research. 2011;55:5572. DOI: 10.3402/fnr.v55i0.5572

55. Horiba T, Katsukawa M, Mita M, Sato R. Dietary obacunone supplementation stimulates muscle hypertrophy, and suppresses hyperglycemia and obesity through the TGR5 and PPARγ pathway. Biochemical and Biophysical Research Communications. 2015;463:846-852. DOI: 10.1016/j.bbrc.2015.06.022

56. Sato R. Nomilin as an anti-obesity and anti-hyperglycemic agent. Vitamins and Hormones. 2013;91:425-439. DOI: 10.1016/B978-0-12-407766-9.00018-3

57. Soda K, Dobashi Y, Kano Y, Tsujinaka S, Konishi F. Polyamine-rich food decreases age-associated pathology and mortality in aged mice. Experimental Gerontology. 2009;44:727-732. DOI: 10.1016/j.exger.2009.08.013

58. Minamisawa M, Suzuki K, Kawai G, Yamaguchi A, Yamanaka S. Functional evaluation of yuzu (Citrus junos) extracts containing limonoids and polyamine for life extension. Journal of Functional Foods. 2017;38:591-600. DOI: 10.1016/j.jff.2017.09.043

59. Neufeld EF. Lysosomal storage diseases. Annual Review of Biochemistry. 1991;60:257-280. DOI: 10.1146/annurev.bi.60.070191.001353

60. Platt FM. Sphingolipid lysosomal storage disorders. Nature. 2014;510:68-75. DOI: 10.1038/nature13476

61. Huang JQ, Trasler JM, Igdoura S, Michaud J, Hanal N, Gravel RA. Apoptotic cell death in mouse models of GM2 gangliosidosis and observations on human Tay-Sachs and Sandhoff diseases. Human Molecular Genetics. 1997;6:1879-1885. DOI: 10.1093/hmg/6.11.1879

62. Abo-Ouf H, Hooper AW, White EJ, Janse van Rensburg HJ, Trigatti BL, Igdoura SA. Deletion of tumor necrosis factor-a ameliorates neurodegeneration in Sandhoff disease mice. Human Molecular Genetics. 2013;22:3960-3975. DOI: 10.1093/hmg/ddt250

63. Wu YP, Proia RL. Deletion of macrophage-inflammatory protein 1 alpha retards neurodegeneration in Sandhoff disease mice. Proceedings of the National Academy of Sciences of the United States of America. 2004;101:8425-8430. DOI: 10.1073/pnas.0400625101

64. Wada R, Tifft CJ, Proia RL. Microglial activation precedes acute neurodegeneration in Sandhoff disease and is suppressed by bone marrow transplantation. Proceedings of the National Academy of Sciences of the United States of America. 2000;97:10954-10959. DOI: 10.1073/pnas.97.20.10954

65. Jeyakumar M, Smith DA, Williams IM, Borja MC, Neville DC, Butters TD, Dwek RA, Platt FM. NSAIDs increase survival in the Sandhoff disease mouse:synergy with N-butyldeoxynojirimycin. Annals of Neurology. 2004;56:642-649. DOI: 10.1002/ana.20242

66. Eisenberg T, Knauer H, Schauer A, Büttner S, Ruckenstuhl C, Carmona-Gutierrez D, Ring J, Schroeder S, Magnes C, Antonacci L, Fussi H, Deszcz L, Hartl R, Schraml E, Criollo A, Megalou E, Weiskopf D, Laun P, Heeren G, Breitenbach M, Grubeck-Loebenstein B, Herker E, Fahrenkrog B, Fröhlich KU, Sinner F, Tavernarakis N, Minois N, Kroemer G, Madeo F. Induction of autophagy by spermidine promotes longevity. Nature Cell Biology. 2009;11:1305-1314. DOI: 10.1038/ncb1975

67. Li L, Rao JN, Bass BL, Wang JY. NF-jB activation and susceptibility to apoptosis after polyamine depletion in intestinal epithelial cells. American Journal of Physiology-Gastrointestinal and Liver Physiology. 2001;280:G992-G1004

68. Matsumoto M, Kurihara S, Kibe R, Ashida H, Benno Y. Longevity in mice is promoted by probiotic-induced suppression of colonic senescence dependent on upregulation of gut bacterial polyamine production. PLOS ONE. 2011;6:e23652. DOI: 10.1371/journal.pone.0023652

69. Franceschi C, Capri M, Monti D, Giunta S, Olivieri F, Sevini F, Panourgia MP, Invidia L, Celani L, Scurti M, Cevenini E, Castellani GC, Salvioli S. Inflammaging and anti-inflammaging: A systemic perspective on aging and longevity emerged from studies in humans. Mechanisms of Ageing and Development. 2007;128:92-105. DOI: 10.1016/j.mad.2006.11.016

70. García-Villalba R, Giménez-Bastida JA, García-Conesa MT, Tomás-Barberán FA, Carlos Espín J, Larrosa M. Alternative method for gas chromatography-mass spectrometry analysis of short-chain fatty acids in faecal samples. Journal of Separation Science. 2012;35:1906-1913. DOI: 10.1002/jssc.201101121

71. Kurihara S, Suzuki H, Oshida M, Benno Y. A novel putrescine importer required for type 1 pili-driven surface motility induced by extracellular putrescine in Escherichia coli K-12. Journal of Biological Chemistry. 2011;286:10185-10192. DOI: 10.1074/jbc.M110.176032

72. Liou AP, Paziuk M, Luevano JM, Machineni S, Turnbaugh PJ, Kaplan LM. Conserved shifts in the gut microbiota due to gastric bypass reduce host weight and adiposity. Science Translational Medicine. 2013;5:178ra41. DOI: 10.1126/scitranslmed.3005687

73. Honda K. Clostridia and the mucosal immune system. Journal of Intestinal Microbiology. 2013;27:187-196

74. Dao MC, Everard A, Aron-Wisnewsky J, Sokolovska N, Prifti E, Verger EO, Kayser BD, Levenez F, Chilloux J, Hoyles L, MICRO-Obes Consortium, Dumas ME, Rizkalla SW, Doré J, Cani PD, Clément K "Akkermansia muciniphila and improved metabolic health during a dietary intervention in obesity: relationship with gut microbiome richness and ecology". Gut. 2015;65:426-436. DOI: 10.1136/gutjnl-2014-308778. ISSN 1468-3288. PMID 26100928

75. Derrien M, Belzer C, de Vos WM.. Akkermansia muciniphila and its role in regulating host functions. Microbial Pathogenesis. 2016;106: 1096-1208. DOI:10.1016/j.micpath.2016.02.005. PMID 26875998

76. Barrett JC, Hansoul S, Nicolae DL, Cho JH, Duerr RH, Rioux JD, Brant SR, Silverberg MS, Taylor KD, Barmada MM, Bitton A, Dassopoulos T, Datta LW, Green T, Griffiths AM, Kistner EO, Murtha MT, Regueiro MD, Rotter JI, Schumm LP, Steinhart AH, Targan SR, Xavier RJ, NIDDK IBD Genetics Consortium, Libioulle C, Sandor C, Lathrop M, Belaiche J, Dewit O, Gut I, Heath S, Laukens D, Mni M, Rutgeerts P, Van Gossum A, Zelenika D, Franchimont D, Hugot JP, de Vos M, Vermeire S, Louis E, Belgian-French IBD Consortium, Wellcome Trust Case Control Consortium, Cardon LR, Anderson CA, Drummond H, Nimmo E, Ahmad T, Prescott NJ, Onnie CM, Fisher SA, Marchini J, Ghori J, Bumpstead S, Gwilliam R, Tremelling M, Deloukas P, Mansfield J, Jewell D, Satsangi J, Mathew CG, Parkes M, Georges M, Daly MJ. Genome-wide association defines more than 30 distinct susceptibility loci for Crohn's disease. Nature Genetics. 2008;40:955-962. DOI: 10.1038/ng.175

77. Atarashi K, Tanoue T, Shima T, Imaoka A, Kuwahara T, Momose Y, Cheng G, Yamasaki S, Saito T, Ohba Y, Taniguchi T, Takeda K, Hori S, Ivanov II, Umesaki Y, Itoh K, Honda K. Induction of colonic regulatory T cells by indigenous Clostridium species. Science. 2011;331:337-341. DOI: 10.1126/science.1198469

78. Atarashi K, Tanoue T, Oshima K, Suda W, Nagano Y, Nishikawa H, Fukuda S, Saito T, Narushima S, Hase K, Kim S, Fritz JV, Wilmes P, Ueha S, Matsushima K, Ohno H, Olle B, Sakaguchi S, Taniguchi T, Morita H, Hattori M, Honda K. Treg induction by a rationally selected mixture of Clostridia strains from the human microbiota. Nature. 2013;500:232-236. DOI: 10.1038/nature12331

79. Arpaia N, Campbell C, Fan X, Dikiy S, van der Veeken J, deRoos P, Liu H, Cross JR, Pfeffer K, Coffer PJ, Rudensky AY. Metabolites produced by commensal bacteria promote peripheral regulatory T-cell generation. Nature. 2013;504:451-455. DOI: 10.1038/nature12726

80. Fukuda S, Toh H, Hase K, Oshima K, Nakanishi Y, Yoshimura K, Tobe T, Clarke JM, Topping DL, Suzuki T, Taylor TD, Itoh K, Kikuchi J, Morita H, Hattori M, Ohno H. Bifidobacteria can protect from enteropathogenic infection through production of acetate. Nature. 2011;469:543-547. DOI: 10.1038/nature09646

81. Maslowski KM, Vieira AT, Ng A, Kranich J, Sierro F, Yu D, Schilter HC, Rolph MS, Mackay F, Artis D, Xavier RJ, Teixeira MM, Mackay CR. Regulation of inflammatory responses by gut microbiota and chemoattractant receptor GPR43. Nature. 2009;461:1282-1286. DOI: 10.1038/nature08530

82. Koh A, De Vadder F, Kovatcheva-Datchary P, Bäckhed F. From dietary fiber to host physiology: short-chain fatty acids as key bacterial metabolites. Cell. 2016;165:1332-1345. DOI: 10.1016/j.cell.2016.05.041

83. Schulthess J, Pandey S, Capitani M, Rue-Albrecht KC, Arnold I, Franchini F, Chomka A, Ilott NE, Johnston DGW, Pires E, McCullagh J, Sansom SN, Arancibia-Cárcamo CV, Uhlig HH, Powrie F. The short chain fatty acid butyrate imprints an antimicrobial program in macrophages. Immunity. 2019;50:432-445.e7. DOI: 10.1016/j.immuni.2018.12.018

84. Sun, Wu W, Chen L, Yang W, Huang X, Ma C, Chen F, Xiao Y, Zhao Y, Ma C, Yao S, Carpio VH, Dann SM, Zhao Q, Liu Z, Cong Y. Microbiota-derived short-chain fatty acids promote Th1 cell IL-10 production to maintain intestinal homeostasis. Nature Communications. 2018;9:3555. DOI: 10.1038/s41467-018-05901-2

85. Qin J, Li R, Raes J, Arumugam M, Burgdorf KS, Manichanh C, Nielsen T, Pons N, Levenez F, Yamada T, Mende DR, Li J, Xu J, Li S, Li D, Cao J, Wang B, Liang H, Zheng H, Xie Y, Tap J, Lepage P, Bertalan M, Batto JM, Hansen T, Le Paslier D, Linneberg A, Nielsen HB, Pelletier E, Renault P, Sicheritz-Ponten T, Turner K, Zhu H, Yu C, Li S, Jian M, Zhou Y, Li Y, Zhang X, Li S, Qin N, Yang H, Wang J, Brunak S, Doré J, Guarner F, Kristiansen K, Pedersen O, Parkhill J, Weissenbach J, MetaHIT Consortium, Bork P, Ehrlich SD, Wang J. A human gut microbial gene catalogue established by metagenomic sequencing. Nature. 2010;464:59-65. DOI: 10.1038/nature08821

Chapter-13

Nondestructive Assessment of Citrus Fruit Quality and Ripening by Visible–Near Infrared Reflectance Spectroscopy

1. Introduction

Citrus fruit are grown commercially in more than 50 countries around the world and are major commodities in the international trade [1, 2]. In Europe, the exceptional characteristics met by some of these produces have granted them the Protected Geographical Indication (PGI), such as the lemons (*Citrus limon* (L.) Osbeck) of Menton in France, Sorrento, Amalfi and Syracuse, and the Sicilian blood orange (*Citrus* × *sinensis*) in Italy, the "Algarve Citrus" in Portugal, or the "Valencianos Citrus" in Spain.

As non-climacteric, citrus fruit are only harvested at their optimal edible ripening stage, and are required to meet the expectations of the current consumer who demands for fruit not only with the best appearance, flavor, and nutritional properties, but that also comply with safety, traceability, and the sustainability of the cultural practices used. Like any other commodity, citrus fruit are subjected to worldwide standard specifications within the value chain [3] on their quality attributes (QA). Additionally, there are also adjustments to these requirements on quality and commercial ripening indices, that arise from the respective PGI normative of each commodity, growing regions and destination markets [4]. The main external quality attributes (EQA) accounted for citrus fruit are general appearance, size, weight, and color. Among the internal quality attributes (IQA), soluble solids content (SSC), titratable acidity (TA), juiciness, maturity index (MI; MI = SSC/TA), and the absence of internal defects are the most relevant. Although firmness is not defined quantitatively, it represents an important IQA, since it is a limiting factor regarding postharvest handling, transport and shelf-life, fruit being expected to maintain a good consistency through the whole supply chain.

Once fruit attain the expected IQA, additional factors will condition the harvest of citrus fruit: orchard yield and size, ripening variability, harvest cost, storage conditions, market prices and consumers' demand. Although dependent on the country, producers are provided with three options to handle these constrains: (i) immediate harvest and marketing; (ii) immediate harvest and cold-storage; or (iii) delayed fruit harvest. Opting for immediate harvest may result in minimal organoleptic quality and low prices, whereas postponing it until favorable market conditions, risks fruit drop, decay and spoilage caused by extreme weather events, pests and diseases [5]. To prevent some of of these consequences, producers resort to the regular use of pesticides, which increase the production costs and impact negatively the environment [6, 7]. Cold-storage is used in some of the major citrus producing countries, such as Spain or South Africa, and require very strict conditions to avoid fruit loss caused by chilling and/or freezing injury [8]. Both, cold-storage and harvest delay may lead to adverse alterations in the citrus-like flavor, and thus fruit quality deterioration, even if MI or SSC remains acceptable for marketing [9]. In all cases, fruit become more susceptible to the occurrence of physiological disorders that cause internal and/or external defects. Among the most typical physiological disorders registered through the supply chain of citrus fruit, there is the section drying, the rind breaking disorder (RBD), the rind pitting disorder (RP), freezing damage, and granulation, as reported for tangerine (*Citrus tangerine* Tanaka [10], 'Nules Clementine' mandarin (*Citrus* × *clementina*) [11], 'Marsh' grapefruit (*Citrus* × *paradisi* Macfad.) [12], sweet lemons (*Citrus limettioides* Tan.) [13], and 'Honey' pomelo (*Citrus maxima* Merr.) [14], respectively. These disorders are difficult to sort out by visual inspection at harvest, but lead to posterior fruit deterioration, limiting their quality, shelf-life, price and acceptance by consumers. In fact, there are strict standards for fruit sorting and grading, which require the detection of some of these disorders, throughout the supply chain, as established by the California Department of Food and Agriculture (CDFA). For exemple, it is not permitted to sell oranges (*Citrus sinensis* (L.) Osbeck) if, generally, more than %15 of fruits per batch have considerable freezing damage [15].

Therefore, the ripening of citrus fruit at harvest is a major determinant of their final quality after the whole postharvest handling processes, the occurrence of storage disorders, and the produce shelf-life span [16]. It also affects the rate of fruit loss between the tree and the consumers' home. Thus, the management and the decision capacity of the optimal harvest date (OHD) is a critical step in the supply chain. The current approach followed by producers and packinghouses to establish it and therefore, to decide on the harvest, is to collect small fruit sets from the various orchards by the beginning of each variety harvest season, and to use them to determine QA through standard methods, that in most cases are destructive, subjective and very time-consuming.

However, all QA vary greatly inside the same orchard, either in terms of absolute values and/or in terms of spatial and temporal distribution, and even in the same tree. This has been shown in citrus orchards of 'Shiranuhi' mandarin (*C. unshiu* × *C. sinensis*) × *C. reticulata* [17], 'Ortanique' (*Citrus reticulata* Blanco x *Citrus sinensis* (L) Osbeck) [18], mandarin (*Citrus reticulata* Blanco) [19], and 'Newhall' and 'Valencia Late' orange [20]. Multiple factors, such as the level of sunlight exposure and the associated fruit temperature on the tree, fruit yield and size, tree vigor and age, rootstocks, site-specific nutritional requirements and micro topographies within the orchard, are reportedly associated to this variability [21, 22, 23, 24]. Furthermore, the location of the orchards and their edaphoclimatic conditions, as well as the cultural practices also induce variability on the fruit maturation process, leading to different levels of QA and different ripening rates observed for the same cultivar at different sites [20, 21]. Consequently, the number of tested fruits with the standard methods is seldom statistically representative of the orchard, leading to the sub-representation of the effective ripening stage of the fruit within and between orchards, which results in a limited assessment of their ripening, heterogeneous fruit quality, a deficient OHD management and a weak traceability in the citrus supply chain [25, 26, 27].

Overall, there is the need to upgrade the management and the sustainability of citrus fruit supply chain with smart and nondestructive technologies that allow a fast, objective, accurate and extensive assessment of fruit QA and ripening on-tree and in the following postharvest, to replace conventional methods. Their aim would be to deliver the best produce to the markets, and contribute to reduce the current level of food loss around the globe, that involves a large portion of fruit and vegetables [28, 29, 30]. Considering how much of the world's population lacks food security, and the importance of these commodities in the provision of essential nutrients and vitamins, which could prevent malnutrition, that kind of technologies would comply with the sustainable development goals (SDGs) proposed by the Food and Agriculture Organization (FAO), International Fund for Agricultural Development (IFAD), and the World Food Programme (WFP), in the 2030 Sustainable Development Agenda, which supports a global commitment to end poverty, hunger and malnutrition by 2030, creating a #ZeroHunger world [31, 32].

The large number of reports published in the past two decades, show an active, and highly motivated research concerning the development of various nondestructive technologies for the assessment of quality and ripening parameters of a wide variety of fruit, including citrus [16, 33, 34, 35, 36, 37]. These techniques are used on inline sorting systems, on the bench or in the field and come in many forms, prices and commercial brands. Among them, the visible–near infrared reflectance spectroscopy (Vis–NIRS), is conceivably one of the most suitable and advanced nondestructive technologies currently used to monitoring several horticultural produces. It has been implemented in applications ranging from the inline automated grading systems, assessing up to 10–12 fruit per second, to handheld units suitable for field use, operating in full sunlight and varying ambient temperature [38, 39]. Additionally, it continues to grow stronger as a major investigation topic worldwide, with a major potential for improvement and contribution to the state of the art of precision agriculture and agronomic systems management [40].

This chapter comprises a brief explanation of Vis-NIRS fundamentals and a review of the various reports on its application published since 2012. Reports published before 2012 were already covered in the last review by [41] and will not be repeated here, with a few exceptions that represent relevant breakthroughs in the area. It will further attempt a critical evaluation on the limiting issues that need further research, to implement it as an effective nondestructive method to assess these commodities' quality and optimal ripening.

The authors invite the reader to complement this chapter with some of the most outstanding reviews published throughout the years, by the main researchers working on the subject (but not only in citrus). These reviews comprise the principles of the technique, its various methods and the listing of fruit and the respective QA for which it has provided calibration models [41, 42, 43, 44, 45], the overview on the publications and main research groups in the field [40], various recommendations for future research activity in the area regarding the adequate experimental design and the reporting requirements [38], as well as the current real-life applications available on the market that seem to comply with the warranted robustness for the technology to be integrated in the supply chain of many crops, including citrus [38, 39].

2. Fundamentals of visible–near infrared reflectance spectroscopy (Vis–NIRS)

In this review we will adopt the most common definition that Vis–NIRS covers the wavelength range 400–2500 nm of the electromagnetic spectrum. The lower limit is consensual, since it is the onset of the visible range, but the upper limit is mainly defined by the spectral response of the most common spectrometers. It comprises the visible (Vis) region (400–750 nm), the more penetrative short wave NIR (SWNIR), or Herschel region (750–1100 nm), and the near infrared region (750–2500 nm) of the spectrum [38, 39, 45]. The NIR radiation was discovered by Friedrich Wilhelm Herschel in 1800, and was first used in agricultural applications to measure the moisture in grain in the late 1960s [45]. The first Vis–NIRS application was commercialized in Japan in 1989 to sort peaches based on SSC in an automated grading line, but the research on its principles, applications and on the development of new customized systems, have only followed some decades later, being quite active nowadays [38, 39, 40].

2.1 Interaction of radiation with the fruit

When a light beam from the sun or a tungsten lamp, hits a fruit or any other sample, the incident radiation may be specularly reflected, absorbed or transmitted, and the relative contribution of each phenomenon depends on the chemical constitution and physical parameters of the sample (Figure 1) [46]. The spectral distribution of the radiation that penetrates the product change through wavelength dependent scattering and absorption processes. The photons that enter the fruit may emerge through multiple scattering in the tissue. Light emerging on the same side of incidence is described as *diffuse reflection*, while light emerging on the opposite side is described as *diffuse transmition*. Both diffuse modes may be understood in a general sense as 'transmitted', according to the initial description. The emerging diffuse light is collected by a spectrometer, originating the term *diffuse reflection spectroscopy*. The spectral features depend on the chemical composition of the product, as well as on its light scattering properties which are related to the sample microstructure. Fruit and vegetables are turbid media, in which scattering events dominate over absorption in the visible (400–750 nm), and particularly in the SWNIR and NIR ranges of the electromagnetic spectrum (750–2500 nm) [44] (see Figure 1).

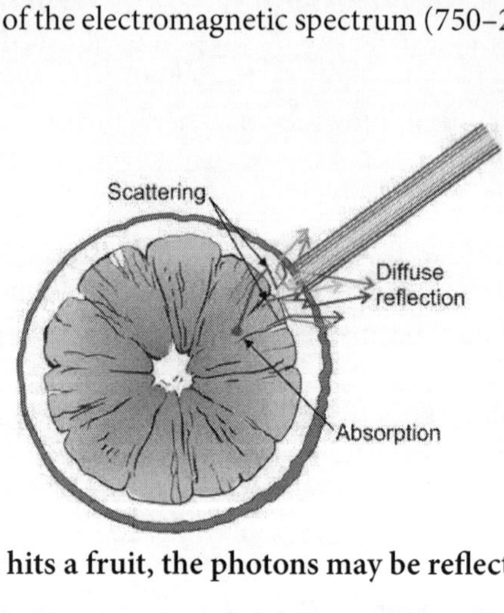

Figure 1. When light hits a fruit, the photons may be reflected at the fruit surface

(specular reflection) or enter the fruit tissue. In the latter case, a succession of scattering events takes place, where the photons change direction. Some of them reemerge, originating diffuse reflection (or transmission, not represented), and the other are eventually absorbed.

In thin rind fruit most of light interaction takes place on the flesh and the skin has mainly a modulation effect upon the spectra. In most citrus, however, most of light interaction occurs in the thick rind and few photons probe the flesh. Thus, the assessment of IQA depends on the interplay between pulp and skin biochemistry and their optical properties [47, 48, 49].

Absolute quantification of diffuse reflected light (for example, as spectral radiance [$W \cdot sr^{-1}\ m^{-2}\ Hz^{-1}$]) is of little use, because it depends obviously on the characteristics of the light source. The calculation of reflectance avoids that subjectivity, since it normalizes the absolute measurement of the sample's reflection by that of a reference material, usually a near perfect reflector ('white') in the wavelength range under study. It should be stressed, however, that even the reflectance depends on the collection geometry (solid angle of collection, viewed area, etc.). Common choices for the reference material include Spectralon or Teflon, with nearly 100% reflection in the Vis-NIR. Reflectance Rλ is calculated according to the Eq. (1) presented below.

Rλ=Sλ−DλRefλ−Dλ×100,E1

where S stands for the Sample counts, D for the Dark counts, Ref for the Reference counts and λ is the wavelength. Here, *counts* refer to the digitized output of the spectrometer, which are proportional to the spectral radiance measured in a specific geometry. The dark counts are obtained with the spectrometer closed and represent the electronic noise, which must be subtracted from the sample and reference measurements.

2.2 'Point' and imaging measurements

Vis-NIRS is most commonly applied on specific 'points' of the fruit, by observing a small area, which produce an average spectrum for that specific site. This is called *point measurements*. But Vis-NIRS can also be applied on extensive sections across the fruit, through *multispectral* and *hyperspectral* measurements, which create an image of the measured sections for each wavelength band [44]. The main difference between multi- and hyperspectral modes are the number of wavebands used. Multispectral imaging uses a set of filters and a common digital camera to deliver typically no more than ten bands, while hyperspectral cameras merge imaging and spectral separation in the optical hardware to produce hundreds of contiguous wavebands. Another way to look into hyperspectral images is to think that it yields the reflectance spectrum for each spatial position of a sample (*i.e.*, for each pixel of the image) [44]. Both techniques, although costly, have been shown to successfully assess several IQA, diseases and defects in several fruit, including citrus fruit [50, 51]. Yet, extensive investigation is needed to allow both the acquisition and image processing software to be implemented in real-time systems. Thus, this chapter will only address the systems that perform 'point' measurements, based on their much wider spread, cost-effective and friendly use through the supply chain, and particularly under field conditions. Further information on the principles and applications of multispectral and hyperspectral Vis-NIRS technologies could be found in the reviews by [44, 50].

2.3 Instrumentation and measurement setup

There are currently, a large variety of both commercial and lab-made customized Vis-NIRS systems, with various shapes, sizes, and prices, that operate in many spectral ranges, and have reportedly allowed the assessment of several QA, the content of critical compounds and the diagnostic and/or prediction of disorders in a wide sort of fruits, including citrus (Tables 2–4). Nevertheless, most of the current commercial fruit applications of Vis-NIRS are based on the use of silicon-based spectrophotometers comprising the Vis-SWNIR region (400–1100 nm), because of their accessible prices and the larger light penetration depth in this band, in comparison to the significantly more expensive InGaAs-based devices (900–2500 nm), that do not add too much value to the quality assessment procedure [39].

From handheld, benchtop, to inline automated grading system, all Vis-NIRS devices comprehend the following fundamental components: an optical spectrometer, a light source (usually a tungsten halogen light bulb) and collection optics (optical fibers, lenses, integration spheres, dedicated probes). The current spectrometers typically include a connection for an optical fiber, an entrance slit (that defines the spectral resolution), a diffraction grating to separate the light into its spectral components, mirrors for collimation and focusing, and a light sensing device that is usually a one-dimensional CCD (Charge-Coupled Device) or CMOS (Complementary Metal-Oxide Semiconductor).

The Vis-NIR spectra may be acquired according to three principal geometrical configurations, as depicted in Figure 2: the reflectance mode (a), the transmittance mode (b), and in the interactance mode (c). The reflectance mode is susceptible to receive specularly reflected light, which may be a disadvantage, since only a fraction of the collected photons probes the fruit interior. In the transmittance mode the photons probe necessarily the fruit interior; however, the optical signal may be weak and noisy. The interactance mode is a tradeoff between the two previous modes: by using a contact probe it avoids specularly reflected photons and receives only those traveling through the fruit flesh. Also, the distance between light injection and collection is small, insuring a good optical signal. However, this is also a disadvantage, since the probing depth into the fruit pulp is shallow.

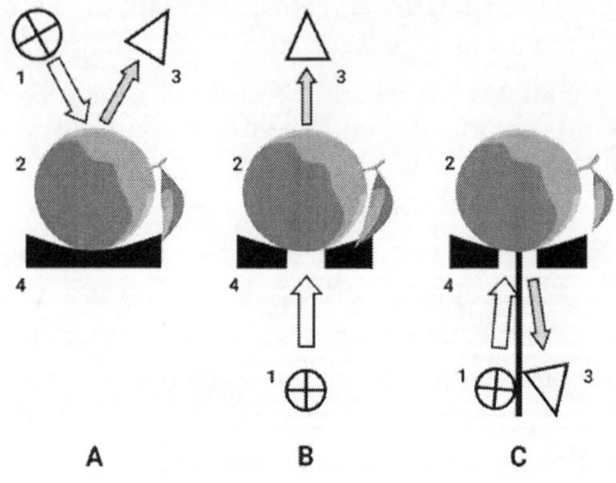

Figure 2. Setup for the acquisition of Vis-NIR reflectance spectra in (A) reflectance, (B) transmittance, and (C) interactance modes. Based on [45].

The choice of the geometry is thus of the utmost importance for obtaining good results, and should account for the fruit and the assessed QA. The penetration of NIR radiation into fruit tissue decreases exponentially with the depth, which is quite critical in thick rind fruit such as citrus [50]. Furthermore, the choice of the detection mode might be influenced by the spectral range used, as report by [48], in which both interactance and reflectance modes produced similar models to assess the SSC of 'Sunkist' navel oranges in the Vis-SWNIR range, but the participation of Vis region degraded this assessment in the transmittance mode. in general, to detect the internal defects, the transmittance mode should be chosen, while the other two modes are quite reliable regarding other IQA (Tables 2–4).

2.4 Typical Vis-NIRS spectrum and its interpretation

All studies on the use of Vis-NIRS to assess the fruit QA start by acquiring the reflectance (R) spectra, which is then converted to the respective absorbance log(1/R) spectra (Figure 3). The main spectral differences observed in a wide variety of fruit are in the visible region, namely in the 400–750 nm range. This is due to changes in the pigments' content through ripening, namely chlorophylls, carotenoids and anthocyanins present on the fruit rind [38]. In fruit that change from green to yellow/orange/red colors through ripening, the spectral information on the pigments' absorption range, may provide accessory indirect correlations with IQA such as firmness, as found in 'Rocha' pear (*Pyrus communis* L.) [53]. This, however, is not so clear in citrus fruit because their color change do not correlate with their maturity and depends on the orchards' location climate [6]. Otherwise, the pattern of the absorption spectra in the NIR range is quite similar among the various fruit species, although position and magnitude of the peaks are fruit specific, even among citrus fruit varieties [41]. The magnitude of the peaks and minima are also dependent on the acquisition mode used, but in general the same features are present and the landscape of the spectra is similar among the same fruit as reported for 'Sunkist' oranges [48].

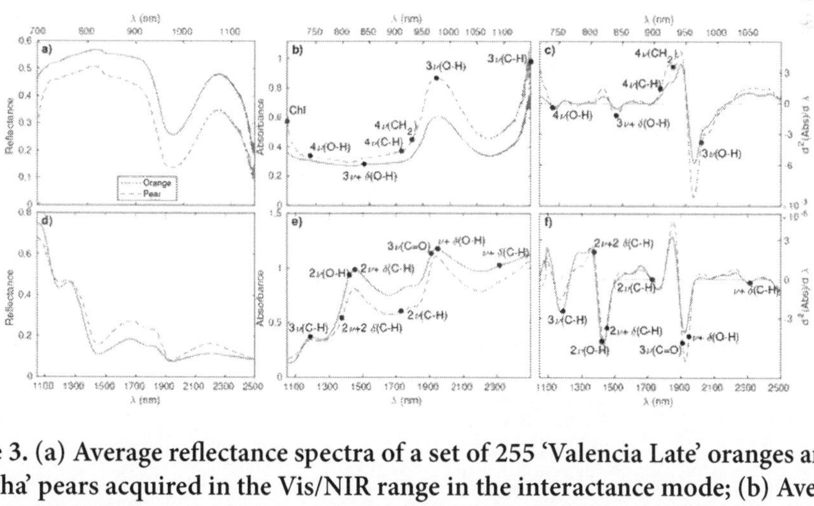

Figure 3. (a) Average reflectance spectra of a set of 255 'Valencia Late' oranges and 239 'Rocha' pears acquired in the Vis/NIR range in the interactance mode; (b) Average absorbance spectra of the same set of fruit. The nominal positions of the most important absorption bands are indicated in the curves. The number is the order of the transition, ν stands for stretching vibration, δ for bending vibrations and the sum

indicates combination bands (for example, 3ν+δ(O–H) represents the combination band of the second overtone of stretching with the fundamental bending in O–H); (c) Savitzky–Golay [52] filter of second derivative order applied to the absorbances. The bands are again indicated; (d) to (f) Same as in (a) to (c) but in the NIR range, with the spectra acquired in reflectance mode.

The spectra in the NIR range convey information mainly related with vibrational bands (stretching and bending) of the relevant functional organic groups, such as O–H, C–H, C–O and C=O. The compilation of the main wavebands present in citrus fruit, namely, O-H and C-H vibration absorptions are presented graphically in Figure 3. These groups exist in all fruit organic molecules, but variations associated with water and storage reserves may induce slight changes in the spectra that may be related with the IQA. Vibration states are quantized and the transitions between states are said to be fundamental or overtones. The fundamental transitions (corresponding to a fundamental band) refer to the transition from the ground state to the first excited state, and take place mainly in the infrared range, that is above 2500 nm. In this range, the absorption peaks are distinguishable and correlate directly to specific compounds, allowing a better assessment of organic compounds such as vitamin C, citric acid or sucrose, as reported in 'Valencia' orange [54]. In contrast, the overtone bands correspond to transitions to higher excited states, with a large number falling in the NIR range. For example, the first overtone band corresponds to the transition from the ground state to the second excited state. A very crude approximation is that the n-th overtone frequency is close to (n-1) times the fundamental frequency. Thus, a general rule is that overtones have higher frequencies and lower amplitudes than the fundamental. For example, the fundamental frequency for O–H stretching (1ν) is around 2700 nm, which means that is beyond the range of the most common NIR dispersive-type spectrometers. However, the overtones are within their instrumental range: 2ν at 1420 nm (first overtone, strong intensity band), 3ν at 970 nm (2nd overtone, medium intensity) and 4ν at 750 nm (3rd overtone, low/very low intensity band). The quoted values are only indicative of a typical band central value. Indeed, the vibrations are dependent on the chemical environment, which results in a frequency spread of the bands. Finally, it is important to refer the combination bands. These correspond to the superposition of vibration motions. For example, the fundamental bending mode (1δ) of water at 6300 nm (infrared range) may combine with the fundamental stretching mode at 2700 nm (1ν), to generate a combination band around 1900 nm [ν+δ (O–H)]. Having in mind that the fruit tissue is composed by many different organic molecules, it is easy to understand that the spectral landscape of fruit NIR reflectance is a continuum, due to band superposition, as previously shown by [54], when comparing NIR and medium infrared spectroscopies (MIR), to assess several compounds in 'Valencia' oranges. Summarizing, the NIR spectra of a fruit contains mainly overtones and combination bands of stretching and bending vibrations of the main functional organic groups of relevant organic compounds regarding the fruit IQA, such as O–H and C–H. The large number of possible vibrations and corresponding bands originates a spectral landscape with very broad and unspecific features, from which it is nevertheless possible to retrieve useful information. For instance, [54] obtained better prediction of fructose and reducing sugars when using NIRS than MIRS.

The typical Vis–NIR and NIR reflectance and absorption spectra of 'Valencia Late' orange and 'Rocha' pear are depicted in Figure 3 [55]. Both fruit spectra were relatively flat from 700 to 910, followed by strong water absorption peaks around 970, 1450 and 1940 nm. However, the C-H bands may distort slightly the water peaks, and the analysis of this distortion conveys more information than the main peaks alone. It is from these patterns associated with the OH and CH vibrations that it is possible to retrieve the information about sugars. Even the water bands by themselves may convey information about the sugars, because the concentration of sugars and water are interdependent [41, 56, 57, 58].

In the Vis–NIR range the most prominent feature is the 3ν(O–H) peak at ~ 975 nm. This peak is reported in the literature in the range 980–960 nm [57], but the actual location depends on multiple factors: (i) the degree of OH bonding; (ii) the temperature; (iii) the presence of other close bands. In other words, within different chemical environments, the OH group will peak at different wavelengths. Effect of (iii) is more clearly observed in Figure 3c. Indeed, the smooth 975 nm peak observed in absorbance has a more fine structure disclosed upon derivation. Thus, the peak 4ν(CH_2) at 930 nm is actually coalescing with the main water peak, causing a depression in the 2nd derivative left positive peak. The form of this overlap is a source of information about all organic compounds content, namely sugars, acids, proteins, etc.

A minor feature, but consistently observed in most fruit, is the slight inflection around 840 nm, which is caused by the band 3ν+δ(O–H). In this case it is more clearly observable in the reflectance spectrum, Figure 3a. In this discussion is important to have in mind that second derivation of a symmetric peaks yields a negative peak at the same position, with two lateral smaller positive peaks. This is clearly observed for the 840 nm feature in the 2nd derivative plot, with the negative peak coinciding with the 3ν+δ(O–H) absorption wavelength. On the contrary, the peaks of the 3ν(O–H) features do not coincide in the absorbance and 2nd derivative plots, which clearly indicates spectral overlap.

Similar curves are observed in other cultivars. For example, in 'Newhall' orange [56] the same structure for the second derivative plot was observed, although a different technique was used, namely the Norris derivative [59].

Concerning the NIR spectra, those from the oranges show three main peaks at 1190, 1450 and 1940 nm, whose origin may be traced to the 3ν(C–H), 2ν(O–H) and ν+δ(O–H) bands, respectively. However, satellite bands overlap, as in the Vis/NIR case. The most 'pure' peak is the first, around 1200 nm, corresponding to the 3ν(C–H) band. As is the 840 nm band, absorbance and 2nd derivative peaks coincide. The other two main peaks are more complex blends of two or more bands. For example, the second peak around 1450 nm, although dominated by the stronger 2ν(O–H) band, has contributions from 2ν+δ(C–H) and 2ν+2δ(C–H). Consequently, the 2nd derivative feature associated with this mix is more complex. The same could be said about the third peak.

Furthermore, due to several causes, the various peaks, even when they coincide among different fruit, may present different levels of importance, signified by their infrared values, with the various IQA. For instance, the combination band of OH reported at 839 nm correlated highly with SSC in 'Rocha' pear samples, but not in 'Valencia Late' oranges [55].

3. Chemometrics

As it has been mentioned earlier, Vis–NIR spectra of fresh fruits tend to be composed by a large superposition of absorption bands. The presence of a substantial amount of water in fresh fruit has a big impact in the spectra, dominating most of the spectral landscape. Therefore, the signals corresponding to the absorption bands of key chemical compounds such as sugars and acids, become masked by water and are only discernible as weak fluctuations in the spectrum. Given the complex interplay between the multiple absorption bands and their weak amplitudes, most of the times the linear relation between chemical compound concentration and absorption (Lambert–Beer law) is almost lost. In order to be able to extract information about chemical concentrations from this type of spectrum, we have to look for relationships' patterns between different wavelengths. This is done resorting to multivariate statistical techniques, that when applied in the field of analytical chemistry is called Chemometrics. This research field can be considered as a subset of the broader area of Machine Learning, and pursues the same goals, i.e., infer critical information from high dimensional data. There is a vast literature on this subject for those who wish to learn more about Chemometrics. Here are some suggestions for introductory and advanced levels [60, 61, 62, 63, 64]. In this section, it is presented a brief introduction to the scientific language used in this area, with the sole objective of familiarize the reader with the main concepts that are often presented in the literature. In Vis-NIRS of fresh fruit, the input data are spectra, i.e. one-dimensional arrays of values, each one corresponding to the intensity of light (diffusively reflected or absorbed) at a specific wavelength. Each spectrum Xn (n=1,... number of samples) represents a measurement or sample and each point xi (i=1,..., number of variables) of the spectrum is usually referred to as input variables, spectral features or simply as 'wavelengths'. The macroscopic properties or QA features that are obtained through laboratory testing (e.g. SSC, firmness, etc.) are commonly defined as target variables Yn or simply attributes. Chemometrics consists on the application of mathematical/statistical methods that allow mapping the spectral features xi into the target variables Y. These methods can be subdivided into two broad subcategories: unsupervised and supervised. In the former type of method, only the input variables xi are used and, the main purpose is to find, for example, trends within the data, clusters that can be used for classification or other general characteristics of the data set. On the other hand, supervised methods use both input and target variables and can be used for classification tasks (e.g. discriminate between different fruit sub-species or origins) or for quantitative (regression) prediction of attributes that have continuous distributions (e.g. SSC, firmness, TA, etc.).

3.1 Spectral pre-processing and outlier detection

In order to implement the best calibration model possible to predict the expected QA, often the spectral data has to be preprocessed before being used. Preprocessing techniques are used to remove irrelevant information (noise, systematic errors and faulty samples) that can degrade the performance of the numerical algorithm used to develop the calibration model. Several preprocessing methods have been created for this purpose and reviews, such as [61, 65], present a wider scope of these techniques. A brief summary of the most commonly used methods is presented in Table 1. For Vis-NIRS, the most common forms of spectral preprocessing can be stacked into two groups: scatter corrections (SC) and derivative techniques (DT). Scatter-corrective methods are used to remove the influence of scattered light that can contaminate the diffuse reflectance spectra. The rationale behind SC techniques is to remove effects that are unrelated to the chemical composition of the samples and that just depend on the measurement geometry or samples morphology. On the other hand, DT are designed to improve signal to noise ratios, eliminate systematic baseline biases and enhance spectral variations. Another type of preprocessing commonly mentioned in the literature is outlier detection. This process consists in identifying and removing from the data set, samples that are very different from the rest of the samples. These outliers can be for example, reflectance spectra that were defectively acquired or fruit with odd properties. The idea is to remove these samples from the data set, using some pre-defined metric in order to feed the model only with the most representative samples in the data set that lead to a correct mapping of the attribute being predicted by the calibration model constructed.

Table 1. Most common spectral preprocessing techniques.

Technique	Description
MSC (Multiplicative Scattering Correction) [66]	Each spectrum Xn is regressed against a reference spectrum (usually the mean spectrum, Xm) using a least square method. Then, the corrected spectrum is computed based on these regression coefficients. There are more sophisticated variants of this method (e.g. Extended MSC [67] that try to correct for additional additive effects).
SNV (Standard Normal Variate) [68]	Each individual spectrum Xn is normalized to have zero mean and unit variance. In some cases, this can also appear as sample standardization.
Smooth derivatives (Savitzky–Golay (SG) [52] and Norris-Williams (NW) [59]	The spectrum derivative in relation to the wavelength is computed. Usually the first derivatives remove baseline effects and the second derivative remove baseline and linear trends in the spectrum. Smooth derivatives are computed using an averaging multi point window in order to be more robust against spurious noise. SG and NW algorithms are the most common methods to compute these derivatives.
De-trending	This process consists in fitting a polynomial (usually 1st or 2nd order) to the spectrum and subtract it from the signal. This provides baseline correction.
Scaling	Scaling is a sort of umbrella under with we can find multiple types of data manipulations. The most common ones are: baseline subtraction, sub-sampling, normalization on columns (all spectral features in the data set are scaled between a max and min values) and normalization on rows (the individual spectral features are scaled between min and max).

3.2 Clustering

In Vis–NIR spectra of fresh fruit, the common spectrum is often described by smooth mounds and soft depressions. This means that adjacent wavelengths can be highly correlated. Therefore, in order to reduce redundancy of information provided by neighbour features xixi+1xi+2… (also called co-linearity), sometimes it is beneficial to restrict the number of used input variables. This is often called dimensionality reduction and is very important for the right operation of certain calibration models. The simplest way to deal with this problem is by using sub-sampling, where a certain number of spectral features are discarded, e.g. every 3rd or 5th point in the spectra. To deal with this problem of dimensionality reduction, some 'clever' algorithms were introduced, the most common being the Principal Component Analysis (PCA) algorithm, the Hierarchical Clustering Analysis (HCA) and K-Means. the latest two methods can be used for classification tasks (mapping a cluster to a class) and for outlier detection as well. If samples are too far apart from the defined clusters (according to some metric such as the Euclidean distance or the Mahalanobis distance), then this suggests that it might be an outlier.

3.3 Classification and regression models

As we mentioned earlier, depending on the problem at hand, we might need to implement a classification or a regression model for our data. Multiple Linear Regression (MLR) is perhaps one of the most straight forward methods to implement. It expands the application of simple linear regression to the multivariate case by linearly combining them. Due to its simplicity, this method has some drawbacks, namely an inefficient applicability in the cases of high co-linearity in the data, and when the number of features in the data set is higher than the number of samples. This is often the case of fresh fruit Vis–NIRS datasets and hence its applicability has been limited. One way of overcoming these limitations, is to use a dimensionality reduction method, such as PCA and then perform MLR on these lower dimensional components. This workflow is known as Principal Component Regression (PCR). Partial Least Square Regression (PLS) is without question the most widely used method to create calibration models to predict the most QA of fresh fruit (Tables 2–4). As opposed to PCA, the PLS algorithm takes into account the covariance between input xi and target Yi variables. In the same spirit as PCA, PLS also projects the data into a latent space, but this time the components are defined along the direction of maximum variance between xi and Yi. These components are called latent variables (also named factors by some researchers), are built in order to model the target variable, and their number is what defines the quality of the PLS model. In general, a low number of latent variables usually lead to more robust predictions, but that might not always be the case. A variant of PLS named PLS Discriminant Analysis (PLS-DA) can be used to deal with classification scenarios when the target variables Yi are not continuous (e.g. 0, 1 for fruits without and with defects). The models mentioned so far are can be described as linear because they rely on a linear combination of multivariate solutions. Besides the easiness of implementation, they are also classically appreciated in Chemometrics because they are easy to interpret in terms of feature importance, i.e., after fitting the model to the data we can back-trace some parameters (e.g. regression coefficients) and find what wavelengths or spectral bands better contributed to the prediction. In turn, this allows inferring information about the chemical concentrations and can be used to identify biological and metabolic behaviors.

Table 2. Overview of applications of Vis-NIRS to measure the quality attributes of citrus fruit by benchtop devices. List of symbols in Table 5.

IQA	Unit	Range (nm)	Mode	Stat Method	Val	RMSEP	R^2	RPD	Ref
'Marsh' grapefruit									
Suc	g/kg	400–2500	R	PLS	I	35	0.94	3.76	[69]
Gluc	mg/g					18	0.77	1.80	
TPh	gGAE/kg					0.37	0.59	1.03	
DPPH	%					5.8	0.85	2.45	
TotCar	g/kg					2.9	0.80	1.71	
VitC	g/kg	400–700				0.20	0.95	3.33	
Chl a	g/kg					0.79	0.87	2.31	
Chl b	g/kg					2.8	0.94	3.61	
'Star Ruby' grapefruit									
SSC	%	850–2500	R	PLS	E	0.31	0.90	2.94	[70]
TA	g/100 mL					0.07	0.83	2.46	
MI	SSC/TA					0.45	0.81	2.32	
BrimA	SSC-TA					0.43	0.86	2.68	
'Trovita' orange									
RateOleoc		400–1000	R	PLS	I	0.008	0.97	—	[71]
DegOleoc						0.005	0.97	—	
'Valencia' orange									
SSC	%	900–1800	R	PLS	E	0.58	0.69	1.87	[72]
VitC	mg/mL	1050–2000	I			8.01	0.43	1.33	
SSC	%	850–2500	R	PLS	E	0.28	0.93	3.57	[70]
TA	g/100 mL					0.02	0.93	3.88	
MI	SSC/TA					0.61	0.96	4.92	
BrimA	SSC-TA					0.006	0.96	3.96	
Firm	N	1000–2500	R	PLS	E	6.22a	0.85	—	[73]
Pectin	%	1000–2500				5.04a	0.49	—	
VitC	mg/L	1000–2500	R	PLS	I	95	0.50	—	[54]
CitAc	g/L					3.8	0.56	—	
Suc	g/L					12	0.56	—	
Gluc	g/L					6.7	0.67	—	
Fruct	g/L					5.0	0.66	—	
TotSug	g/L					12	0.64	—	
RedSug	g/L					20	0.67	—	
'Shatian' pomelo									
WC	%	400–1700	T	PLS	I	0.49	0.74	—	[74]

IQA	Unit	Range (nm)	Mode	Stat Method	Val	RMSEP	R^2	RPD	Ref
Granul						0.096	0.97	—	
'Honey' pomelo									
Granul		400–1700	T	PCA-NN	I	0.97–0.98b	—	—	[14]
sweet lemon									
Freeze	Binary	400–1100	T	SIMCA	I	>92%b	—	—	[13]
Injury				PCA-NN		100%b	—	—	
				SVM		>92%b	—	—	

[a] SEP values.

[b] This value corresponds the accuracy of the classification.

Table 3. Overview of applications of Vis-NIRS to measure the QA of citrus fruit by portable spectrometers/systems under laboratory conditions. List of symbols in Table 5.

IQA	Unit	Range (nm)	Mode	Stat Method	Val	RMSEP	R^2	RPD	Ref
'Marsh' grapefruit									
RindPitt		400–2500	R	PLS	E	5E-4	0.89	—	[12]
DM	%	400–850				0.30c	0.88	—	
'Ehime Kashi28' mandarin									
SSC	%	750–1050	I	PLS	I	0.59	0.71	—	[75]
'Imperial' mandarin									
SSC	%	720–950	I	MPLS	E	0.6–0.8d	0.5-0.8	1.2-3.4	[76]
DM	%					0.6e	0.9e	—	
SSC	%	720-950	I	MPLS	E	0.4–0.5f	0.6-0.9	—	[77]
						0.4–0.5g	0.3-0.7	—	
						0.8–6.8h	0.0-0.8	—	
'Nanfeng' mandarin									
TA	g/100 mL	400–1040	T	PLS	I	0.09	0.41	—	[78]
SSC	%					0.65	0.85	—	
VitC	mg/100 mL					2.80	0.64	—	
TA	g/100 mL			BPNN		0.09	0.42	—	
SSC	%					0.64	0.86	—	
VitC	mg/100 mL					2.70	0.64	—	
'Nules Clementine' mandarin									
WL	g	900–1700	I	PLS	I	2.76	0.83	2.34	[11]
DMC	%					0.92	0.92	3.57	
Sucrose	mg/g DW					24	0.77	1.4	

IQA	Unit	Range (nm)	Mode	Stat Method	Val	RMSEP	R^2	RPD	Ref
Glucose	mg/g DW					11	0.88	2.86	
Fructose	mg/g DW					12	0.90	3.23	
TotSug	mg/g DW					31	0.90	3.15	
RBD	binary	450–1000	I	PLS	E	0.36	0.37	—	[79]
Orange									
SSC	%	600–1100	R	PLS	I	0.73	0.35	—	[80]
		600–950				0.60	0.66	—	
		600-1100B		VABPLS		0.60	0.67	—	
'Gannan' orange									
SSC	%	820–950	I	PLS	I	0.40	0.77	—	[81]
				SVM		0.32	0.84	—	
SSC	%	600–1100	T	PLS	I	0.55	0.85	2.57	[82]
		685-983[a]		SPAPLS		0.57	0.84	2.51	
Navel orange									
SSC	%	580-970[b]	T	PPSOPLS	I	0.43	0.79	—	[83]
'Valencia Late' orange									
SSC	%	400–1000	R	PLS	I	0.33	0.92	—	[84]
TA	mg/L					0.07	0.73	—	
BrimA	SSC-TA					0.37	0.84	—	
MI	SSC/TA					1.03	0.76	—	
'Page' tangelo									
TA	c.a./100 g	400–1000	R	PLS	I	0.03	0.77	—	[85]
SSC	%					0.40	0.72	—	

[a] Selected wavelengths in this range.

[b] Selected wavelength bands in this range.

[c] Cross validation.

[d] Combinations of cal. and val. Sets.

[e] Calibration.

[f] Combinations of cal. and val. sets (harvest day).

[g] Combinations of cal. and val. sets (orchard location).

[h] Combinations of cal. and val. sets (season).

Table 4. Overview of applications of Vis-NIRS to measure the quality attributes of citrus fruit by portable systems on-tree. List of symbols in Table 5.

IQA	Unit	Range (nm)	Mode	Stat Method	Val	RMSEP	R^2	RPD	Ref
'Clemenvilla' mandarin									
TA	% acid citric	1600–2400	R	MPLS	I	0.15	0.56	—	[86]
SSC	%					0.77a	0.50	—	
MI	SSC/TA					1.46a	0.53	—	
pH						0.11a	0.61	—	
MPF	N					7.3a	0.45	—	
PerTh	mm					0.54a	0.51	—	
TA	% acid citric	1600–2400	R	LOCAL	I	0.13a	0.76	2.08	[87]
SSC	%					0.71a	0.57	1.51	
MI	SSC/TA					1.13a	0.79	2.14	
pH						0.11a	0.74	1.91	
BrimA	SSC-TA					0.70a	0.75	2.0	
5 cultivars of orange									
SSC	%	450–1100	I	PLS	E	0.86	0.56	—	[88]
'Newhall' orange									
SSC	%	680–1100	I	PLS	I/E	1/1.1 ± .2	.5/.3 ± .1	1.4/1.1 ± .1	[56]
Juice pH						.2/.27 ± .04	.5/.4 ± 0.1	1.4/1.1 ± 0.1	
'Powell Summer' orange									
SSC	%	500–1690	R	LOCAL	I	0.62	0.74	—	[89]
MPF	N					4.64	0.74	—	
PerTh	mm	1100–1690				0.58	0.68	—	
Juice Weight	g					18	0.68	—	
SSC	%	1600–2400	R	LOCAL	I	0.80	0.81	2.23	[87]
MI	SSC/TA					3.56	0.65	1.69	
BrimA	SSC-TA					0.85	0.82	2.33	
'Clemenvilla' mandarin and 'Powell Navel' orange									
JuiceMass	g	1600–2400	R	LOCAL	I	0.92a	0.69	1.77	[90]
PerTh	mm					28a	0.72	1.86	
SSC	%	1600–2400	R	LOCAL		0.86a	0.78	2.09	[87]
pH						0.15a	0.72	1.93	

IQA	Unit	Range (nm)	Mode	Stat Method	Val	RMSEP	R^2	RPD	Ref
TA	% acid citric					0.14a	0.84	2.43	
MI	SSC/TA					2.98a	0.77	2.08	
BrimA	SSC-TA					0.84a	0.78	2.15	

[a] Corrected to bias.

Table 5. List of abbreviatures used in Tables 2-4.

Colums	Rows' content
IQA: Internal Quality Attributes	β Car: β carotene; **BrimA**: Brix minus Acids (SSC-k(TA)); **Chl a**: chlorophyll a; **Chl b**: chlorophyll b; **CitAc**: Citric Acid; **DegOleoc**: Degree of oleocellosis; **DMC**: Dry Matter Content; **DPPH**: 2,2-diphenyl-1-picrylnhydrazyl; **DM**: dry matter; **Firm**: Firmness; **Fruct**: Fructose; **Gluc**: Glucose; **Granul**: Granulation; **JuiceVol**: Juice Volume; **MPF**: Maximum Penetration Force; **MI**: Maturation Index; **Pectin**: Pectin Content; **PerTh**: Pericarp Thickness; **RateOleoc**: Rate of oleocellosis; **RBD**: Rind Breakdown disorder; **RedSug**: Reduced Sugars; **RindPitt**: Rind Pitting; **SSC**: Soluble Solid Content; **Suc**: Sucrose; **TA**: Titratable Acidity; **TAO**: total antioxidant capacity; **TotCar**: Total Carotenoids; **TotSug**: total sugars; **TPh**: Total Phenolic content; **VitC**: Vitamin C; **WC**: Water Content; **WL**: Weight Loss
Mode: mode of acquisition	R (Reflectance), I (interactance) or T (transmittance)
StatMethod: Statistical method employed	**iPLS**: Interval PLS; **LAR**: Least Angle Regression; **LOCAL**: Patented algorithm for PLS regression from a subset of proximal calibration samples; **LS-SVM**: Least squares support vector machines; **MLR**: Multiple linear regression; **MPLS**: Modified PLS **PCA-NN**: Principal components analysis combined with artificial neural networks; **PCA-BPNN**: back propagation neural network (BPNN) based on principal component analysis (PCA) **PLS**: Partial Least Squares; **PPSOPLS**: piecewise particle swarm optimization (PPSO) algorithm based PLS **SIMCA**: Soft independent modelling of class analogy; **SPAPLS**: Succesive Projections algorithm coupled with PLS; **SVM**: Support vector machines; VABPLS: variable adaptive boosting partial least squares
Val: Model validation type	I (internal), CV (cross validation only), E (external)
RMSEP: Root Mean Square Error of Prediction	
R^2: Coefficient of determination	
RPD: Residual predictive deviation	

In the last couple of decades, non-linear models imported from other areas of Machine Learning have begun to permeate Chemometrics, and given its high use case in the literature, Support Vector Machines (SVM) is one of the most popular. The strategy of this model consists in searching for boundaries that separate two cluster or classes. The algorithm tries to find the best boundary between classes by maximizing a distance margin between neighbor samples. It has the advantage that it can use kernel tricks to transform the data points into another mathematical space, where these boundaries are easier to establish. SVMs were initially used for classification tasks, but have been extended to deal with regression problems as well (SVR). SVR has been used successfully for many datasets, and the most often mentioned drawback is the complexity of its optimization task. Another popular type of non-linear models that is often used for classification and regression problems is Neural Networks (NN). These represent a wide class of algorithms with many types of architectures and are derived from the field of Artificial Intelligence. In recent years, classical NN architectures such as the Multi-Layer Perceptron has been increasingly substituted by more modern architectures developed for Deep Learning. The most promising of these NN are the so called Convolutional Neural Networks that have been very successful in image recognition tasks.

3.4 The quality of a calibration model

Independent of the type of model that is used for prediction or classification, the important thing is to find how well it performs on the desired data. To assess the quality of the predictions made by the calibration models, several metrics are often used. In a recent review by [38], the author makes a case for the uniformization of the report of error metrics in future publications. In what follows these recommendations are highligted. The partitioning of the data for model development is very important. The data set should always be split into two sub-sets, called train and test sets. The train set, as its name suggests, is used for the calibration model development and, once the main hyper-parameters have been established, the model is used to predict the test set and assess its performance. Model development can be done with the full train set using a cross-validation strategy or by further splitting it into calibration and tuning (or assessment or validation) sets. As a note of caution, it is important to mention that for different areas of Machine Learning the names given to these data splits can vary and that can lead to some confusion. Otherwise, the test set should be derived from a different distribution from that of the train, in which case it is named as external validation set [45]. For example, data from two consecutive harvest seasons are used as train set, while the test set uses data from a third season. A similar situation can be envisaged by using a train set collected from different orchards or origins, than that used for validation. Yet, given the large amount of time invested in acquiring this type of data, multi-seasonal or multi-orchards test data sets are not often found in the literature. In contrast, laboratory models with homogeneous fruit sets are abundant (Tables 2–4). Currently, the most common procedure when constructing and validating Vis-NIRS models for the various fruit QA is to separate a fraction of the available samples as train/calibration set (usually 80%), and the remaining as test/validation set (usually 20%). Furthermore, the validation samples are typically chosen as the best possible representation of the whole set and within the variation range of the train set. This has been applied even when the models comprehend several species and/or cultivars, orchard locations and harvest years, which are mixed in the calibration and validation sets [90]. All studies included in Tables 2–4 that used this approach, were labeled as internal in the validation column. Internal validation does not ensure the success of a continuous monitoring application, which is a dynamic and open process, particularly, if one aims to use the Vis-NIRS in real world applications, being in inline grading systems or handheld devices. Once the model is applied to the test set and a final prediction is made, one can assess how well the model performs by computing several metrics. If the model was developed for regression, the most used are the root mean squared error (RMSE), bias, the slope, the coefficient of determination (R^2), and the standard deviation ratio (SDR) or ratio of performance to deviation (RPD) or residual standard deviation (RSD). If the model is developed for classification, the advised metrics are accuracy (ACC), F1 score and receiver operating characteristic (ROC) curve. For completeness, these metrics are often computed not only for the test set, but also for the calibration, and tuning sets as well. The comparison between calibration (C), tuning (CV) and test set (P) error metrics allows to understand how well the model generalizes, i.e. how the information learned by the model during training transposes to the final external validation dataset.

4. Prediction of quality attributes

Vis-NIRS combined with various chemometric methods has produced calibration models to predict simultaneously multiple QA of various citrus species and varieties, which are presented in Tables 2–4. These attributes range from fruit size, weight and color [90] to SSC, MI, external and internal defects, several compounds such as sugars, acids, pigments and antioxidants. As expected, these models address predominantly several varieties of orange and mandarin, but also grapefruit, lime, pomelo, sweet lemon and tangelo. The spectral ranges used cover the whole Vis-NIRS range. The majority of the Vis-NIRS calibration models were obtained from samples collected and assessed under controlled conditions in the laboratory, after fruit temperature equilibration, either with benchtop or handheld devices (Tables 2 and 3). Despite the large market availability of the latter, presenting different levels of portability, spectral ranges, sizes, and prices, only a few studies have focused on its application to assess the quality and ripening of oranges and mandarins on-tree (Table 4), perhaps due to the complexities involved under field conditions, and the performance deterioration of calibration models, in spite of the spectral range used [86, 87, 88, 89, 90]. Nevertheless, the QA assessed on-tree (Table 4) comprise fruit mass and size, color parameters [86, 89, 90], pericarp thickness, SSC, TA, firmness, MI, juice pH and mass, and BrimA index, which measures the balance between sweetness and acidity as described by [12]. Noteworthy, the majority of the calibration models exhibited R2<0.8, despite the range used, and did not include external validation, except for [56, 88].

Grading lines equipped with Vis-NIR sensors are now commercially available from various companies, to assess both the EQA and IQA of citrus fruit [38, 39, 40]. Unfortunately, the scientific evidence about the accuracy of these systems is very scarce, due to the 'industrial secrecy'. Nevertheless, there are a few cases of partnership among the industrial sector and the research groups to know the real applicability and their performance in assessing citrus fruit QA by such equipment [91], in most cases still in the prototype stage, as reported by [92, 93]. Valero [91] evaluated the performance of a customized NIR equipment installed underneath the fruit conveyor to sort oranges and mandarins in a Spanish packinghouse. This system working in a transmittance mode in the 650–970 nm spectral range, only provided calibration models that could discriminate between low and high values of SSC for both mandarin (R = 0.76–0.86; SEP ~ 0.9 °Brix; RPD ~ 0.74) and orange (R = 0.87; SEP ~ 0.7 °Brix; RPD < 1.5). No acceptable models were obtained for TA in neither species. Miller and Zude [92] evaluated the performance of a Vis-NIRS system to assess the SSC of 'Indian River' red grapefruit and 'Honey' tangerine from Florida in a sorting inline prototype, with R^2 ranging from 0.15 to 0.67. The prototype percent correct classification averaged 85% for SSC at 10 °Brix and 79% for an 11 °Brix setpoint in the second-year of tests. Otherwise, [93] reported on the development and laboratory testing of the nondestructive citrus fruit quality monitoring prototype system, which consisted of a light detection and ranging (LIDAR) and Vis-NIRS sensors installed on an inclined conveyor for mimicking real-time fruit size and SSC measurement respectively, during harvest. Laboratory tests in 'Valencia' orange revealed that the system was applicable for instantaneous fruit size ($R^2 = 0.91$) and SSC ($R^2 = 0.677$, SEP = 0.48 °Brix) determination.

The various Vis-NIRS calibration models presented in this chapter, show different levels of accuracy, prediction and robustness for the various attributes, SSC being the most successfully IQA assessed at all spectral ranges, independently of the devices used (Tables 2–4). Both juice pH and vitamin C also seem easily assessed by devices operating in the Vis-SWNIRS range devices [78, 94, 95], but TA has been shown to require wavelengths range > 1000 nm [13, 96]. Additionally, calibration models for firmness have been difficult to obtain, although there are a few exceptions reported for several orange varieties, in the reflectance mode and in the ranges 500–1690 nm [89], and 1000–2500 nm [73]. Of course, the calibration models for specific compounds, such as sugars, acids or antioxidants, require in most cases the longer NIR spectral range [12, 54, 69].

Both external and internal defects in citrus have been successfully predicted by Vis-NIRS, such is the case of the section drying in tangerine (transmittance; 780 and 960 nm) [10], oleocellosis in 'Trovita' orange (reflectance; 400–1000 nm) [71], the freeze damage in sweet lemon (half-transmittance; 400–110 nm) [13], the rind pitting in 'Marsh' grapefruit (reflectance; 400–2500) [12] or the granulation in 'Shantian' pomelo (transmittance; 400–700) [74]. However, [97] was not able to predict the external disorder RBD in 'Nules Clementine' mandarin (interactance; range 450–1000 nm).

Among the chemometrics methods used to construct the calibration models, PLS is still the main linear regression technique used, and the one to produce the best models for the widest number of QA (Tables 2–4). However, there are some exceptions, regarding the use of non-linear techniques, which were shown to deliver calibration models with equivalent or even better prediction and accuracy for several QA than PLS. Among these, there is the WT-LSSVR, BP-NN and LS-SMV that provided models with higher prediction capacity for SS in 'Gannan' orange [81], SSC, TA and vitamin C in Nanfeng mandarin [78] and in 'Newhall' orange [96], respectively. The LOCAL algorithm has also shown to produce better models than MPLS for firmness and juice mass in 'Powell Summer Navel' orange [89], and in 'Clemenvilla' mandarin [90]. SIMCA, SVM and particularly PCA-ANN, also allowed to assess with a total accuracy >98% the freeze damage in sweet lemon [13] and PCA-GRNN allowed the assessment of the granulation in 'Honey' pomelo at a classification accuracy (CA) >95% [14].

Independently of the chemometrics technique used, for the majority of the models presented in Tables 2–4, the train and test fruit sets were chosen from the same batch, orchards or seasons. Even when the whole data set comprised fruit from several orchards, harvest season or citrus varieties, the usual approach was to randomly choose 80% of the whole data set to construct the calibration model and 20% to validate it, as reported by [90, 97]. A truly stringent external validation is thus required to have a realistic idea of the models' performance in orchard and/or cargo batch monitoring. External validation means validation through a dataset with a different origin (spatial or temporal) relatively to the datasets used in calibration. Nevertheless, there are some clear examples of this approach, such as previously reported for mandarin [76, 79, 97, 98, 99], orange [12, 56, 70, 72, 73, 79, 88, 97], and grapefruit [12, 70]. Without the effective external validation, it is not possible to know exactly how well these models would work in real conditions due to the large variability within the trees, orchards, sites and harvest seasons. Yet, a certain degree of deterioration of the initial model prediction is expected, which would warrant further attention. This has been reported by [56, 77, 88]. Yet, there is space and potential for improvement beyond the 'proof of concept', if one aims to use these devices on the daily routines of the orchards' management. This has been suggested by [56, 78] through model recalibration using a few fruits from the new harvest season/orchard, or by achieving a strong degree of robustness by constructing a multi-seasonal and multi-orchard model as reported by [89] for 'Newhall' orange, which will be much more advantageous when assessing the ripening of fruit on-tree.

5. Future research and perspectives

Vis-NIRS has been incorporated by a large number of companies in commercial applications to be used on inline, benchtop and handheld systems. However, there are several topics regarding the full potential and limitations of this technology that require attention and further research in order to provide the consistency warranted by the daily basis routines of the citrus supply chain when assessing fruit quality and ripening. Firstly, all researchers engaged in this area should report their results in a uniform way, particularly in what respects the obtained models' metrics [38]. This would allow a better understanding on the effective advances and contributions made by each study. Other models' metrics parameters, such as the prediction gain, may also be useful, as reported by [100]. Secondly, the calibration models' robustness must be addressed and solved through a stringent multi-year, multi-cultivar and multi-orchard validation, such as previously reported by [56]. The usual approach of validating calibration models with a random fraction of the total available data set, even when the models comprehend several varieties, orchard sites and harvest years does not ensure the success of a continuous monitoring application and delivers unrealistic performance metrics [76, 77]. The usual recalibration and spiking approaches used to improve the initial calibration models with a few fruits from independent data sets that will be then assessed, assume that those fruits used to recalibrate/update the model constitute a faithful representation of the new population and are common techniques in various commercial devices, for inline and benchtop systems. However, this becomes quite difficult to apply if one aims to monitor the on-tree fruit ripening evolution through time, for the fruit sampled in the first weeks cannot represent those to be measured in the last weeks of the harvest season. Thirdly, there is a large potential for models' improvement, by using the non-linear techniques of machine learning, and those of deep learning. Fourthly, there is much to understand on the effect of the rind in the assessment of the pulp IQA in citrus fruit, since the NIR radiation hardly gets to the fruit pulp, and both biochemical and optical properties have a major role to play in the spectral data acquired [47, 49, 95, 101, 102]. Fifthly, the calibration models should be able to predict attributes that are closer to the organoleptic evaluation of the fruit. It is the case of BrimA index, a better indicator of fruit sweetness that the SSC, which was satisfactorily predicted in orange, grapefruit and mandarin [12, 87]. Finally, the handheld devices must really be tested under field conditions, if one aims to assess the fruit on-tree, which is essential for the OHD decision.

6. Conclusions

The usefulness of Vis-NIRS combined with different chemometric techniques in the supply chain of citrus fruit is already quite extensive and growing, similarly to many other commodities. In this chapter the authors only addressed the classic spectral 'point' measurements, but it is quite clear that both inline, benchtop and handheld devices are used to assess nondestructively multiple QA in various citrus species and cultivars, with a clear predominance of orange and mandarin. Among these attributes, there are both EQA and IQA, as well as defects caused by various factors, such as physiological disorders. The devices available on the market are from various brands, operate in various ranges and present a wide variety of prices. Aside from the "proof of concept" made by many studies, that the authors tried to comprise as much as possible in this chapter, there are several issues that still need to be addressed by researchers, a major one being the need for a stringent external validation of the calibration models, in order to assure robustness and to fulfill with the essential requirements to include this technology in the daily routines of these crop supply chain. This is of the utmost importance when considering the assessment of fruit ripening on-tree to determine the optimal harvest date for each orchard, or sections of the orchard. This is highly significant based on the determinant effect of producing and harvesting the fruit at its best ripening stage, thus assuring the best quality throughout the whole postharvest and shelf-life. As a concluding remark, it is very important to add that these devices are of medium and high cost, and that are not the kind of technology to 'set and forget', as reiterated by [39], which demands not only for a budget to acquire the systems, but also to maintain them, and to keep the continuous update and improvement of the calibration models, that in most cases need the selling company assistance. Thus, there must be a cost–benefit that both the producers and packinghouses have to meet through the added commercial value to citrus fruit by these systems, and the consumer willingness to pay for fresh fruit graded in terms of IQA such as sugar content, acidity and nutraceutical properties.

References

1. FAO - Citrus fruit fresh and processed statistical bulletin, 66 pp. 2017. Available from: http://www.fao.org/publications [Accessed: 2020-10-20]

2. Liu Y, Heying E, Tanumihardjo SA. History, Global Distribution, and Nutritional Importance of Citrus Fruits. Comprehensive Reviews in Food Science and Food Safety. 2012; 11(6): 530–545. doi.org/10.1111/j.1541-4337.2012.00201.x

3. Codex Alimentarius - International Food Standards [Internet]. 2020 Available from: http://www.fao.org/fao-who-codexalimentarius/codex-texts/list-standards/En/ [Accessed: 2020-12-07]

4. ado J, Rodrigo MJ, Zacarías, L. Maturity indicators and citrus fruit quality. Stewart Postharvest Review. 2014; 2 (2): 1–6

5. adaniya MS. Citrus fruit: Biology, Technology and Evaluation. 1st Edition. Academic Press (Elsevier); San Diego; 2008. 576 p
6. Abbas F, Fares, A. Best management practices in Citrus production. In: Tennant P, Benkeblia N (Eds) Citrus II. Tree and Forestry Science and Biotechnology. 2009; 3 (Special Issue 1):1–11
7. Etebu E, Nwauzoma AB. A review on sweet orange (Citrus sinensis L Osbeck): health, diseases and management. American Journal of Research Communication. 2014; 2 (2): 33–70
8. Ferreyra MLF, Perotti V, Figueroa CM, Garrán S, Anderson PC, Vázquez D, Iglesias AA, Podestá FE. Carbohydrate metabolism and fruit quality are affected in frost-exposed Valencia orange fruit. Physiologia Plantarum 2006; 128(2): 224–236. doi.org/10.1111/j.1399-3054.2006.00744.x
9. Goldenberg L, Yaniv Y, Choi HJ, Doron-Faigenboim, A, Carmi N, Porat R. Elucidating the biochemical factors governing off-flavor perception in mandarins. Postharvest Biology Technology. 2016; 120: 167–179. doi.org/10.1016/j.postharvbio.2016.06.009
10. Peiris KHS, Dull GG, Leffler RG, Burns JK, Thai CN, Kays SJ. Nondestructive detection of selection drying, an internal disorder in tangerine. HortScience. 1998; 33: 310–312
11. Magwaza LS, Opara UL, Terry LA, Landahl S, Cronje PJ, Nieuwoudt H, Mouazen AM, Saeys W, Nicolaï B. Prediction of 'Nules Clementine' mandarin susceptibility to rind breakdown disorder using Vis/NIR spectroscopy. Postharvest Biology and Technology 2012; 74: 1–10. doi.org/10.1016/j.postharvbio.2012.06.007
12. Ncama K, Tesfay SZ, Fawole OA, Opara UL, Magwaza LS. Non-destructive prediction of 'Marsh' grapefruit susceptibility to postharvest rind pitting disorder using reflectance Vis/NIR spectroscopy. Scientia Horticulturae. 2018; 231: 265–271 doi.org/10.1016/j.scienta.2017.12.028
13. Moomkesh S, Mireei SA, Sadeghi M, Nazeri M. Early detection of freezing damage in sweet lemons using Vis/SWNIR spectroscopy. Biosystems Engineering. 2017; 164: 157–170. doi.org/10.1016/j.biosystemseng.2017.10.009
14. Sun X, Xu S, Lu H. Non-Destructive Identification and Estimation of Granulation in Honey Pomelo Using Visible and Near-Infrared Transmittance Spectroscopy Combined with Machine Vision Technology. Applied Sciences. 2020; 10: 5399. doi.org/10.3390/app10165399
15. .USDA - National Agricultural Statistics Service, Publications - 1999 Agricultural Statistics Annual [Internet]. 2018. Available from: https://www.nass.usda.gov/Publications/Ag_Statistics/1999/index.php [Accessed: 2020-12-07]
16. Srivastava S, Sadistap S. Non-destructive sensing methods for quality assessment of on-tree fruits: a review. Food Measure. 2018; 12: 497–526. doi.org/10.1007/s11694-017-9663-6
17. Moon DG, Joa JH, Moon YE, Seong KC, Kim CH, Ahn YK. Plant growth and fruit quality as affected by canopy locations in 'Shiranuhi' mandarin. Horticulture, Environment and Biotechnology. 2011; 52: 443. doi.org/10.1007/s13580-011-0004-0

18. Pires R, Cavaco AM, Antunes MD, Panagopoulos T, Brázio A, Afonso AM, Daniel M, Rosendo ML, Gonçalves I, Guerra R. Ripening Forecast of 'Ortanique' (Citrus reticulata Blanco x Citrus sinensis (L) Osbeck) based on georeferenced internal quality parameters. In: XXIII Meeting of the Spanish Society of Plant Physiology and the XVI Hispano-Portuguese Congress of Plant Physiology, 26–28 June, Pamplona, Spain. 2019. p156

19. Timilsina K, Tripathi KM. Chemical quality attributes of mandarin (Citrus reticulata Blanco) as affected by altitude and fruit bearing position in Kavre, Nepal Archives of Agriculture and Environmental Science. 2019; 4 (3): 319–325. doi.org/10.26832/24566632.2019.0403010

20. Cavaco AM, Cruz SP, Antunes MD, Guerra R, Pires R, Afonso AM, Brázio A, Silva L, Lucas MR, Daniel M, Panagopoulos T. Spatiotemporal modelling of the quality and ripening of two cultivars of "Algarve Citrus" orchards at different edaphoclimatic conditions. Postharvest Biology and Technology. 2021; 172: 111386. doi.org/10.1016/j.postharvbio.2020.111386

21. Barman K, Ahmad MS, Siddiqui MW. Factors affecting the quality of fruit and vegetables: recent understandings. In Siddiqui MW Editor. Postharvest Biology and technology of horticultural crops. Apple Academic Press (Taylor and Francis Group), Oakville; 2015. pp. 1–50

22. Colaço AF, Molin JP. Variable rate fertilization in citrus: a long-term study. Precision Agriculture. 2017; 18:169–191. doi.org/10.1007/s11119-016-9454-9

23. Ohashi Y, Ueyama H. Numerical simulations of winter cold damage to citrus fruits using the WRF model. Advances in Science and Research. 2019; 16:1–6. doi.org/10.5194/asr-16-1-2019

24. Gullo G, Dattola A, Vonella V, Zappia R. Effects of two reflective materials on gas exchange, yield, and fruit quality of sweet orange tree Citrus sinensis (L.) Osb. European Journal of Agronomy. 2020; 118:126071. doi.org/10.1016/j.eja.2020.126071

25. Canavari M, Centonze R, Hingley M, Spadoni R. Traceability as part of competitive strategy in the fruit supply chain. British Food Journal. 2010; 112 (2):171–186. doi.org/10.1108/00070701011018851

26. Kondo N. Automation on fruit and vegetable grading system and food traceability. Trends in Food Science & Technology. 2010; 21 (3): 145–152. doi.org/10.1016/j.tifs.2009.09.002

27. Haleem A, Khan S, Khan MI. Traceability implementation in food supply chain: A grey-DEMATEL approach. Information Processing in Agriculture. 2019; 6 (3): 335–348. doi.org/10.1016/j.inpa.2019.01.003

28. Conrad Z, Niles MT, Neher DA, Roy ED, Tichenor NE, Jahns L. Relationship between food waste, diet quality, and environmental sustainability. PLoS ONE. 2018;13(4): e0195405. doi.org/10.1371/journal.pone.0195405

29. Muth MK, Birney C, Cuéllar A, Finn SM, Freeman M, Galloway JN, Gee I, Gephart J, Jones K, Low L, Meyer E, Read Q, Smith T, Weitz K, Zoubek S. A systems approach to assessing environmental and economic effects of food loss and waste interventions

in the United States. Science of the Total Environment. 2019; 685:1240–1254. doi.org/10.1016/j.scitotenv.2019.06.230

30. Yahaya SM, Mardiyya AY. Review of Post-Harvest Losses of Fruits and Vegetables. Biomedical Journal of Scientific and Technical Research. 2019; 13(4):10192–10200. http://dx.doi.org/10.26717/BJSTR.2019.13.002448

31. FAO, IFAD and WFP. Achieving Zero Hunger: the critical role of investments in social protection and agriculture. 2015. Available from: http://www.fao.org/3/a-i4951e.pdf [Accessed: 2020-12-10]

32. Steensland A, Zeigler M. Productivity in Agriculture for a Sustainable Future. In: Campos H, Editor. The Innovation Revolution in Agriculture. Springer, Cham. 2021 p.33–69 http://doi-org-443.webvpn.fjmu.edu.cn/10.1007/978-3-030-50991-0_2

33. García-Ramos FJ, Valero C, Homer I, Ortiz-Cañavate J, Ruiz-Altisent M. Non-destructive fruit firmness sensors: a review. Spanish Journal of Agricultural Research. 2005; 3(1): 61–73

34. Nicolaï BM, Defraeye T, Ketelaere B, Herremans E, Hertog MLATM, Saeys W, Torricelli A, Vandendriessche T, Verboven P. Nondestructive Measurement of Fruit and Vegetable Quality. Annual Review of Food Science and Technology. 2014; 5:285–312 doi.org/10.1146/annurev-food-030713-092410

35. Beghi R, Buratti S, Giovenzana V, Benedetti S, Guidetti R. Electronic nose and visible-near infrared spectroscopy in fruit and vegetable monitoring. Reviews in Analytical Chemistry. 2017; 20160016 doi.org/10.1515/revac-2016-0016

36. Cubero S, Lee WS, Aleixos N, Albert F, Blasco J. Automated Systems Based on Machine Vision for Inspecting Citrus Fruits from the Field to Postharvest–a Review. Food and Bioprocess Technology. 2016; 9:1623–1639. doi.org/10.1007/s11947-016-1767-1

37. Kittiyanpunya C, Phongcharoenpanich C, Krairiksh M. Dual-Frequency Sensor for Thick Rind Fruit Quality Assessment. IEEE Transactions on Instrumentation and Measurement. 2020; 69 (10):7617–7626. https://ieeexplore.ieee.org/document/9057693

38. Walsh KB, Blasco J, Zude-Sasse M, Sun X. Visible-NIR 'point' spectroscopy in postharvest fruit and vegetable assessment: The science behind three decades of commercial use. Postharvest Biology and Technology 2020; 168: 111246 doi.org/10.1016/j.postharvbio.2020.111246

39. Walsh KB, McGlone VA, Han DH. The uses of near infra-red spectroscopy in postharvest decision support: A review. Postharvest Biology and Technology 2020; 163: 111139 doi.org/10.1016/j.postharvbio.2020.111139

40. Cattaneo TMP, Stellari A Review: NIR Spectroscopy as a Suitable Tool for the Investigation of the Horticultural Field. Agronomy. 2019; 9: 503. doi:10.3390/agronomy9090503

41. Magwaza LS, Opara UL, Nieuwoudt H, Cronje PJR, Saeys W, Nicolaï B. NIR Spectroscopy Applications for Internal and External Quality Analysis of Citrus Fruit - a Review. Food and Bioprocess Technology. 2012; 5:425–444

42. Wang H, Peng J, Xie C, Bao Y, He Y. Fruit Quality Evaluation Using Spectroscopy Technology: A Review. Sensors. 2015; 15: 11889–11927 doi:10.3390/s150511889

43. Torricelli A, Contini D, Dalla Mora A, Martinenghi E, Tamborini D, Villa F, Tosi A, Spinelli L. Recent advances in Time-Resolved NIR spectroscopy for nondestructive assessment of fruit quality. Chemical Engineering Transactions. 2015; 44: 43–48 doi: 10.3303/CET1544008

44. Lu R, Van Beers R, Saeys W, Li C, Cen H. Measurement of optical properties of fruits and vegetables: A review. Postharvest Biology and Technology. 2020; 159:111003. doi.org/10.1016/j.postharvbio.2019.111003

45. Nicolaï BM, Beullens K, Bobelyn E, Peirs A, Saeys W, Theron KI, Lammertyn J. Nondestructive measurement of fruit and vegetable quality by means of NIR spectroscopy: a review. Postharvest Biology and Technology 2007; 46:99–118

46. Tuchin VV. Tissue optics and photonics: Light-tissue interaction. Journal of Biomedical Photonics & Engineering. 2015; 1(2): 98–135. doi: 10.18287/jbpe-2015-1-2-98

47. Sun C, Van Beers R, Aernouts B, Saeys W. Bulk optical properties of citrus tissues and the relationship with quality properties. Postharvest Biology and Technology. 2020; 163: 111127. doi.org/10.1016/j.postharvbio.2020.111127

48. Wang A, Hu D, Xie L. Comparison of detection modes in terms of the necessity of visible region (VIS) and influence of the peel on soluble solids content (SSC) determination of navel orange using VIS-SWNIR spectroscopy. Journal of Food Engineering. 2014; 126:126–132 doi.org/10.1016/j.jfoodeng.2013.11.011

49. Sun C, Aernouts B, Van Beers R, Saeys W. Simulation of light propagation in citrus fruit using monte carlo multi-layered (MCML) method. Postharvest Biology and Technology. 2021; 291: 110225. doi.org/10.1016/j.jfoodeng.2020.110225

50. Arendse E, Fawole OA, Magwaza LS, Opara UL. Non-destructive prediction of internal and external quality attributes of fruit with thick rind: A review. Journal of Food Engineering. 2018; 217: 11–23

51. Teerachaichayut S, Ho HT. Non-destructive prediction of total soluble solids, titratable acidity and maturity index of limes by near infrared hyperspectral imaging. Postharvest Biology and Technology. 2017; 133: 20–25 http://dx.doi.org/10.1016/j.postharvbio.2017.07.005

52. Savitzky A, Golay MJE. Smoothing and Differentiation of Data by Simplified Least Squares Procedures. Analytical Chemistry. 1964; 36 (8): 1627–1639 doi: 10.1021/ac60214a047

53. Cavaco AM, Pinto P, Antunes MD, Silva JM, Guerra R. 'Rocha' pear firmness predicted by a Vis/NIR segmented model. Postharvest Biology and Technology. 2009; 51(3): 311–319 doi.org/10.1016/j.postharvbio.2008.08.013

54. Borba KR, Spricigo PC, Aykas DP, Mitsuyuki MC, Colnago LA, Ferreira MD. Non-invasive quantification of vitamin C, citric acid, and sugar in 'Valencia' oranges using infrared spectroscopies. Journal of Food Science and Technology. 2020 doi.org/10.1007/s13197-020-04589-x

55. Castañeda-Loaiza, V, Magalhães, T, Cavaco AM and Guerra, R, 2021. Comparison of Vis/NIR interactance and NIR reflectance spectroscopies for the assessment of internal quality attributes of 'Rocha' pear (Pyrus communis L.) and 'Valencia Late' oranges (Citrus sinensis (L.) Osbeck). Under preparation

56. Cavaco AM, Pires R, Antunes MD, Panagopoulos T, Brázio A, Afonso AM, Silva L, Lucas MR, Cadeiras B, Cruz SP, Guerra R. Validation of short wave near infrared calibration models for the quality and ripening of 'Newhall' orange on tree across years and orchards. Postharvest Biology and Technology. 2018; 141: 86–97. doi.org/10.1016/j.postharvbio.2018.03.013

57. Gauglitz G, Vo-Dinh T., editors. Handbook of spectroscopy, 2003. WILEY-VCH Verlag GmbH & Co. KGaA, Weinheim

58. Golic M, Walsh K, Lawson P. Short-Wavelength Near-Infrared Spectra of Sucrose, Glucose, and Fructose with Respect to Sugar Concentration and Temperature. Applied Spectroscopy 2003; 57(2): 139–145

59. Norris KH, Williams PC. Optimization of mathematical treatments of raw near-infrared signal in the measurement of protein in hard red spring wheat. I. Influence of particle size. Cereal Chemistry.1984(2); 61: 158–165

60. Chau F-T, Liang Y-Z, Gao J, Shao X-G, Winefordner J, editors. Chemometrics: From Basics to Wavelet Transform. Chemical Analysis Series, Vol. 164. John Wiley & Sons Inc.; 2004. 316p. doi: 10.1002/0471454745

61. Mark H, Workman J. Chemometrics in Spectroscopy. 1st ed. Cambridge: Academic Press; 2007. 558p. doi: 10.1016/B978-0-12-374024-3.X5000-4

62. Brereton RG. Chemometrics for Pattern Recognition. John Wiley & Sons, Ltd; 2009. 512p. doi: 10.1002/9780470746462

63. Sun D-W, editor. Infrared Spectroscopy for Food Quality Analysis and Control. Elsevier Inc.; 2009. 448p. doi: 10.1016/B978-0-12-374136-3.X0001-6

64. Franca AS, Nollet L. editors. Spectroscopic Methods in Food Analysis. CRC Press; 2018. 663p. doi: 10.1201/9781315152769

65. Rinnan A, van der Berg F, Engelsen SB. Review of the most common pre-processing techniques for near-infrared spectroscopy, Trends in Analytical Chemistry. 2009; 28 (10): 1201–1222. doi.org/10.1016/j.trac.2009.07.007

66. Martens H, Jensen SA, Geladi P. Multivariate linearity transformations for near infrared reflectance spectroscopy. In: OHJ Christie (Editor), Proc. Nordic Symp. Applied Statistics, Stokkland Forlag, Stavanger, Norway. 1983. pp. 205–234

67. Martens H, Stark E. Extended multiplicative signal correction and spectral interference subtraction: New preprocessing methods for near infrared spectroscopy. Journal of Pharmaceutical and Biomedical Analysis. 1991; 9(8): 625–635. doi.org/10.1016/0731-7085(91)80188-F

68. Barnes RJ, Dhanoa MS, Lister SJ. Standard normal variate transformation and de-trending of near-infrared diffuse reflectance spectra. Applied Spectroscopy. 1989; 43:772. doi.org/10.1366/0003702894202201

69. Olarewaju OO, Magwaza LS, Nieuwoudt H, Poblete-Echeverría C, Fawole OA, Tesfay SZ, Opara UL. Model development for non-destructive determination of rind biochemical properties of 'Marsh' grapefruit using visible to near-infrared spectroscopy and chemometrics. Spectrochimica Acta Part A: Molecular and Biomolecular Spectroscopy. 2019; 209: 62–69. doi:10.1016/j.saa.2018.10.027

70. Ncama K, Opara UL, Tesfay SZ, Fawole OA, Magwaza LS. Application of Vis/NIR spectroscopy for predicting sweetness and flavour parameters of 'Valencia' orange (Citrus sinensis) and 'Star Ruby' grapefruit (Citrus x paradisi Macfad). Journal of Food Engineering. 2017; 193: 86–94. http://dx.doi.org/10.1016/j.jfoodeng.2016.08.015

71. Zheng Y, He S, Yi S, Zhou Z, Mao S, Zhao X, Deng L. Predicting oleocellosis sensitivity in citrus using VNIR reflectance spectroscopy. Scientia Horticulturae. 2010; 125 (3): 401–405 doi:10.1016/j.scienta.2010.04.036

72. Magwaza LS, Opara UL, Terry LA, Landahl S, Cronje PJR, Nieuwoudt HH, Hanssens A, Saeys W, Nicolaï BM. Evaluation of fourier transform-NIR spectroscopy for integrated external and internal quality assessment of Valencia oranges. Journal of Food Composition and Analysis. 2013; 31: 144–154. http://dx.doi.org/10.1016/j.jfca.2013.05.007

73. Bizzani M, Flores DWM, Colnago LA, Ferreira MD. Non-invasive spectroscopic methods to estimate orange firmness, peel thickness, and total pectin content. Microchemical Journal. 2017; 133: 168–174. http://dx.doi.org/10.1016/j.microc.2017.03.039

74. Xu S, Lu H, Ference C, Qiu G, Liang X. Rapid Nondestructive Detection of Water Content and Granulation in Postharvest "Shatian" Pomelo Using Visible/Near-Infrared Spectroscopy. Biosensors. 2020; 10, 41. doi.org/10.3390/bios10040041

75. Li Y, Jin G, Jiang X, Yi S, Tian X. Non-destructive determination of soluble solids content using a multi-region combination model in hybrid citrus. Infrared Physics & Technology. 2020; 104: 103138 doi.org/10.1016/j.infrared.2019.103138

76. Guthrie JA, Walsh KB, Reid DJ, Liebenberg CJ. Assessment of internal quality attributes of mandarin fruit. 1. NIR calibration model development. Australian Journal of Agricultural Research. 2005; 56: 405–416

77. Guthrie JA, Reid DJ, Walsh KB. Assessment of internal quality attributes of mandarin fruit. 2. NIR calibration model robustness. Australian Journal of Agricultural Research.. 2005; 56: 417–426

78. Xudong S, Hailiang Z, Yande L. Nondestructive assessment of quality of Nanfeng mandarin fruit by a portable near infrared spectroscopy. International Journal of Agricultural and Biological Engineering. 2009; 2(1): 65–71

79. Magwaza LS, Landahl S, Cronje PJR, Nieuwoudt HH, Mouazen AM, Nicolaï BM, Terry LA, Opara UL. The use of Vis/NIRS and chemometric analysis to predict fruit defects and postharvest behaviour of 'Nules Clementine' mandarin fruit. Food Chemistry. 2014; 163: 267–274 doi: http://dx.doi.org/10.1016/j.foodchem.2014.04.085

80. Li P, Li S, Du., Jiang L, Liu X, Ding S, Shan Y. A simple and nondestructive approach for the analysis of soluble solid content in citrus by using portable visible to near-infrared spectroscopy. Food Science & Nutrition. 2020; 00:1–10. doi.org/10.1002/fsn3.1550

81. Liu Y, Gao R, Hao Y, Sun X, Ouyang A. Improvement of Near-Infrared Spectral Calibration Models for Brix Prediction in 'Gannan' Navel Oranges by a Portable Near-Infrared Device. Food and Bioprocess Technology. 2012; 5: 1106–1112. doi.org/10.1007/s11947-010-0449-7
82. Tian X, Li J, Yi S, Jin G, Qiu X, Li Y. Nondestructive determining the soluble solids content of citrus using near infrared transmittance technology combined with the variable selection algorithm. Artificial Intelligence in Agriculture. 2020; 4: 48–57 doi.org/10.1016/j.aiia.2020.05.001
83. Song J, Li G, Yang X, Liu X, Xie L. Rapid analysis of soluble solid content in navel orange based on visible-near infrared spectroscopy combined with a swarm intelligence optimization method. Spectrochimica Acta Part A: Molecular and Biomolecular Spectroscopy. 2020; 228: 117815 doi: 10.1016/j.saa.2019.117815
84. Jamshidi B, Minaei S, Mohajerani E, Ghassemian H. Reflectance Vis/NIR spectroscopy for nondestructive taste characterization of Valencia oranges. Computers and Electronics in Agriculture. 2012; 85: 64–69
85. Antonucci F, Pallottino F, Paglia G, Palma A, D'Aquino S, Menesatti P. Non-destructive Estimation of Mandarin Maturity Status Through Portable VIS-NIR Spectrophotometer. Food and Bioprocess Technology. 2011; 4:809–813 doi.org/10.1007/s11947-010-0414-5
86. Sánchez MT, De la Haba MJ, Pérez-Marín D. Internal and external quality assessment of mandarins on-tree and at harvest using a portable NIR spectrophotometer. Computers and Electronics in Agriculture. 2013; 92: 66–74
87. Torres I, Sánchez MT, De la Haba MJ, Pérez-Marín D. LOCAL regression applied to a citrus multispecies library to assess chemical quality parameters using near infrared spectroscopy. Spectrochimica Acta Part A: Molecular and Biomolecular Spectroscopy. 2019; 217: 206–214 doi.org/10.1016/j.saa.2019.03.090
88. Zude M, Pflanz M, Kaprielian C, Aivazianc, BL. NIRS as a tool for precision horticulture in the citrus industry. Biosystems Engineering. 2008; 99: 455–459. doi.org/10.1016/j.biosystemseng.2007.10.016
89. Sánchez MT, De la Haba MJ, Serrano I, Pérez-Marín D. Application of NIRS for nondestructive measurement of quality parameters in intact oranges during on tree ripening and harvest. Food Analytical Methods. 2013; 6: 826–837. doi.10.1007/s12161-012-9490-7
90. Torres I, Pérez-Marín D, De la Haba MJ, Sanchéz MT. Developing universal models for the prediction of physical quality in citrus fruits analysed on-tree using portable NIRS sensors. Biosystems Engineering. 2017; 153: 140–148. http://dx.doi.org/10.1016/j.biosystemseng.2016.11.007
91. Valero C, Ortiz-Cañavate J, Riquelme MT, Homer I. Evaluation of Commercial Equipment for On-Line and Laboratory Quality Sorting of Citrus. Frutic Chile. "8th Fruit, Nut, and Vegetable Production Engineering Symposium". 2009
92. Miller WM, Zude M. NIR-based sensing to identify soluble solids content of Florida citrus. Applied Engineering in Agriculture. 2004; 20: 321–327

93. Yamakawa M, Khot LR, Ehsani R, Kondo N. Real-time nondestructive citrus fruit quality monitoring system: development and laboratory testing. Agricultural Engineering International: CIGR Journal. 2012; 14(3): 117–124

94. Gómez AH, He Y, Pereira AG. Non-destructive measurement of acidity, soluble solids and firmness of Satsuma mandarin using Vis/NIR-spectroscopy techniques. Journal of Food Engineering. 2006; 77: 313–319. doi.org/10.1016/j.jfoodeng.2005.06.036

95. Jamshidi B, Minaei S, Mohajerani E, Ghassemian H. Prediction of Soluble Solids in Oranges Using Visible/Near-Infrared Spectroscopy: Effect of Peel. International Journal of Food Properties. 2014; 17(7): 1460–1468. doi.org/10.1080/10942912.2012.717332

96. Liu C, Yang SX, Deng L. A comparative study for least angle regression on NIR spectra analysis to determine internal qualities of navel oranges. Expert Systems with Applications. 2015; 42: 8497–8503. http://dx.doi.org/10.1016/j.eswa.2015.07.005

97. Magwaza LS, Opara UL, Cronje PJR, Landahl S, Nieuwoudt HH, Mouazen AM, Nicolaï BM, Terry LA. Assessment of rind quality of 'Nules Clementine' mandarin during postharvest storage: 1. Vis/NIRS PCA models and relationship with canopy position. Scientia Horticulturae. 2014;165: 410–420. doi.org/10.1016/j.scienta.2013.09.035

98. Miyamoto K, Kitano Y. Non-Destructive Determination of Sugar Content in Satsuma Mandarin Fruit by near Infrared Transmittance Spectroscopy. Journal of Near Infrared Spectroscopy. 1995. 3: 227–237. doi: 10.1255/jnirs.73

99. 99.Ou AS, Lin S, Lin T, Wu S, Tiarn M. Studies on the determination of quality-related .constituents in Ponkan Mandarin by near infrared spectroscopy. Journal of the Chinese Agricultural Chemical Society. 1997. 35: 462–474

100. Passos D, Rodrigues D, Cavaco AM, Antunes MD, Guerra R. Non-Destructive Soluble Solids Content Determination for 'Rocha' Pear Based on VIS-SWNIR Spectroscopy under 'Real World' Sorting Facility Conditions. Sensors. 2019; 19:5165. doi.org/10.3390/s19235165

101. Wang A, Wang C, Xie L. Influence of the peel on predicting soluble solids content of navel oranges using visible and near-infrared spectroscopy. Transactions of the ASABE. 2016; 59(1): 31–37 doi.10.13031/trans.59.11152

102. Sun C, Aernouts B, Saeys W. Effects of harvest time, fruit size and cultivar on the bulk optical properties of Satsuma mandarin. Postharvest Biology and Technology. 2020; 111412 doi.org/10.1016/j.postharvbio.2020.111412

Chapte-14

Electrochemical Applications for the Antioxidant Sensing in Food Samples Such as Citrus and Its Derivatives, Soft Drinks, Supplementary Food and Nutrients

1. Introduction

Free radicals occur when an atom or molecule contains one or more unpaired electrons in its outermost orbitals [1]. Basically, three main factors play a role in the formation of free radicals. i) The atoms or molecules can become radical as a result of the fragmentation of covalently bonded molecules exposed to high-energy electromagnetic waves or high temperatures. ii) A molecule that does not have a radical feature experience an electron loss and radicals are formed by leaving unpaired electrons in its outer orbital. iii) A radical is formed when a molecule that does not have a radical property receives an electron from outside and has an unpaired electron in its outer orbital [1, 2]. These unshared electrons as known radicals are highly unstable, transforming them into high-energy and very efficient chemical species. The most active free radicals in biological systems are those based on oxygen and are commonly referred to as reactive oxygen species (ROS) with pathological [3]. This family group includes superoxide radical ($O_2^{\cdot}$), singlet oxygen, nitroxide (NO), hydroxyl radical ($OH^{\cdot}$), and hydrogen peroxide (H_2O_2) which is not itself radical but causes the formation of radical [1]. Besides, we can classify the causes of free radicals in two groups as endogenous or exogenous [1, 2]. Cigarettes, air pollution, alcohol, radiation, heavy organic solvents and pesticides are among exogenous sources, while enzymes, proteins, oxidative stressors, and heavy metals are endogenous sources [1, 4].

Free radicals cause the greatest damage to human health on basic cellular components such as lipids, proteins and nucleic acids [1, 5]. Therefore, these radicals lead to immune deficiency, hypertension and even important diseases such as cancer, neurodegenerative diseases, heart disease, and atherosclerosis [1, 2]. Also, studies are revealing that radicals disturb the homeostatic balance [6]. To scavenge these drawbacks effects of radicals, which are extremely important for human health, the human body needs antioxidants obtained from the body or nutrition to fulfill biological activities such as survival and healthy life. Antioxidants can be defined as molecules that usually contain phenolic functional groups in their structure and prevent the formation of free radicals that damage the cell or by scavenging existing radicals [3]. The functional task of antioxidants is that they act as shields in the body and neutralize them by donating their electrons with the s-free radicals. Thus, radicals found in a rather unstable structure do not become a threat to human health by transforming into a more stable structure reacted with antioxidants. Moreover, many different equivalent antioxidant expressions are used in antioxidant quantification in food samples. The leading ones are the expressions of "total antioxidant capacity (TAC)", "antioxidant activity (AA)", and "antioxidant capacity (AC)". The total amount of antioxidants is expressed by measurement units such as equivalent trolox, rutin, ascorbic acid, and quercetin, etc.

Antioxidants are mainly obtained via natural and synthetic [7]. The first of these, natural antioxidants, are molecules synthesized by the organism or obtained from food sources. Natural antioxidants produced by the organism are the most important source for human health. Many factors affect the production process of this natural antioxidant. The most important of these is the age of the person. As a person gets older, the amount of natural antioxidants produced by his organism decreases day by day. For this reason, there is a greater need for the natural antioxidants found in foods for older people. The importance of healthy food sources, especially organic-based foods, is increasing day by day. Also, such nutrients should be accessible to all segments of society.

Important dietary flavonoid sources are fruits especially citrus fruits such as oranges, apples, grapes, mandarins, berries lemons, limes and their derived products as well as juices [8]. In general, citrus fruits contain pectin, sugar, carotenoid pigments, vitamins (A, B1, and C), and; organic acids such as ascorbic acid and citric acid, minerals and a number of active phytochemicals such as flavonoids and coumarins, as naringenin, naringin, hesperidin, neohesperidin, hesperetin, rutin, narirutin and tangeretin [9]. For example; polyphenol antioxidants such as flavanols (epicatechin, catechin), phenolic acids (caffeic acid and gallic acid), anthocyanins (e.g., malvidin-3-glucoside), oligomeric and polymeric proanthocyanidins, flavonols (myricetin, quercetin, and their glycosides), and many others polyphenols exist in wine, especially in red wine [10]. Flavonoids have an important role in scavenging reactive oxygen species, which can counteract lipid oxidation, decrease peroxide formation in vivo, and improve activity of the body's antioxidant enzyme. Citrus flavonoids such as naringin, naringenin, and hesperidin have antioxidant activity [11]. Naringenin is a flavonoid, particularly a flavanone, found in citrus fruits especially oranges and grape fruits and in vegetable's such as tomatoes and their preparations. The pharmacological and biological properties of phytoestrogen naringenin and its derivatives include, anticancer, anti-inflammatory, antiulcer, antifibrotic, diastolic, antioxidant and skin protective effects [8]. Also, citrus species are a rich source of flavanone glycosides such as hesperidin and narirutin, which have anticancer, antioxidant, antiobesity and anti-inflammatory activities [12].

Secondly, the antioxidant group is synthetic, that is a molecule that is obtained as a result of chemical reactions and is generally used as food preservatives [13]. Synthetic antioxidants such as butylated hydroxytoluene (BHT), butylated hydroxyanisole (BHA), and tertiary butylhydroxyquinone (TBHQ) also extend the shelf life of foods [14]. However, natural antioxidants that can be taken from foods are less risky in terms of human health since synthetic antioxidants can have toxicity even if they are very little, they require high costs and have less capacity than natural antioxidants. Due to this reason, the investigations of foods types that can contain high levels of antioxidants in different types of endemic, organic and traditional food samples have been remarkably increased recently.

For antioxidant content and amount analyzes, oxygen radical absorbance capacity (ORAC) and radical-arrest antioxidant parameter (TRAP), ferric thiocyanate (FTC), Trolox equivalent antioxidant capacity (ABTS/TEAC), cupric ion (Cu^{2+}) reduction antioxidant capacity (CUPRAC), iron ion reducing antioxidant capacity (FRAP), DPPH radical scavenging activity determination and Folin–Ciocalteu methods are the most widely preferred as analytical methods [15, 16, 17]. Furthermore, to evaluate and characterize the antioxidant substances in food samples, various analytical methods such as high-pressure liquid chromatography (HPLC) combined with different detection, gas chromatography, micellar electrokinetic capillary chromatography, capillary electrophoresis includes different detection systems and UV–visible spectrophotometry have been used [18, 19, 20]. However, these classical methods have great shortcomings for fully validated analyzes such as long pre-treatment, need for too much solvent, expensive equipment, long analysis time. They do not provide the necessary procedures for green chemistry, especially due to the use of too much solvent and too much waste in antioxidant analyses. For these reasons, scientists have turned to alternative methods for antioxidant quantification in food samples. Especially in recent years, they have focused on electrochemical techniques which are fast, inexpensive, reliable, non-pre-treatment, and environmentally friendly in the analysis of drugs, pesticides, metal ions and organic molecules such as antioxidants, vitamins and nucleic acid [21, 22, 23].

In this chapter, the applicability, sensitivity and reliable maintenance of electrochemical methods, which have attracted great attention in food and food samples, have been examined for the analysis of antioxidants. Moreover, which types of electrochemical methods are used and what advantages they provide have been investigated for the antioxidant sensing in food samples. It also describes the classification of each used in electrochemical methods by working electrode type, dynamic operating range, the limit of detection (LOD), measurement limit (LOQ), sample type, and standard analyte, etc. While many articles referenced for determining antioxidants by electrochemistry have gained momentum in the literature in the last two decades, we focused our study on the studies conducted in the last 4 years.

2. Electrochemistry

Electrochemistry is the branch of science which is investigating the physical and chemical changes coming from the interaction of the material with electrical factors such as current, potential, and electron charge. Electroanalytical chemistry is based on measuring the electrical properties of solutions containing analytes and switching to quantification using measured electrical signals a collection of electrochemical methods. Moreover, electroanalytical measurement methods are based on two basic points: potentiometric (static methods) and potentiostatic (dynamic methods). Electrode systems in both methods are immersed in the solution containing the analyte, called the electrochemical cell. Potentiostatic methods are widely used for routine analysis because they are less costly, high sensitive, and selective and have wider potential application areas than other electroanalytical methods. The basic principle of these methods is to measure the current that occurs during the oxidation or reduction of the analyte in the chemical reaction.

Electrochemical methods began with the Czech chemist Jaroslav Heyrovsky, discovering the basis of polarography in 1922 and took an important place among the analytical methods. Especially, since the 1980s, it has been possible to develop electrodes that have been modified mechanically or chemically with improved technology. In modification processes, polymers, organic ligands, inorganic clays, phthalocyanines and nanoparticles have been commonly used for the detection of electroactive substances in very small volume complex samples such as biological, environmental and human bodies. In the last twenty years, even very small quantities of substances that are electroactive have been additionally analyzed at high precision, selective by electrochemical methods by carbon-based or modified electrodes have wonderful properties. Electroanalytical methods have also an important place in quantification as well as in obtaining details such as determination, adsorption, reaction rate and equilibrium constants of the number of electrons transferred in the reduction or oxidation electrode reactions. In short, electroanalytical methods provide details on direct or indirect quantitative and qualitative analysis of electroactive species such as antioxidants, drugs, pesticides, etc.

2.1 Voltammetric application for the determination of antioxidant capacity

Voltammetry is a potentiostatic assay based on the recording of the peak current at controlled potential variation by the oxidation or reduction which enables qualitative and quantitative analysis by means in electrochemical reactions. Over the last two decades compared to other electroanalytical techniques, voltammetry has been intensely curious in all the electroanalytical methods due to their are used to analyze numerous compounds by anodic or cathodic scanning and to investigate their conceptual basis of electro-mechanism. There are four voltammetric techniques including cyclic (CV), linear (LSV), differential (DPV), and square (SWV) are commonly used to determination of antioxidant-type compounds.

Voltammetric techniques are an alternative analytical method, proved to have an excellent correlation compare with another conventional analytical process, for a while to study the AC in various food and beverage samples. They can be a benefit to characterize which species compounds have a greater contribution to the antioxidant capacity present for the real samples in terms of quantitative and qualitative by controlled the half-wave peak potential, peak current and the electron transfer number in reaction. The antioxidant capacity is related to the peak currents of oxidation species caused by hydroxyl groups (–OH) and antioxidant species contains many hydroxyl groups. They commonly give an electro-oxidation broad peak at a range of 400 mV- 600 mV depend on pH. So that, almost all antioxidant substances have electro-activity compounds and their peak current and peak potential provide quantitative and qualitative details, respectively. Further, the voltammetric techniques allow investigating the electrochemical behavior of antioxidant agents and interaction with oxygenated species.

Voltammetric methods have gained an important place among determinations of the antioxidant capacity in the last decade. Moreover, due to their great superiority, the use of complex samples such as food and beverages they have become widespread and widely found in the literature. Among these electroanalytical methods, square wave stripping, different pulse stripping, and cyclic voltammetric techniques are the most commonly preferred for the analysis of antioxidants by accuracy and precision. From past to these days, the compounds used as standard agents for the evolution of the AC by studies electrochemical methods are apigenin, ascorbic acid, caffeine, catechin, chlorogenic chrysin, p-coumarin acid, eugenol, fisetin, gallic acid, kaempferol, luteolin, morin, quercetin, rutin, t-resveratrol, Trolox and Malvidin-3-glucoside. As far as we have examined the literature, scientists have however preferred ascorbic acid, caffeic acid, gallic acid, catechin, rutin and quercetin which are often used as antioxidant standard substances due to excessive availability of these substances in food and drink. The chemical structures of some antioxidant molecules are given in Figure 1.

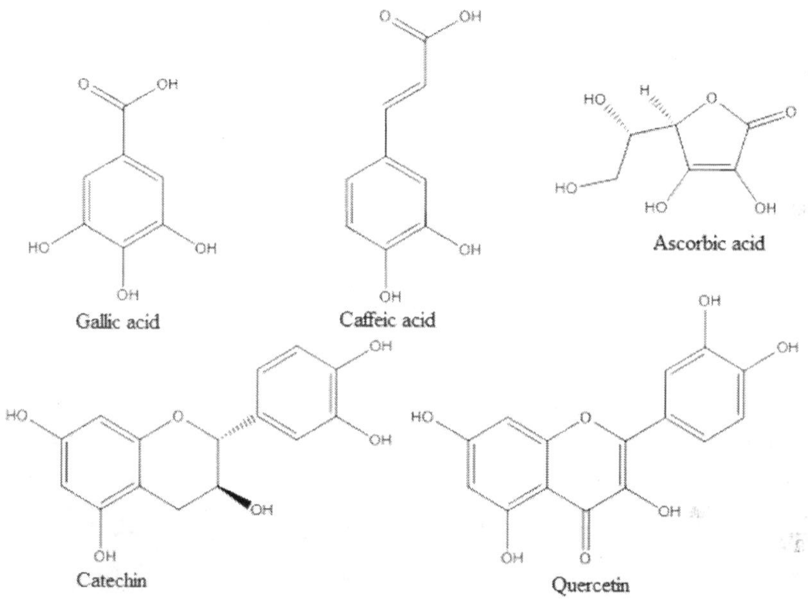

Figure 1. Molecular formulas of commonly used antioxidants.

2.1.1 Cyclic voltammetric technique

Cyclic voltammetry (CV) is usually the first experiment in the electrochemical operation of a compound in biological materials as nature samples to get in details about the electro-behaviors. In particular, to study the thermodynamics, kinetic, electron transfer, substance transfer type, and as well as quantitative determinations of oxidation or reduction processes can be carried out by cyclic voltammetric technique. In addition to taking a single measurement with CV, sequential multiple measurements can be taken. The most common applications of cyclic voltammetry are additionally electro-polymerization, electrochemical characterization, and the design of modified electroanalytical systems. Two types of cyclic voltammograms can be obtained as irreversible or reversible, depending on the chemical components of the target molecules. In reversible voltammetry, there is a difference of about 59 mV between the reduction and oxidation peak potentials (Figure 2).

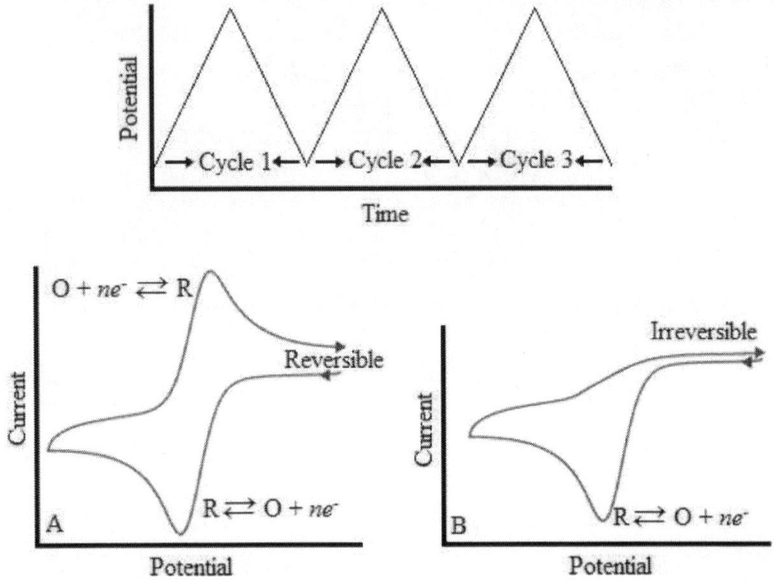

Figure 2. Potential-excitation signal and voltammograms for the cyclic voltammetry in details.

During the past years, cyclic voltammetry has been used as an alternative to existing methods to evaluate the antioxidant sensing in natural samples such as teas, biological fluids, beverage juices plants, foods and beverage juices on different working electrodes. The most using parameter is peak current because of its proportional to the concentration of the antioxidants. Peak current heights also provide quantitative information about the amount of antioxidant capacity in food samples. The carbon-based working electrodes such as glassy carbon electrode (GCE), carbon paste electrode (CPE), screen printed carbon electrode (SPCE), and modified electrodes (Nanoparticle/GCE, Nanoparticle/CPE, Fe_3O_4/GCE) have been widely preferred in electrochemical measurements for the analysis of total antioxidant capacity (TAC). Peak current and peak potential values of standard substances such as ascorbic acid, caffeic acid, catechin, coumarin, gallic acid, morin, quercetin and rutin were commonly taken care of for the evaluation of TAC. The amount of antioxidants in food samples is generally given as equivalent gallic acid, equivalent value quercetin, etc.

Even though the CV method raises doubts about sensitivity, it also has great advantages. Quick, simple, low detection limit, cheaper and easier application are summarized as great advantages. Interferences effect on antioxidant capacity by a non-antioxidant agent to reducing TAC and non-selective to a family of molecules between carotenoids and polyphenols unless the electrode is modified are drawbacks properties. Despite all of these disadvantages, CV attracts a great deal of attention among analytical methods, and a large number of studies deal with CV are also being undertaken. A large part of the work done up to day time to determine the antioxidant capacity by the CV method is summarized in Table 1. Table 1 includes the type of working electrode, working range, the limit of determination (LOD), the limit of quantification (LOQ), measurement parameter, standard compound and food sample.

Table 1. Evaluation of antioxidant capacity by CV technique.

Method	Electrode	Analyte	Linear range	LOD	LOQ	Samples	Optimum pH	Peak Potentials	Interferences	Ref
CV	Pt electrode	juglone (5-hydroxy-1,4,-naphthoquinone)	—	—	—	walnut	—	2.1 V	—	[24]
CV	GCE	polyphenols	—	—	—	Black Tea Samples *Camellia sinensis*	pH 7.0 (PBS)	+ 0.5 V	Catechin, gallic acid	[25]
CV	Graphite Paste Electrode	Rutin (RT)	200–1000 μM	89,4 μM		pharmaceutical sample (Captopril)	pH 4.0 (PBS)	+ 0.44 V	—	[26]
CV	Nanotuned Gold Nanoparticles and Solvothermally Reduced Graphene modified GCE (GCE/EAuNPs4 / rGO/Naf.)	sinapic acid (SA)	20 μM - 200 μM	33.43 (±0.21) nM		Human urine samples	pH 7.6 (PBS)	0.47 V	L-cystine, glycine, alanine, serum albumin, uric acid, citric acid, ascorbic acid, and urea	[27]
CV	glassy carbon electrode	caffeic acid, chlorogenic acid, quercetin, gallic acid, (+)-catechin, ascorbic acid	—	—	—	Apricot pomace extracts	pH 4 acetate buffer	0.51 V	—	[28]
						black currant pomace extracts		0.54 V		
						Grape pomace extracts.		0.48 V		
CV	glassy carbon disc electrode	Tannins	—	—	—	wine solution	—	—	—	[29]
CV	glassy carbon electrode	polyphenols and flavonoids	—	—	—	Venezuelan propolis	pH 7.00 (PBS)	−0.90 V (cathodic) −0.75 V (anodic)	—	[30]

CV	carbon paste electrode	Trolox	—	1.9 μM	0.6 μM	red wine, coffee and green tea	pH 7.0 (PBS)	—	—	[31]
CV	Carbon Paste Electrode (CPE)	Quinizarin (H2Qz)	0–36 μM	3.129 ± 1.200 μM	10.429 ± 1.133 μM	—	pH 7.00 (Aqueous)	—	Anthrarufin (H2Arf), Chrysazine (H2Cz), Anthraflavin (H2Afv)	[32]
CV	glassy carbon electrode (GCE)	polyphenols	—	—	—	Malaysian honey	pH 7 (PBS)	—	glucose and fructose	[33]
CV	ZnO nanoflowers modified carbon paste electrode	p-nitrophenol (p-NP)	0.1–1 μM	0.08 μM	—	Astragalus membranaceus	pH 7.0 (PBS)	—	—	[34]
CV	carbon nanotube (CNT)-carboxymethylcellulose (CMC) electrode MWCNT-CMC/Au	Curcumin	1.0–48 μM	0.21 μM	—	Real samples	pH 6.0 citric acid	0.30 V	—	[35]
CV	GCE	polyphenols, tannins, flavonoids, and sterols/triterpènes.	—	—	—	*Thymus vulgaris*	pH 7 (PBS)	—	—	[36]
CV	carbon screen printed electrode (cSPE)	Ethoxyquin (EQ)	20–100 mM	7.5 mM	20.0 mM	Salmon Samples	pH 3.5 ammonium formate buffer	+0.45 V	BHA, BHT, diphenylamine, and ascorbic acid (AA)	[37]
CV	carbon nanotube (CNT)-carboxymethylcellulose (CMC) electrode	monohydroxycinnamic acid	1.0–194 μM	0.071 μM	—	real food samples	pH 6.0 citric acid	—	—	[38]

CV	CPE	catechol, (CAT)	30.0–540 μM	2.47 μM	8.24 μM	wine and food samples	pH 7.4 (PBS)	0.24 and 0.46 V	—	[39]
		4-ethylcatechol (4-EC)	10.0–350 μM	0.282 μM	0.339 μM					
		4-ethylguaiacol (4-EG)	1.00–210 μM	0.111 μM	0.371 μM					
	CPME-CNT	CAT	30.0–540 μM	1.37 μM	4.58 μM					
		4-EC	10.0–350 μM	0.184 μM	0.613 μM					
		4-EG	1.00–120 μM	0.106 μM	0.353 μM					
	CPME-AB	CAT	30.0–540 μM	1.85 μM	6.16 μM					
		4-EC	0.20–350 μM	0.0863 μM	0.288 μM					
		4-EG	1.00–120 μM	0.0937 μM	0.312 μM					
CV	HP-ZnO/GCE	Gallic Acid (GA)	0.1–130 μM	0.02 μM	—	Wine sample	pH 3.0 (PBS)	+0.59 V	catechol (CT), dopamine (DA), caffeic acid (CA), morin (MR), hydroquinone (HQ), uric acid (UA), ascorbic acid (AA), ferulic acid (FA)	[40]

CV	glassy carbon electrode/ poly(3,4-ethylenedioxythiophene)-gold nanoparticles-sinusoidal voltage (GC/ PEDOT-AuNPs-SV)	caffeic acid (CA)	10 μM - 1 mM	4.24 (±0.12) μM	—	juice samples (like peaches and apple juices)	pH 7 (PBS)	—		[41]
CV	carbon electrodes	piperine	5 mM	—	—		pH 1.2 $HClO_4$			[42]
CV	Poly(3,4-ethylenedioxythiophene)- tyrosinase PEDOT-Tyr	Caffeic acid (CA)	10–300 μM	4.33 μM	14.43 μM	Wines and beers	0.1 M H2SO4	0.22 V	—	[43]
CV	glassy carbon electrode (GCE)	Catechin	0.1 mM	—	—	grape skin and seed	pH 3.6 tartaric acid buffer	483 mV	—	[44]
		Caffeic acid						445 mV		
		Gallic acid						472 mV		
		Oenin chloride						652 mV		
		Rutin						260 mV		
CV	Single Walled Carbon Nanotubes modified Screen Printed Carbon Electrodes (SWCNT-SPCE)	Catechin	0.1 mM	—	—	grape skin and seed	pH 3.6 tartaric acid buffer	132 mV	—	[44]
		Caffeic acid						139 mV		
		Gallic acid						122 mV		
		Oenin chloride						377 mV		
		Rutin						201 mV		

CV	Carbon nanofibers CNF	Caffeic acid (CA)	0.1–40 μM	3,23 nM	10,77 nM	Active Detox, DVR-Stem Glycemo, and green tea	pH 3.6 (PBS)	—	uric acid, ferulic acid, vanillic acid, gallic acid, and catechol	[45]
CV	Graphene/Neutral Red -GCE	UA	0.5–50 μM	0.076 μM	—	human urine and blood serum sample	—	—	urine and blood serum samples	[46]
CV	PE-DOT(poly(3,4-ethylenedioxythiophene) /GCE	UA	6–100 μM	7 μM	23 μM	milk sample	pH 6.6 (PBS)	—	A-lactose, L-aspartic acid, L-glutamic acid, L-histidine	[47]
		AA	30–500 μM	45 μM	149 μM			—		
CV	ZnO-graphene/ ITO	UA	5-80 μM	5.0 μM	—	—	1 M H_2SO_4	—	AA	[48]
CV	GOx-chitosan / Co3O4/Au-graphene transistors (GOx-CHIT/ Co3O4 modified SGGT)	UA	0,3-3 μM	0.1 μM	—	real tear samples	PBS	—	AA, Fructose, Xylose, Mannose	[49]

2.1.2 Square wave voltammetric technique

Square wave voltammetry (SWV) can be used to perform a faster experiment than other voltammetric techniques. Commonly when the scanning speeds of other techniques are of 1–10 mV/second or more, in the square wave voltammetry a scanning speed is used at 1 V/second. Thus, the target molecule can be analyzed more quickly by SWS. The square wave voltammetry can combine with the stripping technique. Thus, a stripping voltammetric technique was developed to determine electroactive substances at high sensitive enables in ultra-trace concentration levels. Especially, ultra-trace target substances in complex samples can be analyzed by combining the technique with the enrichment stripping process. The working principle of the stripping technique is the same as square wave voltammetry and only two new parameters are more applied as the accumulation time and the accumulation potential (Figure 3).

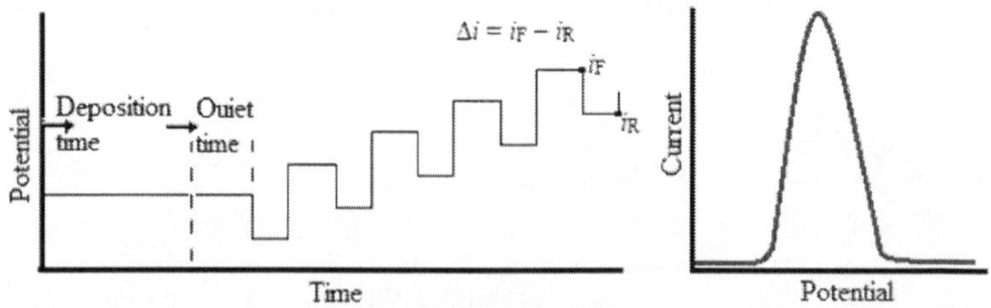

Figure 3. Potential-excitation signal and voltammogram for the square wave stripping voltammetry in details.

Nowadays SWV and square wave stripping voltammetry (SWSV) are frequently applied to deduce compounds such as drugs, heavy metals, pesticides and antioxidants, etc. in numerous specimen types because they have excellent analytical sensitivity and selectivity. Furthermore, SWV and its derivate combined technique can be applied for simultaneous determination of compounds which are close oxidation or reduction peak potentials like paracetamol, ascorbic acid, uric acid and dopamine. In the last decade, SWV and SWSV have been more effective in determining antioxidant substances in the complex matrix samples and are superior compared with analytical methods especially spectrophotometric to evaluate quantification and qualification. It is one of the most important electroanalytical methods for the determination of antioxidants since it is a wide working range, low detection limits, easy to apply, cheap and non-pretreatment. Furthermore, they have been successfully analyzed the phenols in food samples which is called a type of important antioxidant such as o-phenylenediamine, p-chlorophenol, p-aminophenol hydroquinone, pyrocatechol and phenol, etc. At the same time, various antioxidant substances such as gallate, gallic acid, quercetin and caffeine were easily studied in food or beverage samples at high precision, accuracy and selective on the carbon-based electrode. Besides, at nM concentration of antioxidant substances comparable to chromatographic techniques have been determined by modified electrodes which are increasing conductivity accurately and selectively in tea samples. Evaluation of antioxidant capacity by SWV or SWSV techniques in the last 4 years are summarized in Table 2 according to the type of working electrode, working range, the limit of detection (LOD), quantity limit (LOQ), measurement parameter, standard composition and food sample.

Table 2. Evaluation of antioxidant capacity by SWV or SWSV.

Method	Electrode	Analyte	Linear range	LOD	LOQ	Samples	Optimum pH	Peak Potentials	Interferences	Ref
SWV	glassy carbon electrode modified with graphite/bismuth (III) oxide (Gr/Bi_2O_3/GCE)	Ellagic acid (EA)	—	0.07 nM	0.21 nM	walnut and pomegranate	pH 3.0 (BRB)	0.6 V	inorganic ions (Na^+,K^+,Ca^{2+}, Cl^-, SO_4^{2-}, CO_3^{2-}), Glucose, Fructose, Eugenol, Capscasin	[50]
SWV	CPE/PAG	quercetin (QRT)	0.099–1.090 μM	0.029 μM	—	crude natural fruits (orange, apple and onion)	pH 6.0 (PBS)	0.18–0.22 V (ox and red)	Aspartic acid (ASP), Gallic acid (GAL), Sucrose (SUC) and Tartaric acid (TAC)	[51]
		rutin (RT)		0.058 μM				0.31–0.30 V (ox and red)		
SWV	$TPCo_3O_4$&SWCNT@CPE	α-lipoic Acid	2–100 μM	0.37 μM	—	dietary supplements	pH 6 (BRB)	—	Vitamins (vitamin C, B2, and B6), possible ingredients in LA pharmaceutical formulations	[52]
SWV	pencil graphite electrode	naringenin (NGN)	75 nM-0,1 mM	44 nM	0,111 μM	Citrus juice, fruits and peel	pH 4.00 (KHPT)	—	—	[8]
SWV	Single-Walled Carbon Nanotube Modified Glassy Carbon Electrode (SWCNT/GCE)	Quercetin (QCT)	0.01–100 μM	0.007 μM	—	tea samples (tea:green, basil and black)	pH 5.0 (PBS)	—	—	[53]
SWV	Untreated boron doped diamond electrode (BDDE)	Sesamol	0.2 mM–1.0 mM	85 nM	—	tahini halva samples	pH 2.0 H_2SO_4	—	Cu^{2+}, Pb^{2+}, Cd^{2+}, Mg^{2+}, Ca^{2+}, K^+, Cl^-,and ascorbic acid and catechol, glucose, and fructose	[54]

Method	Electrode	Analyte	Linear range	LOD	LOQ	Samples	Optimum pH	Peak Potentials	Interferences	Ref
SWV	immobilization (in solution) of laccase onto the activated carboxylic groups of carboxymethyl-botryosphaeran (CMB). (CBPE-CMB/LCE) biosensor	Quercetin (QCT)	0.0498–0.794 μM	0.026 μM	—	Red wine Green tea Apple juice Lemon juice	pH 6.0 (PBS)	0.23 V	epinephrine, dopamine, paracetamol, guaiacol and catechol, uric acid and inorganic ions (Ca^{2+},Mn^{2+},Fe^{2+}, Zn^{2+},SO_4^{2-} and NO_3^-)	[55]
SWV	CNF-ZnO modified glassy carbon electrode (CNF-ZnO-GCE)	Silymarin	2–123 nM	1 nM	—	Human serum samples and urine samples were	pH 7.0 (PBS)	+0.20 V	NO_3^-,Na^+, Cu^{2+}, K^+, 4-nitrophenol, rutin, dopamine, caffeic acid, luteolin, tetracycline, hydrogen peroxide, glucose, ascorbic acid, epinephrine, uric acid, and quercetin	[56]
SWV	gold nanoparticle/ graphene quantum dots (AuNP/GQD) nanozyme–modified screen-printed carbon electrode (AuNP/GQDs/ SPCE)	Quercetin	0,1 nM - 1 mM	0,033 nM	0,1 nM	Human plasma	pH 5 (BRB)	—	glucose, sucrose, ascorbic acid, riboflavin, phenylalanine, L-tryptophan, L-tyrosine, bisphenol A, lysine, uric acid, and two metal ions such as Na + and Co2+	[57]
SWV	SWCNTs-SPCE	Polyphenols (caffeic acid, gallic acid, catechin and malvidin-3-glucoside	—	—	—	Wine samples	pH 3.6	—	—	[10]

Method	Electrode	Analyte	Linear range	LOD	LOQ	Samples	Optimum pH	Peak Potentials	Interferences	Ref
SWV and AdSV	boron-doped diamond electrode (CPT-BDDE)	5-O-Caffeoylquinic acid (5-CQA)	2.8 µM - 0,17 mM	0.4 µM	—	Food&beverage samples (vanilla-enriched instant coffee, vanilla sugar, cola soft drink)	0.1 M HNO3	0.68 V	caffeic acid, p-coumaric acid, gallic acid, ferulic acid, sinapic acid, and syringic acid, K^+, Na^+, Ca^{2+}, Mg^{2+}, Zn^{2+}, Cu^{2+}, Fe^{3+}, NO_3^-, Cl^- and SO_4^{2-}	[58]
		vanillin (VAN)	3.3 µM - 0,33 mM	0.38 µM				1.15 V		
		caffeine (CAF)	0.52 µM - 0,21 mM	0.15 µM				1.50 V		
SWASV	Gold disk	Cu(II) and TBHQ	5.66–113.37 µg/kg 4.76–92.40 mg/kg	0.351 µg/kg 1.13 mg/kg	—	—	pH 2 (BRB)	—	—	[59]

2.1.3 Differential pulse voltammetric technique

Differential pulse voltammetric technique (DPV) is one of the most widely used for the analysis of both organic and inorganic species. Pulse voltammetry techniques were proposed by Baker and Jenkin in 1952 as a more sensitive measurement electroanalytical method. Differential pulse voltammetry techniques can be used to determine up to 10^{-8} M concentration of the target agents. The peak current (I_p) is a function of the concentration for the electroactive species and is linear as $Ip = f(C)$. Also, it is possible to analyze substances not only quantitative analysis but also qualitative analysis with pulse technique. The peak currents are related to the concentration of the substance whereas the peak potential values are related to the selectivity. Thus, simultaneous determinations of the substances have been studied by DPV on bare or modified electrodes (Figure 4).

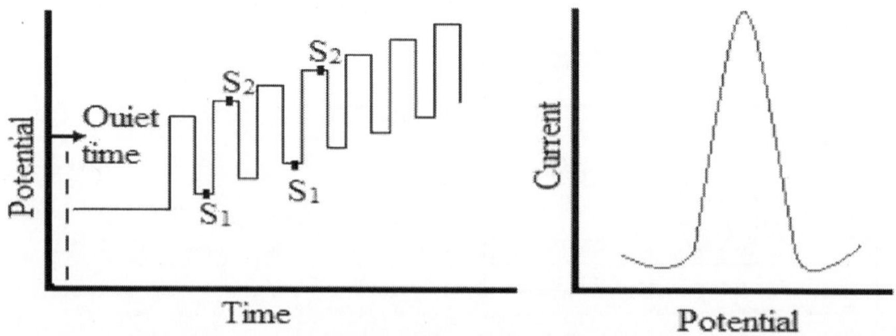

Figure 4. Potential-excitation signal and voltammogram for the differential pulse stripping voltammetry in details.

Nowadays, quite a lot of DPV studies can be found in the literature for the very sensitive detection of heavy metal, drug, pesticide, antioxidant agent and inorganic/organic species on numerous bare and modified working electrodes. Besides, DPV is one of the most important candidates to determine the trace amount of target agents in analytical methods due to its high sensitivity and selective. Also, it can be applied to complex samples as biological and food samples such as blood and serum, beverages. Especially, DPV has an important place among antioxidant determination methods because of these advantages and the availability of low concentration.

In recent years, DPV has been used frequently in determining the total antioxidant capacity without any pretreatment of solid and liquid food samples. The complex matrix such as biological and food samples contain very dense different types of substances. For this reason, despite it is indeed very difficult to selectively and precisely determine the antioxidant capacity in some complex matrixes; DPV is the most applicable method for such species. There are also plenty of studies were published which deal with chlorogenic acid, caffeic acid, p-coumaric acid, quercetin, gallic acid and ferulic acid, etc. as illustrating the antioxidant properties were determined by DPV on bare or modified electrodes based on carbon nanomaterials. Several applications, based commonly on the used as a determination of antioxidant capacity are given in Table 3.

Table 3. Evaluation of antioxidant capacity by DPV or DPSV.

Method	Electrode	Analyte	Linear range	LOD	LOQ	Samples	Optimum pH	Peak Potentials	Interferences	Ref
DPV	GCE/PoPD/Pt	Rosmarinic acid (RA)	3 μM – 7 μM	0.9 μM	—	Melissa officinallis, Rosmarinus officinals	pH 2 H_2SO_4	0,63 V	phenolic compounds-caffeic acid, ascorbic acid, coumaric acid, 2,5 dihydroxybenzoic acid, chlorogenic acid, rutin, and gallic acid	[60]
		protocatechuic acid (PCA)	2 μM – 70 μM	0.8 μM	—			0,53 V		
DPV	ZrO2NPs-AuNPs-DES/CPE	Caffeic acid (CA)	0.22–55 μM	25 nM	—	green tea and fruit juices	pH 3 (BRB)	—	—	[61]
DPV	Fluorine doped graphene oxide/GCE	Caffeic acid (CA)	0.5–100.0 μM	0.018 μM		wine	pH 2.65 (BRB)	—	p-coumaric acid, hydroquinone, trans-ferulic acid, gallic acid, glucose, and ascorbic acid	[11]
DPV	CuO nano-rice/GCE	UA	1–160 μM	1.2 μM	—	real samples of dopamine injection, human serum, and urine samples	pH 7 (PBS)	—	Glucose, Fructose, Galactose	[62]
		DA	1–150 μM	0.42 μM	—			—		
DPV	Poly(DPA)/SiO_2@Fe_3O_4/CPE	UA	1.2–8.2 μM	0.4 μM	1.2 μM	Fresh human serum samples	pH 7.0 (PBS)	0.3 V	Sucrose, DA, AA, Glucose, Folic acid	[63]
DPV	Carbon paste modified with Bi decorated multiwalled carbon nanotubes and cetrimonium bromide (CTAB)	Caffeic acid (CA)	0.06–500 μM	0.157 nM	1.910 nM	Coconut water, coffee, tea	pH 7.0 (PBS)		AA, UA, FA, Trp, Mor, GA, Glucose and FoA	[64]

Method	Electrode	Analyte	Linear range	LOD	LOQ	Samples	Optimum pH	Peak Potentials	Interferences	Ref
DPV	Bimetallic $CoFeSe_2$ nanosphere in functionalized carbon nanofibers $CoFeSe_2$/f-CNF	Caffeic acid (CA)	0.01–263.96 μM	0.002 μM	—	Red wine samples by	pH 7.0 (PBS)	0.21 V	catechol (CC), hydroquinone (HQ), epinephrine (EP), dopamine (DA), uric acid (UA), and ascorbic acid (AA)	[65]
DPV	Carbon/iron-based active catalyst f-MWCNTs/a-$NaFeO_2$	Caffeic acid (CA)	0.1–17.2 μM	0.002 μM	0.0068 μM	coffee, green tea, red wine	pH 7.0 (PBS)	—	catechol (CT), gallic acid (GA), ascorbic acid (AA), hydroquinone (HQ), and uric acid (UA)	[66]
DPV	N-doped carbon quantum dots/ hexagonal porous copper oxide decorated multiwall carbon nanotubes N-CQD/HP-Cu_2O/ MWCNT/GCE	Caffeic acid (CA)	0.05–43 μM	0.004 μM	—	red wine samples	pH 7.0 (PBS)	—	dopamine (DA), catechol (CC), ascorbic acid (AA), uric acid (UA) and epinephrine (EP)	[67]
DPV	Ce-TiO_2/carbon nanotube composite Ce-TiO_2/CNTs	Caffeic acid (CA)	0.001 -10 μM	0.0003 μM		caffeic acid tablets samples	pH 6.0 (PBS)	—	Cl^-, Br^-, $SO4^{2-}$, $NO3^{-}$, $H_2PO_4^-$, Na^+, K^+, Mg^{2+} and Al^{3+}, glucose, L-serine, uric acid, urea, oxalic acid, glycine, alanine, L-cysteine, L-tyrosine, L-glutamic acid, and guanidine acid	[68]

Method	Electrode	Analyte	Linear range	LOD	LOQ	Samples	Optimum pH	Peak Potentials	Interferences	Ref
DPV	MOF-818 metal–organic framework-reduced graphene oxide/ multiwalled carbon nanotubes composite MOF-818/ RGO/MWCNTs/ GCE	caffeic acid (CA), chlorogenic acid (CGA), and gallic acid (GA)	0.2–7 μM 7–50 μM	5,2 nM	—	human serum and urine samples	pH 3.0 (PBS)	—	Na^+, K^+, SO_4^{2-}, Cl^-, 40-fold glutamic acid, glycine, glucose, sucrose, urea, ascorbic acid, uric acid, and equal concentration of baicalein, luteolin, and vanillic acid	[69]
DPV	Fe_3O_4 @ZIF-4 nano-hybrid on a glassy carbon electrode (GCE) (Fe_3O_4 /GCE, ZIF- 4/ GCE)	p-coumaric acid (CA)	0.50–12.00 μM	0.18 μM	0.60 μM	orange juices samples	pH 4 (BRB)	0.71 V	anions, cations and other polyphenols such as SO_4^{2-}, NO_3^-, Cl^-, Fe^{3+}, Fe^{2+}, Zn^{2+}, Ni^{2+}, Cu^{2+}, Mg^{2+}, Ca^{2+}, K^+, Na^+, Li^+ ions, citric acid, glucose, catechin and quercetin	[70]
DPV	graphene modified screen-printed electrode	Melatonin	—	0.03 mg/L	—	food supplements	pH 7.4	0.268 (±0.014) V	—	[71]
DPV	multi-walled carbon nanotubes modified carbon paste electrode (MWCNTs/CPE).	quercetin (QU)	—	1.96 nM	—	Orange juice	pH 2.0 (BRB)	—	tannic acid (TA)	[72]
DPV	—	polyphenols	—	—	—	Black Tea Samples Camellia sinensis	pH 5.5 (PBS)	+ 0.5 V	Catechin, gallic acid	[25]
DPV	screen printed carbon electrode	gallic acid	0.1–2 mM	23–103 μM	70–310 μM	White wine Green tea Apple juice	pH 5.8; 7; 8 (PBS)	—	caffeic and ascorbic acid	[73]

Method	Electrode	Analyte	Linear range	LOD	LOQ	Samples	Optimum pH	Peak Potentials	Interferences	Ref
DPV	alumina-modified glassy carbon electrode GCE	Caffeic Acid (CA)	0.1–5 µM	0.004 µM	0.01 µM	Tea (Green, Black, Mint, Hibiscus, Rosemary), wine and phytotherapics	0.1 M $HClO^4$	0.519 ± 0.002 V (Green tea) 0.528 ± 0.002 V (Black tea) 0.526 ± 0.001 V (Mint tea) 0.533 ± 0.002 V (Hibiscus tea) 0.508 ± 0.001 V (Rosemary tea) 0.571 ± 0.005 V (Phytotherapic) 0.532 ± 0.001 V (Wine 1) 0.525 ± 0.002 V (Wine 2)	—	[74]
		Gallic Acid (GA)	0.1–5 µM	0.005 µM	0.02 µM					
		Catechin	0.1–5 µM	0.001 µM	0.003 µM					
		Quercetin (QCT)	0.1–15 µM	0.005 µM	0.02 µM					
DPV	nanoporous gold electrodes (NPG)	ascorbic acid (AA)	0.32 to 3.4 mM	63 µM	—	mimic human serum sample of fetal bovine serum	pH 7.4 (PBS)	0.05 V	—	[75]
		uric acid (UA)	0.065 to 1.5 mM	9.0 µM				0.35 V		
DPV	glassy carbon electrode (GCE)	Gallic Acid (GA)	19.92–98.04 ppm	—	—	mango (pulp, peel, and seed)	pH 5 (BRB)	0.445 and 0.550 V (oxidation peaks)	0.05 V	[76]
		Trolox	2.34–472.18 µM							
DPV	CoSe2@rGO modified SPCE	propyl gallate	0.075–460.15 µM	16.35 (±0.46) nM	—	spiked meat samples (Chicken, Beef)	pH 7.0 (PBS)	0.34 V	uric acid (UA), ascorbic acid (AA), dopamine (DA), hydroquinone (HQ), catechol (CT), epinephrine (EP), and norepinephrine (NEP)	[77]

Method	Electrode	Analyte	Linear range	LOD	LOQ	Samples	Optimum pH	Peak Potentials	Interferences	Ref
DPV	Nano-Graphene-platelets (nGp)-Brilliant green (Bg)/ Modified carbon paste electrode (nGp- Bg/MCPE)	Hesperidin (HES)	0.1–7.0 µM 7.0–100.0 µM	50.0 nM	—	Fortified Fruit Juice Samples (lemon juice, orange rind, and peppermint extract that contain HES)	pH 7.5 (PBS)	—	AA, Bioflavonoids (such as quercetin, rutin, naringenin, morin), Inorganic ions (Ca^{2+}, Mg^{2+}, K^{+}, Zn^{+}, Cu^{2+}, Cl^{-}, SO_4^{2-})	[78]
DPV	5-amino-2-mercapto-1,3,4-thiadiazole (p-AMT) on nitrogen-doped carbon sphere (N-CS) modified glassy carbon (GC) electrode p-AMT@N-CS/ GC (1) p-AMT@N-CS/ GC (simultaneous addition) (2)	Gallic Acid (GA)	5–1187 µM (1) 5–128 µM (2)	0.58 µM (1) 0.82 µM (2)	—	Grape juice samples	pH 7.0 (PBS)	+0.06 V (GA)	ascorbic acid (AA), uric acid (UA), catechol (CC), and hydroquinone (HQ), K^{+}, Na^{+}, Mg^{2+}, Cu^{2+}, Ni^{2+}, NO^{3-}, Cl^{-}, sulfate ion (SO_4^{2-}, and glucose	[79]
		Caffeic Acid (CA)	5–2082 µM (1) 5–128 µM (2)	0.143 µM (1) 0.30 µM (2)				+0.14 V (CA)		
DPV	SWCNT-Subphthalocyanine (CS) Hybrid Material modified GCE electrode (CS/ GCE)	catechin	0.1–1.5 µM	13 nM	43 nM	real tea samples (such as green, rosehip fruit, Turkish and Indian black tea)	pH 3 (BRB)	—	metal ions (such as K^{+}, Na^{+}, Li^{+}, Cu^{2+}, Ca^{2+}, Mg^{2+}, Fe^{2+}, Zn^{2+}, Cd^{2+}, Fe^{3+}), rutin, 6-methoxy flavone, gallic acid, caffeic acid, biomolecules (viz. caffeine, ascorbic acid, citric acid and glucose)	[80]

Method	Electrode	Analyte	Linear range	LOD	LOQ	Samples	Optimum pH	Peak Potentials	Interferences	Ref
DPV	Cobalt oxide nanoparticles-modified carbon-paste electrodes (CoO-NPs-CPE)	Gallic Acid (GA)	0,1–1 μM	1.52 μM	—	Red and White Wine	pH 2.0 (PBS)	0.61 V	Metals ions (K+, Cl–, Na+, Fe3+), ascorbic acid and quercetin	[81]
DPV	graphite/chemically modified silica ceramic electrode (SMICl/C)	Quercetin (QRT)	9–102 μM	3.2 μM	—	pharmaceutical “Quercetin”	Ethanol	0.102–0.155 V	—	[82]
			10–100 μM	3 μM			3:2 ethanol/ water	0.561–0.571 V		
			13–95 μM	4.4 μM			4:1 ethanol/ water	0.561–0.592 V		
			0.15–60 μM	0.46 μM			Water	0.134–0.155 V		
DPV	glassy carbon electrode modified with polyaminobenzene sulfonic acid functionalized single-walled carbon nanotubes (f-SWNT) and poly(pyrocatechol violet) (polyPCV/f-SWNT/GCE)	Gallic acid (GA)	0.75–10 10–100 μM	0.12 μM	0.41 μM	Cognac XO Brandy VS Brandy 5-Star	pH 2.0 (BRB)	0.48 V	K^+, Na^+, Mg^{2+}, Ca^{2+}, NO_3^-, Cl^-, and SO_4^{2-} and glucose, rhamnose, sucrose as well as ascorbic acid, phenolic aldehydes (vanillin, syringaldehyde)	[83]
		Ellagic Acids (EA)	0.75–7.5 7.5–100 μM	0.11 μM	0.37 μM			0.63 V		
DPV	Sodium dodecyl sulfate modified carbon composite paste electrode	Curcumin	0,2 - 1 μM 1.5 - 4.5 μM	27 nM	92 nM	Natural food supplement	pH 6.0 (PBS)	—	Na^+,K^+, Mg^{2+}, Zn^{2+}, ascorbic acid, glucose, starch, tyrosine and tartazine	[84]

Method	Electrode	Analyte	Linear range	LOD	LOQ	Samples	Optimum pH	Peak Potentials	Interferences	Ref
DPV	implemented functionalized-MWCNT/Nileblue-composite on carbon paste electrode (fMWNCT/NB/MCPE)	Naringenin (NR)	10.0–50.0 μM 0.9–10.0 μM	0.30 μM	0.93 μM	fruit juices(Grape juice, Tomato juice, Orange juice)	pH 7.0 (PBS)	—	AA, GLU, Na^+, Mg^{2+}, K^+, Ca^{2-}, Cl^-, SO^{4-}	[85]
DPV	vitreous carbon electrode	Trolox	50 μM to 600 μM	43.8 μM	120 μM	Greigia sphacelata fruit (Chupón or Quiscal)	pH 7.4 (PBS)	—	—	[86]
DPV	Screen Printed Carbon Electrodes	Polyphenols	—	—	—	Wine	pH 3.20 Tartaric Acid Solutions	—	—	[87]
DPV	Poly(L-Methionine)/Carbon Nanotube Glassy Carbon Electrode (PLM/MWCNT/GCE)	Gallic acid (GA)	0.004–1.1 μM 1.7–20 μM	3.1 nM	—	green tea, black tea, and red wine samples	pH 2.2 (BRB)	—	Na^+,K^+,Ca^{2+} Mg^{2+}, Zn^{2+}, Cu^{2+}, Ni^{2+}, ascorbic acid, theophylline, caffeine, cysteine, glucose, fructose, sucrose, and glycine	[88]
DPV	pencil graphite electrode	naringenin (NGN)	78,6 nM - 0,182 mM	30,6 nM	102 nM	citrus juice	pH 4.00 (KHPT)	—	—	[8]
DPSV	3D SWCNTs-coumarin hybrid modified glassy carbon electrode (3DSWCNTs-coumarin/GCE)	Quercetin (QCT)	0.25–3 μM	20 nM	66 nM	Tea samples	pH 2.0 (BRB)		ascorbic acid, caffeine, citric acid, l-cysteine, glycine, glucose, Na^+, Mg^{2+} Ca^{2+}, SO_4^{2-}, NO^{3-} and Cl^-, gallic acid, 6-methoxyflavon	[89]

Method	Electrode	Analyte	Linear range	LOD	LOQ	Samples	Optimum pH	Peak Potentials	Interferences	Ref
DPAdSV	unmodified screen-printed carbon electrodes (SPCEs)	Capsaicinoids	0.16 - 16.37 µM	0.05 µM	0.15 µM	fresh chili pepper samples (Meiren chili pepper, Chaotian green chili pepper, Chaotian red chili pepper, Xiaomi green chili pepper, and Xiaomi red chili pepper)	0.10 M HCl	0.40 V	Fe^{3+},Cu^{2+},K^{+},Na^{+}, Ga^{2+},Cl^{-},SO_4^{2-} and glucose, and 100-fold of Mg^{2+}	[90]
DpAdSV	screen-printed carbon electrode modified with single-walled carbon nanotubes (SWCNTs)) and Prussian blue (PB) coated with chitosan	Rutin	0.03 to 0.24 µM 0.25 to 2.0 µM	0.01 µM	—	black tea, coffee and synthetic drink of tea	pH 3.0 (PBS)	0.25 V (ox) 0.096 V (red)	morin and quercetin	[91]
DPCV	molecularly imprinted poly (p-aminobenzene sulphonic acid) on carbon nanodots coated pencil graphite electrode (FA-imp/CNDs/ PGE)	folic acid (FA)	2.2–30.8 ng/mL	2.02 ng/mL		drug tablets and human urine samples	pH 6.2 (PBS)	—	Methotrexate (MTX), folinic acid (FCA), tetrahydrofolic acid (THF), pyridoxine (PYR), and 5-methyltetrahydrofolate (5- THF)	[92]

In amperometric techniques, the current produced during the reduction or oxidation of an electroactive species at a constant potential value that is applied between a working electrode and reference electrode is measured, in this way providing specific quantitative electroanalytical knowledge for the target analyte. Especially, amperometric, which is based on electrical current analysis, is commonly utilized in microchip electrophoresis applications owing to its high sensitivity, it also lets for the determination of electroanalytical active species without derivatization, accomplishing adjustable versatility and selectivity (Table 4).

Table 4. Evaluation of antioxidant capacity by Amperometric technique.

Method	Electrode	Analyte	Linear range	LOD	LOQ	Samples	Optimum pH	Peak Potentials	Interferences	Ref
Amperometry (AMP)	HP-ZnO/GCE	Gallic Acid (GA)	0.1–130 μM	0.02 μM	—	Wine sample	pH 3.0 (PBS)	+0.48 V	catechol (CT), dopamine (DA), caffeic acid (CA), morin (MR), hydroquinone (HQ), uric acid (UA), ascorbic acid (AA), ferulic acid (FA)	[40]
Amperometry (AMP)	Nickel oxide nanoparticles modified glassy carbon electrode (NiO NPs/GCE)	Dopamine (DA)	—	11 μM	—	*Limonia acidissima* natural fruit juice	pH 7.2 (PBS)	0.41 V	—	[93]
amperometry (AMP)	Poly(L-Methionine)/ Carbon Nanotube Glassy Carbon Electrode (PLM/ MWCNT/GCE)	Gallic acid (GA)	0.002–0.1 μM 0.2–12 μM	0.5 nM	—	green tea, black tea and red wine samples	pH 2.2 (BRB)	0.5 V	Na^+,K^+,Ca^{2+} Mg^{2+}, Zn^{2+}, Cu^{2+}, Ni^{2+}, ascorbic acid, theophylline, caffeine, cysteine, glucose, fructose, sucrose, and glycine	[88]

Method	Electrode	Analyte	Linear range	LOD	LOQ	Samples	Optimum pH	Peak Potentials	Interferences	Ref
Amperometry	GCE/PoPD/Pt	Rosmarinic acid (RA)	1 μM – 55 μM	0.5 μM	—	Melissa officinallis, Rosmarinus officinals	pH 2 H_2SO_4	—	—	[60]
		protocatechuic acid (PCA)	1 μM – 60 μM	0.6 μM	—			—		
Chronoamperometry (CA)	Graphite/Lacc–PDA	gallic acid	1–150 μM	0.29 μM	—	Chestnut shell waste extract/TPC		—	—	[94]
		Caffeic acid	1–50 μM	0.14 μM	—			—		
		Rosmarinic acid	1–20 μM	0.09 μM	—			—		
Chronoamperometry (CA)	CuO nano-rice/ GCE	UA	0.83–253 μM	0.83 μM	—	real samples of dopamine injection, human serum and urine samples	pH 7 (PBS)	—	Glucose, Fructose, Galactose	[62]
		DA	0.083–428.8 μM	0.083 μM	—			—		

Ganesh et al., synthesized zinc oxide nanoparticles using mechanochemical synthesis technique. New ZnO nanoparticle as hexagonal prism was investigated by scanning electron microscopy, X-ray diffraction, particle size distribution, ultraviolet–visible spectroscopy, and energy-dispersive X-ray spectroscopic methods. Electrochemical properties of the newly prepared electrode were characterized by using an amperometric method and cyclic voltammetry technique. The prepared electrode has a wide working linear range between 0.1–130 μM with a detection limit of 0.02 μM. Obtained results showed that the prepared electrode has numerous active surface sites, good electronic activity, and surface area. They applied the proposed electrode to the determination of gallic acid in samples as wine successfully [40].

Kumar and coworkers successfully synthesized NiO nanoparticles from natural fruit using an efficient, simple, and low-cost technique. The obtained NiO nanoparticles were investigated with various methods such as FTIR, XRD, TEM, SEM, UV, and PL. XRD studies showed that NiO nanoparticles have cubic geometry. The band of Ni-O bond was shown at 430 cm^{-1}. Photocatalytic properties of the obtained NiO nanoparticles were applied to photodegrade the methylene blue dye. They used the prepared electrode to the determination of dopamine with the LOD of 11 μM [93].

Koçak et al. prepared a new composite electrode using carbon nanotube and poly-l-methionine onto the glassy carbon electrode. Electrochemical properties and surface structure of the prepared electrode were studied using electrochemical impedance spectroscopy and scanning electron microscopy. Electrochemical properties of gallic acid with the proposed electrode were investigated in various techniques such as differential pulse voltammetry, cyclic voltammetry and amperometry. The obtained results of electrochemical studies exhibited that the prepared electrode shows a suitable method of determination for gallic acid in pH 2.2 BR buffer solution. The prepared sensor has a wide working linear range with two linear segments between 4 nM-1.1 μM and 1.7–20.0 μM with LOD of 3.1 nM. They used the prepared new sensor for the detection of gallic acid in various samples as black tea, green tea and wine samples. The experimental results showed that the proposed sensor exhibit high selectivity, reproducibility, stability and catalytic effect [88].

Potentiometry is an electrochemical technique based on measuring the potential difference between two electrodes called working and reference electrodes. The working basis of the potentiometry technique is the potential difference based on the concentration of an analyte in the sample solution relative to a reference electrode (Table 5).

Table 5. Evaluation of antioxidant capacity by potentiometric technique.

Method	Electrode	Analyte	Linear range	LOD	LOQ	Samples	Optimum pH	Peak Potentials	Interferences	Ref
Potentiometry	—	Antioxidant	—	—	—	black and green tea microsuspensions	pH 7.2 (PBS)	—	—	[95]
Potentiometry	GCE	chicoric acid	—	—	—	Echinacea flowers	pH 7.4 (PBS)	—	—	[96]
Potentiometry	POM immobilization on the surface of a glassy carbon electrode	Polyoxometalates (POMs)	—	—	—	—	0.1 M HClO4	—	—	[97]

Table 6. Evaluation of antioxidant capacity by linear sweep voltammetry (LSV).

Method	Electrode	Analyte	Linear range	LOD	LOQ	Samples	Optimum pH	Peak Potentials	Interferences	Ref
LSV	Ionic liquid-rGO-titania-Nafion-GCE	capsaicin	0.03–10 µM	0.0032 µM	—	Korean hot pepper (Chungyang pepper) solution	pH 1.0 (BRB)	0.75 V	—	[98]
LSV)	gold disk electrode	2-tert-butylphenol (2-TBF)	9.12–80.83 µg cm^{-3}	0.67 µg /L	2.22 µg cm^{-3}	mineral and synthetic oils	0,16 M H_2SO_4	—	—	[99]

Table 7. Evaluation of antioxidant capacity by other techniques.

Method	Electrode	Analyte	Linear range	LOD	LOQ	Samples	Optimum pH	Peak Potentials	Interferences	Ref
EI electrochemical index	CPEs (carbon paste electrodes)	TAC	0.105–0.500 µM	40,4 nM	0,105 µM	olive oil samples	pH 7 (PBS)	—	—	[100]
PC peak current			8.02 x 10^{-2} – 0.500 µM	30,5 nM	80,2 nM					
Redox microsensor	Redox measurements	gallic acid	0.2–2 mM	49 µM	148 µM	White wine Serum	pH 5.8	—	—	[73]
			0.1–2 mM	109 µM	331 µM		pH 7			
			0.1–1.5 mM	74 µM	223 µM		pH 8			

3. Conclusion

Electrochemistry is a powerful and versatile analytical technique for the determination of numerous substances such as drugs, pesticides, inorganic, antioxidant-type compounds and electroactive compounds by rapidly possible applications in a lot of fields. Electroanalytical methods besides providing details on quantitative and qualitative of analyte that offer validation parameters such as sensitivity, accuracy and precision, selective and linear working range. Moreover, it is superior to determine the target analyte by electroanalytical methods lack of interferences effect especially in a complex matrix such as biological and food samples contain countless substances. The improvement of simultaneous determination of analytes considerably has been carried out to be applied in biological and environmental systems by the sensitive and selective electrochemistry methods. Because of this, the use of many areas of electrochemistry is widespread.

Nowadays, electrochemical methods, especially voltammetry from medicine to the determination of antioxidants, have made an important place especially in the world of science. Not only analytical chemists but also biology, food engineering and all people who are engaged in food have been used electrochemical methods to determine the antioxidant capacity in plants, tea, beverages, carbonated beverages and solid food samples, etc. Compounds such as ascorbic acid, caffeic, catechin, ascorbic acid, quercetin, gallic acid and coumarin have been widely used as reference standard agents to an evaluation of antioxidant capacity by electrochemical methods have been carried out until today. Due to advances in electronics and computer science have provided significant benefits in terms of electrochemical instrumentation such as accuracy, sensitivity and easy application, the electro-analysis of antioxidant compounds is successfully applied by stripping voltammetric techniques at nM concentration level. The purpose of this review is to show that electroanalytical methods for commonly used antioxidant types may be the best analytical method for the quantitative and qualitative analyte and that they can successfully compete with more conventional methods especially spectrometric methods. Consequently, voltammetric techniques supply that even at low concentrations, the antioxidant capacities of food samples can be determined to be very fast, simple, non-pretreatment and highly sensitive compared to conventional analytical methods. The review presented that the antioxidant capacity of various food samples can be carried out by voltammetric techniques in the estimation in real samples.

References

1. A. Phaniendra, D. B. Jestadi, and L. Periyasamy, "Free Radicals: Properties, Sources, Targets, and Their Implication in Various Diseases," Indian J. Clin. Biochem., vol. 30, no. 1, pp. 11–26, 2015, doi: 10.1007/s12291-014-0446-0
2. K. Rahman, "Studies on free radicals, antioxidants, and co-factors.," Clin. Interv. Aging, vol. 2, no. 2, pp. 219–236, 2007
3. V. Lobo, A. Patil, A. Phatak, and N. Chandra, "Free radicals, antioxidants and functional foods: Impact on human health," Pharmacogn. Rev., vol. 4, no. 8, pp. 118–126, 2010, doi: 10.4103/0973-7847.70902
4. G. S. B. Aseervatham, T. Sivasudha, R. Jeyadevi, and D. Arul Ananth, "Environmental factors and unhealthy lifestyle influence oxidative stress in humans--an

overview.," Environ. Sci. Pollut. Res. Int., vol. 20, no. 7, pp. 4356–4369, 2013, doi: 10.1007/s11356-013-1748-0

5. S. Kumar, "Free Radicals and Antioxidants: Human and Food System," Adv. Appl. Sci. Res., vol. 2, no. 1, pp. 129–135, 2011
6. J. S. Aprioku, "Pharmacology of free radicals and the impact of reactive oxygen species on the testis," J. Reprod. Infertil., vol. 14, no. 4, pp. 158–172, 2013
7. E. M. Atta, N. H. Mohamed, and A. A. M. Abdelgawad, "Antioxidants: an Overview on the Natural and Synthetic Types," Eur. Chem. Bull., vol. 6, no. 8, p. 365, 2017, doi: 10.17628/ecb.2017.6.365-375
8. I. G. David et al., "Voltammetric analysis of naringenin at a disposable pencil graphite electrode-application to polyphenol content determination in citrus juice," Anal. Methods, vol. 10, no. 48, pp. 5763–5772, 2018, doi: 10.1039/c8ay02281j
9. P. J. Divya, P. Jamuna, and L. A. Jyothi, "Antioxidant properties of fresh and processed Citrus aurantium fruit," Cogent Food Agric., vol. 2, no. 1, 2016, doi: 10.1080/23311932.2016.1184119
10. E. F. Newair, P. A. Kilmartin, and F. Garcia, "Square wave voltammetric analysis of polyphenol content and antioxidant capacity of red wines using glassy carbon and disposable carbon nanotubes modified screen-printed electrodes," Electroanalysis, vol. 32, no. 5, pp. 1–9, 2020, doi: 10.1155/2020/8869436
11. V. S. Manikandan, B. Sidhureddy, A. R. Thiruppathi, and A. Chen, "Sensitive electrochemical detection of caffeic acid in wine based on fluorine-doped graphene oxide," Sensors (Switzerland), vol. 19, no. 7, 2019, doi: 10.3390/s19071604
12. D. S. Kim and S. Bin Lim, "Extraction of flavanones from immature Citrus unshiu pomace: process optimization and antioxidant evaluation," Sci. Rep., vol. 10, no. 1, pp. 1–13, 2020, doi: 10.1038/s41598-020-76965-8
13. M. Carocho, P. Morales, and I. C. F. R. Ferreira, "Antioxidants: Reviewing the chemistry, food applications, legislation and role as preservatives," Trends Food Sci. Technol., vol. 71, no. November 2017, pp. 107–120, 2018, doi: 10.1016/j.tifs.2017.11.008
14. .J. Sanhueza, S. Nieto, and A. Valenzuela, "Thermal stability of some commercial synthetic antioxidants," JAOCS, J. Am. Oil Chem. Soc., vol. 77, no. 9, pp. 933–936, 2000, doi: 10.1007/s11746-000-0147-9
15. R. Apak, K. Güçlü, M. Özyürek, and S. E. Karademir, "Novel total antioxidant capacity index for dietary polyphenols and vitamins C and E, using their cupric ion reducing capability in the presence of neocuproine: CUPRAC method," J. Agric. Food Chem., vol. 52, no. 26, pp. 7970–7981, 2004, doi: 10.1021/jf048741x
16. D. Huang, O. U. Boxin, and R. L. Prior, "The chemistry behind antioxidant capacity assays," J. Agric. Food Chem., vol. 53, no. 6, pp. 1841–1856, 2005, doi: 10.1021/jf030723c
17. N. Chaves, A. Santiago, and J. C. Alías, "Quantification of the antioxidant activity of plant extracts: Analysis of sensitivity and hierarchization based on the method used," Antioxidants, vol. 9, no. 1, 2020, doi: 10.3390/antiox9010076

18. J. Karovičová and P. Šimko, "Determination of synthetic phenolic antioxidants in food by high-performance liquid chromatography," J. Chromatogr. A, vol. 882, no. 1, pp. 271–281, 2000, doi: 10.1016/S0021-9673(00)00353-8

19. .M. M. Delgado-Zamarreño, I. González-Maza, A. Sánchez-Pérez, and R. Carabias Martínez, "Analysis of synthetic phenolic antioxidants in edible oils by micellar electrokinetic capillary chromatography," Food Chem., vol. 100, no. 4, pp. 1722–1727, 2007, doi: 10.1016/j.foodchem.2005.10.018

20. R. Apak, M. Özyürek, K. Güçlü, and E. Çapanoɿlu, "Antioxidant activity/capacity measurement. 1. Classification, physicochemical principles, mechanisms, and electron transfer (ET)-based assays," J. Agric. Food Chem., vol. 64, no. 5, pp. 997–1027, 2016, doi: 10.1021/acs.jafc.5b04739

21. R. K. Mishra et al., "Wearable Flexible and Stretchable Glove Biosensor for On-Site Detection of Organophosphorus Chemical Threats," ACS Sensors, vol. 2, no. 4, pp. 553–561, 2017, doi: 10.1021/acssensors.7b00051

22. O. Inam, E. Demir, and B. Uslu, "Voltammetric Pathways for the Analysis of Ophthalmic Drugs," Curr. Pharm. Anal., vol. 16, no. 4, pp. 367–391, 2019, doi: 10.2174/15734129 15666190225163637

23. E. Demir, A. Senocak, M. F. Tassembedo-Koubangoye, E. Demirbas, and H. Y. Aboul-Enein, "Electrochemical Evaluation of the Total Antioxidant Capacity of Yam Food Samples on a Polyglycine-Glassy Carbon Modified Electrode," Curr. Anal. Chem., vol. 16, no. 2, pp. 176–183, 2018, doi: 10.2174/1573411014666180619143729

24. A. Masek, E. Chrzescijanska, M. Latos-Brozio, and M. Zaborski, "Characteristics of juglone (5-hydroxy-1,4,-naphthoquinone) using voltammetry and spectrophotometric methods," Food Chem., vol. 301, no. February, p. 125279, 2019, doi: 10.1016/j.foodchem.2019.125279

25. D. V. Thomaz et al., "Electrochemical study of commercial black tea samples," Int. J. Electrochem. Sci., vol. 13, no. 6, pp. 5433–5439, 2018, doi: 10.20964/2018.06.55

26. D. M. da Silva and M. C. da Cunha Areias, "Rutin as an Electrochemical Mediator in the Determination of Captopril using a Graphite Paste Electrode," Electroanalysis, vol. 32, no. 2, pp. 301–307, 2020, doi: 10.1002/elan.201900145

27. A. Kumar, B. Purohit, K. Mahato, S. Roy, A. Srivastava, and P. Chandra, "Design and Development of Ultrafast Sinapic Acid Sensor Based on Electrochemically Nanotuned Gold Nanoparticles and Solvothermally Reduced Graphene Oxide," Electroanalysis, vol. 32, no. 1, pp. 59–69, 2020, doi: 10.1002/elan.201900406

28. G. S. Vasyliev, V. I. Vorobyova, and O. V. Linyucheva, "Evaluation of Reducing Ability and Antioxidant Activity of Fruit Pomace Extracts by Spectrophotometric and Electrochemical Methods," J. Anal. Methods Chem., vol. 2020, 2020, doi: 10.1155/2020/8869436

29. A. Ricci, G. P. Parpinello, N. Teslić, P. A. Kilmartin, and A. Versari, "Suitability of the cyclic voltammetry measurements and DPPH• spectrophotometric assay to determine the antioxidant capacity of food-grade oenological tannins," Molecules, vol. 24, no. 16, pp. 1–12, 2019, doi: 10.3390/molecules24162925

30. L. G. Mohtar, G. A. Messina, F. A. Bertolino, S. V. Pereira, J. Raba, and M. A. Nazareno, "Comparative study of different methodologies for the determination the antioxidant activity of Venezuelan propolis," Microchem. J., vol. 158, no. May, p. 105244, 2020, doi: 10.1016/j.microc.2020.105244

31. J. Juárez-Gómez, M. T. Ramírez-Silva, D. S. Guzmán-Hernández, M. Romero-Romo, and M. Palomar-Pardavé, "Novel electrochemical method to evaluate the antioxidant capacity of infusions and beverages, based on in situ formation of free superoxide radicals," Food Chem., vol. 332, no. March 2019, p. 127409, 2020, doi: 10.1016/j.foodchem.2020.127409

32. A. K. Rivas-Sánchez, D. S. Guzmán-Hernández, M. T. Ramírez-Silva, M. Romero-Romo, and M. Palomar-Pardavé, "Quinizarin characterization and quantification in aqueous media using UV-VIS spectrophotometry and cyclic voltammetry," Dye. Pigment., vol. 184, no. August 2020, 2021, doi: 10.1016/j.dyepig.2020.108641

33. N. I. Ismail, S. Sornambikai, M. R. A. Kadir, N. H. Mahmood, R. M. Zulkifli, and S. Shahir, "Evaluation of Radical Scavenging Capacity of Polyphenols Found in Natural Malaysian Honeys by Voltammetric Techniques," Electroanalysis, vol. 30, no. 12, pp. 2939–2949, 2018, doi: 10.1002/elan.201800493

34. F. U. Khan et al., "An Astragalus membranaceus based eco-friendly biomimetic synthesis approach of ZnO nanoflowers with an excellent antibacterial, antioxidant and electrochemical sensing effect," Mater. Sci. Eng. C, vol. 118, no. August 2020, p. 111432, 2021, doi: 10.1016/j.msec.2020.111432

35. R. Wada, S. Takahashi, H. Muguruma, and N. Osakabe, "Electrochemical detection of curcumin in food with a carbon nanotube-carboxymethylcellulose electrode," Anal. Sci., vol. 36, no. 9, pp. 1113–1118, 2020, doi: 10.2116/analsci.20P021

36. S. Amamra et al., "Determination of total phenolics contents, antioxidant capacity of thymus vulgaris extracts using electrochemical and spectrophotometric methods," Int. J. Electrochem. Sci., vol. 13, no. 8, pp. 7882–7893, 2018, doi: 10.20964/2018.08.57

37. M. Vandeput et al., "Electrochemical Studies of Ethoxyquin and its Determination in Salmon Samples by Flow Injection Analysis with an Amperometric Dual Detector," Electroanalysis, vol. 30, no. 7, pp. 1293–1302, 2018, doi: 10.1002/elan.201700611

38. R. Wada, S. Takahashi, and H. Muguruma, "New perspective on ECE mechanism of monohydroxycinnamic acid oxidation with carbon nanotube electrode," Electrochim. Acta, vol. 359, 2020, doi: 10.1016/j.electacta.2020.136964

39. C. Kalinke et al., "Voltammetric Electronic Tongue Based on Carbon Paste Electrodes Modified with Biochar for Phenolic Compounds Stripping Detection," Electroanalysis, vol. 31, no. 11, pp. 2238–2245, 2019, doi: 10.1002/elan.201900072

40. K. Ganesh et al., "Synthesis and characterization of hexagonal prism like zinc oxide for electrochemical determination of Gallic acid in wine samples," Int. J. Electrochem. Sci., vol. 14, no. 5, pp. 4769–4780, 2019, doi: 10.20964/2019.05.40

41. D. Bottari et al., "Electrochemical sensing of caffeic acid using gold nanoparticles embedded in poly(3,4-ethylenedioxythiophene) layer by sinusoidal voltage

procedure," Chemosensors, vol. 7, no. 4, pp. 1–14, 2019, doi: 10.3390/chemosensors7040065

42. O. E. Carp, A. Moraru, M. Pinteala, and A. Arvinte, "Electrochemical behaviour of piperine. Comparison with control antioxidants," Food Chem., vol. 339, no. September 2020, p. 128110, 2021, doi: 10.1016/j.foodchem.2020.128110
43. J. J. García-Guzmán et al., "Assessment of the polyphenol indices and antioxidant capacity for beers and wines using a tyrosinase-based biosensor prepared by sinusoidal current method," Sensors (Switzerland), vol. 19, no. 1, 2019, doi: 10.3390/s19010066
44. N. Benbouguerra, T. Richard, C. Saucier, and F. Garcia, "Voltammetric behavior, flavanol and anthocyanin contents, and antioxidant capacity of grape skins and seeds during ripening (Vitis vinifera var. merlot, tannat, and syrah)," Antioxidants, vol. 9, no. 9, pp. 1–19, 2020, doi: 10.3390/antiox9090800
45. A. V. Bounegru and C. Apetrei, "Voltammetric sensors based on nanomaterials for detection of caffeic acid in food supplements," Chemosensors, vol. 8, no. 2, 2020, doi: 10.3390/CHEMOSENSORS8020041
46. C. Liu, Z. Xu, and L. Liu, "Covalent Bonded Graphene/Neutral Red Nanocomposite Prepared by One-step Electrochemical Method and its Electrocatalytic Properties Toward Uric Acid," Electroanalysis, vol. 30, no. 6, pp. 1017–1021, 2018, doi: 10.1002/elan.201700817
47. M. Motshakeri, J. Travas-Sejdic, A. R. J. Phillips, and P. A. Kilmartin, "Rapid electroanalysis of uric acid and ascorbic acid using a poly(3,4-ethylenedioxythiophene)-modified sensor with application to milk," Electrochim. Acta, vol. 265, pp. 184–193, 2018, doi: 10.1016/j.electacta.2018.01.147
48. W. Zhang, X. Jin, H. Chai, G. Diao, and Y. Piao, "3D Hybrids of Interconnected Porous Carbon Nanosheets/Vertically Aligned Polyaniline Nanowires for High-Performance Supercapacitors," Adv. Mater. Interfaces, vol. 5, no. 11, pp. 1–8, 2018, doi: 10.1002/admi.201800106
49. C. Xiong et al., "ZIF-67 derived porous Co3O4 hollow nanopolyhedron functionalized solution-gated graphene transistors for simultaneous detection of glucose and uric acid in tears," Biosens. Bioelectron., vol. 101, no. July 2017, pp. 21–28, 2018, doi: 10.1016/j.bios.2017.10.004
50. A. N. Raja, Annu, K. Singh, and R. Jain, "Ultrasensitive quantification of ellagic acid using Gr/Bi2O3/GCE as voltammetric sensor," Int. J. Electrochem. Sci., vol. 15, pp. 10040–10057, 2020, doi: 10.20964/2020.10.05
51. S. Mbokou Foukmeniok, O. Ilboudo, Y. Karanga, I. Tapsoba, E. Njanja, and I. Tonle Kenfack, "Direct and simultaneous quantification of rutin and quercetin in natural fruits base on purified Arabic Gum modified carbon paste electrode," SN Appl. Sci., vol. 1, no. 5, pp. 1–9, 2019, doi: 10.1007/s42452-019-0413-8
52. B. B. Petković et al., "Easily Prepared Co3O4Doped Porous Carbon Material Decorated with Single-wall Carbon Nanotubes Applied in Voltammetric Sensing of Antioxidant α-lipoic Acid," Electroanalysis, pp. 1–10, 2020, doi: 10.1002/elan.202060290

53. E. Kuyumcu Savan, "Square Wave Voltammetric (SWV) Determination of Quercetin in Tea Samples at a Single-Walled Carbon Nanotube (SWCNT) Modified Glassy Carbon Electrode (GCE)," Anal. Lett., vol. 53, no. 6, pp. 858–872, 2020, doi: 10.1080/00032719.2019.1684514

54. B. Aslışen, Ç. C. Koçak, and S. Koçak, "Electrochemical Determination of Sesamol in Foods by Square Wave Voltammetry at a Boron-Doped Diamond Electrode," Anal. Lett., vol. 53, no. 3, pp. 343–354, 2020, doi: 10.1080/00032719.2019.1650752

55. A. Gomes, G. J. Mattos, B. Coldibeli, R. F. H. Dekker, A. M. Barbosa Dekker, and E. R. Sartori, "Covalent attachment of laccase to carboxymethyl-botryosphaeran in aqueous solution for the construction of a voltammetric biosensor to quantify quercetin," Bioelectrochemistry, vol. 135, p. 107543, 2020, doi: 10.1016/j.bioelechem.2020.107543

56. A. T. E. Vilian et al., "Controllable synthesis of bottlebrush-like ZnO nanowires decorated on carbon nanofibers as an efficient electrocatalyst for the highly sensitive detection of silymarin in biological samples," Sensors Actuators, B Chem., vol. 321, no. June, p. 128544, 2020, doi: 10.1016/j.snb.2020.128544

57. C. Stefanov, C. C. Negut, L. A. D. Gugoasa, and J. (Koos) F. van Staden, "Gold nanoparticle-graphene quantum dots nanozyme for the wide range and sensitive electrochemical determination of quercetin in plasma droplets," Microchim. Acta, vol. 187, no. 11, 2020, doi: 10.1007/s00604-020-04587-y

58. N. Alpar, Y. Yardım, and Z. Şentürk, "Selective and simultaneous determination of total chlorogenic acids, vanillin and caffeine in foods and beverages by adsorptive stripping voltammetry using a cathodically pretreated boron-doped diamond electrode," Sensors Actuators, B Chem., vol. 257, pp. 398–408, 2018, doi: 10.1016/j.snb.2017.10.100

59. A. L. Squissato, E. M. Richter, and R. A. A. Munoz, "Voltammetric determination of copper and tert-butylhydroquinone in biodiesel: A rapid quality control protocol," Talanta, vol. 201, no. April, pp. 433–440, 2019, doi: 10.1016/j.talanta.2019.04.030

60. K. V. Özdokur and Ç. C. Koçak, "Simultaneous Determination of Rosmarinic Acid and Protocatechuic Acid at Poly(o-Phenylenediamine)/Pt Nanoparticles Modified Glassy Carbon Electrode," Electroanalysis, vol. 31, no. 12, pp. 2359–2367, 2019, doi: 10.1002/elan.201900144

61. S. A. Shahamirifard, M. Ghaedi, Z. Razmi, and S. Hajati, "A simple ultrasensitive electrochemical sensor for simultaneous determination of gallic acid and uric acid in human urine and fruit juices based on zirconia-choline chloride-gold nanoparticles-modified carbon paste electrode," Biosens. Bioelectron., vol. 114, no. January, pp. 30–36, 2018, doi: 10.1016/j.bios.2018.05.009

62. K. Krishnamoorthy, V. Sudha, S. M. Senthil Kumar, and R. Thangamuthu, "Simultaneous determination of dopamine and uric acid using copper oxide nano-rice modified electrode," J. Alloys Compd., vol. 748, pp. 338–347, 2018, doi: 10.1016/j.jallcom.2018.03.118

63. Y. Veera Manohara Reddy, B. Sravani, S. Agarwal, V. K. Gupta, and G. Madhavi, "Electrochemical sensor for detection of uric acid in the presence of ascorbic

acid and dopamine using the poly(DPA)/SiO2@Fe3O4 modified carbon paste electrode," J. Electroanal. Chem., vol. 820, no. March, pp. 168–175, 2018, doi: 10.1016/j.jelechem.2018.04.059

64. V. Erady et al., "Carbon paste modified with Bi decorated multi-walled carbon nanotubes and CTAB as a sensitive voltammetric sensor for the detection of Caffeic acid," Microchem. J., vol. 146, no. November 2018, pp. 73–82, 2019, doi: 10.1016/j.microc.2018.12.023
65. M. Sakthivel, S. Ramaraj, S. M. Chen, B. Dinesh, H. V. Ramasamy, and Y. S. Lee, "Entrapment of bimetallic CoFeSe2 nanosphere on functionalized carbon nanofiber for selective and sensitive electrochemical detection of caffeic acid in wine samples," Anal. Chim. Acta, vol. 1006, pp. 22–32, 2018, doi: 10.1016/j.aca.2017.12.044
66. R. Nehru, Y. F. Hsu, and S. F. Wang, "Electrochemical determination of caffeic acid in antioxidant beverages samples via a facile synthesis of carbon/iron-based active electrocatalyst," Anal. Chim. Acta, vol. 1122, pp. 76–88, 2020, doi: 10.1016/j.aca.2020.05.001
67. G. Muthusankar et al., "N-doped carbon quantum dots @ hexagonal porous copper oxide decorated multiwall carbon nanotubes: A hybrid composite material for an efficient ultra-sensitive determination of caffeic acid," Compos. Part B Eng., vol. 174, no. March, p. 106973, 2019, doi: 10.1016/j.compositesb.2019.106973
68. X. Long, "Ce-TiO2/carbon Nanotube Composite Modified Glassy Carbon Electrode for Electrochemical Detection of Caffeic Acid," Int. J. Electrochem. Sci., vol. 14, pp. 7832–7841, 2019, doi: 10.20964/2019.08.13
69. Y. Yan, X. Bo, and L. Guo, "MOF-818 metal-organic framework-reduced graphene oxide/multiwalled carbon nanotubes composite for electrochemical sensitive detection of phenolic acids," Talanta, vol. 218, no. April, p. 121123, 2020, doi: 10.1016/j.talanta.2020.121123
70. A. Şenocak, "Fast, Simple and Sensitive Determination of Coumaric Acid in Fruit Juice Samples by Magnetite Nanoparticles-zeolitic Imidazolate Framework Material," Electroanalysis, vol. 32, no. 10, pp. 2330–2339, 2020, doi: 10.1002/elan.202060237
71. A. Miccoli et al., "Sensitive electrochemical detection method of melatonin in food supplements," Rev. Chim., vol. 69, no. 4, pp. 854–859, 2018, doi: 10.37358/rc.18.4.6215
72. M. Mosleh, S. M. Ghoreishi, S. Masoum, and A. Khoobi, "Determination of quercetin in the presence of tannic acid in soft drinks based on carbon nanotubes modified electrode using chemometric approaches," Sensors Actuators, B Chem., vol. 272, no. March, pp. 605–611, 2018, doi: 10.1016/j.snb.2018.05.172
73. M. Badea et al., "Electrochemical strategies for gallic acid detection: Potential for application in clinical, food or environmental analyses," Sci. Total Environ., vol. 672, pp. 129–140, 2019, doi: 10.1016/j.scitotenv.2019.03.404
74. A. P. Lima, W. T. P. dos Santos, E. Nossol, E. M. Richter, and R. A. A. Munoz, "Critical evaluation of voltammetric techniques for antioxidant capacity and activity: Presence of alumina on glassy-carbon electrodes alters the results," Electrochim. Acta, vol. 358, 2020, doi: 10.1016/j.electacta.2020.136925

75. T. A. Silva, M. R. K. Khan, O. Fatibello-Filho, and M. M. Collinson, "Simultaneous electrochemical sensing of ascorbic acid and uric acid under biofouling conditions using nanoporous gold electrodes," J. Electroanal. Chem., vol. 846, no. January, p. 113160, 2019, doi: 10.1016/j.jelechem.2019.05.042

76. J. Hoyos-Arbeláez, L. Blandón-Naranjo, M. Vázquez, and J. Contreras-Calderón, "Antioxidant capacity of mango fruit (Mangifera indica). An electrochemical study as an approach to the spectrophotometric methods," Food Chem., vol. 266, no. March, pp. 435–440, 2018, doi: 10.1016/j.foodchem.2018.06.044

77. S. M. Chen et al., "Determination of the antioxidant propyl gallate in meat by using a screen-printed electrode modified with CoSe 2 nanoparticles and reduced graphene oxide," Microchim. Acta, vol. 185, no. 11, 2018, doi: 10.1007/s00604-018-3048-3

78. G. Manasa, R. J. Mascarenhas, A. K. Bhakta, and Z. Mekhalif, "Nano-graphene-platelet/Brilliant-green composite coated carbon paste electrode interface for electrocatalytic oxidation of flavanone Hesperidin," Microchem. J., vol. 160, no. PB, p. 105768, 2021, doi: 10.1016/j.microc.2020.105768

79. D. R. Kumar, M. S. Sayed, M. L. Baynosa, and J. J. Shim, "5-Amino-2-mercapto-1,3,4-thiadiazole coated nitrogen-doped-carbon sphere composite for the determination of phenolic compounds," Microchem. J., vol. 157, no. May, p. 105023, 2020, doi: 10.1016/j.microc.2020.105023

80. A. Şenocak, T. Basova, E. Demirbas, and M. Durmuş, "Direct and Fast Electrochemical Determination of Catechin in Tea Extracts using SWCNT-Subphthalocyanine Hybrid Material," Electroanalysis, vol. 31, no. 9, pp. 1697–1707, 2019, doi: 10.1002/elan.201900214

81. C. O. Chikere, E. Hobben, N. H. Faisal, P. Kong-Thoo-Lin, and C. Fernandez, "Electroanalytical determination of gallic acid in red and white wine samples using cobalt oxide nanoparticles-modified carbon-paste electrodes," Microchem. J., vol. 160, no. PB, p. 105668, 2021, doi: 10.1016/j.microc.2020.105668

82. M. O. Onizhuk, O. S. Tkachenko, A. V. Panteleimonov, V. V. Varchenko, K. Belikov, and Y. V. Kholin, "Electrochemical oxidation of quercetin in aqueous and ethanol-water media with the use of graphite/chemically modified silica ceramic electrode," Ionics (Kiel)., vol. 24, no. 6, pp. 1755–1764, 2018, doi: 10.1007/s11581-017-2320-6

83. G. Ziyatdinova et al., "Simultaneous voltammetric determination of gallic and ellagic acids in cognac and brandy using electrode modified with functionalized SWNT and poly(pyrocatechol violet)," Food Anal. Methods, vol. 12, no. 10, pp. 2250–2261, 2019, doi: 10.1007/s12161-019-01585-6

84. C. Raril, J. G. Manjunatha, and G. Tigari, "Low-cost voltammetric sensor based on an anionic surfactant modified carbon nanocomposite material for the rapid determination of curcumin in natural food supplement," Instrum. Sci. Technol., vol. 48, no. 5, pp. 561–582, 2020, doi: 10.1080/10739149.2020.1756317

85. G. Manasa, R. J. Mascarenhas, A. K. Bhakta, and Z. Mekhalif, "MWCNT/Nileblue Heterostructured Composite Electrode for Flavanone Naringenin Quantification in Fruit Juices," Electroanalysis, vol. 32, no. 5, pp. 939–948, 2020, doi: 10.1002/elan.201900573

86. C. Aravena-Sanhueza, Felipe; Pérez-Rivera, Mónica; Castillo-Felices, Rosario; Mundaca-Uribe, Rodolfo; Bustos, Mario Aranda ; Peña Farfal, "Determination Of Antioxidant Capacity (Orac) Of Greigia Sphacelata And Correlation With Voltammetric Methods," J. Chil. Chem. Soc., Vol. 3, Pp. 4925–4928, 2020, Doi: Http://Dx.doi.org/10.4067/S0717-97072020000204925

87. P. Rocha, Â. Vilas-Boas, N. Fontes, D. Geraldo, and F. Bento, "Evaluation of Polyphenols in Wine by Voltammetric Techniques with Screen Printed Carbon Electrodes," Electroanalysis, vol. 32, no. 1, pp. 159–165, 2020, doi: 10.1002/elan.201900392

88. Ç. C. Koçak, Ş. U. Karabiberoğlu, and Z. Dursun, "Highly sensitive determination of gallic acid on poly (L-Methionine)-carbon nanotube composite electrode," J. Electroanal. Chem., vol. 853, no. October, 2019, doi: 10.1016/j.jelechem.2019.113552

89. A. Şenocak, B. Köksoy, E. Demirbaş, T. Basova, and M. Durmuş, "3D SWCNTs-coumarin hybrid material for ultra-sensitive determination of quercetin antioxidant capacity," Sensors Actuators, B Chem., vol. 267, pp. 165–173, 2018, doi: 10.1016/j.snb.2018.04.012

90. W. Lyu et al., "A simple and sensitive electrochemical method for the determination of capsaicinoids in chilli peppers," Sensors Actuators, B Chem., vol. 288, no. February, pp. 65–70, 2019, doi: 10.1016/j.snb.2019.02.104

91. E. Nagles, J. Penagos-Llanos, O. García-Beltrán, and J. Hurtado, "Determination of Rutin in Drinks Using an Electrode Modified with Carbon Nanotubes-Prussian Blue," J. Anal. Chem., vol. 73, no. 5, pp. 504–511, 2018, doi: 10.1134/S1061934818050064

92. S. Güney, "Electrochemical synthesis of molecularly imprinted poly(p-aminobenzene sulphonic acid) on carbon nanodots coated pencil graphite electrode for selective determination of folic acid," J. Electroanal. Chem., vol. 854, no. October, p. 113518, 2019, doi: 10.1016/j.jelechem.2019.113518

93. M. S. S. Kumar et al., "Multifunctional applications of Nickel oxide (NiO) nanoparticles synthesized by facile green combustion method using Limonia acidissima natural fruit juice," Inorganica Chim. Acta, vol. 515, no. September 2020, p. 120059, 2021, doi: 10.1016/j.ica.2020.120059

94. L. C. Almeida et al., "Electrochemical deposition of bio-inspired laccase-polydopamine films for phenolic sensors," Electrochim. Acta, vol. 319, pp. 462–471, 2019, doi: 10.1016/j.electacta.2019.06.180

95. K. Brainina, N. Stozhko, M. Bukharinova, E. Khamzina, and M. Vidrevich, "Potentiometric method of plant microsuspensions antioxidant activity determination," Food Chem., vol. 278, no. November 2018, pp. 653–658, 2019, doi: 10.1016/j.foodchem.2018.11.098

96. H. Yi, Y. Cheng, Y. Zhang, Q. Xie, and X. Yang, "Potentiometric and UV-Vis spectrophotometric titrations for evaluation of the antioxidant capacity of chicoric acid," RSC Adv., vol. 10, no. 20, pp. 11876–11882, 2020, doi: 10.1039/d0ra01248c

97. Y. Tanaka, T. Hasegawa, T. Shimamura, H. Ukeda, and T. Ueda, "Potentiometric evaluation of antioxidant capacity using polyoxometalate-immobilized electrodes," J.

Electroanal. Chem., vol. 828, no. May, pp. 102–107, 2018, doi: 10.1016/j.jelechem.2018.09.024

98. D. H. Kim, S. Nam, J. Kim, and W. Y. Lee, "Electrochemical determination of capsaicin by ionic liquid composite-modified electrode," J. Electrochem. Sci. Technol., vol. 10, no. 2, pp. 177–184, 2019, doi: 10.5229/JECST.2019.10.2.177

99. J. Chýlková, L. Janíková, R. Šelešovská, and J. Mikšíček, "New voltammetric method for rapid determination of phenolic antioxidant 2-tert-butylphenol in synthetic oils using gold electrode," Eur. Food Res. Technol., vol. 150, pp. 1651–1654, 2019, doi: 10.1007/s00706-019-02417-3

100. P. J. Juarez-Luna, S. Mendoza, and A. Cardenas, "Comparison of electrochemical methods using CUPRAC, DPPH, and carbon paste electrodes for the quantification of antioxidants in food oils," Anal. Methods, vol. 11, no. 45, pp. 5755–5760, 2019, doi: 10.1039/c9ay01921a

Index

V

W

X

Z